Materials

François SORRENTINO,
Denis DAMIDOT
and Charles FENTIMAN

CaO–SiO$_2$–Al$_2$O$_3$–Fe Oxides Chemical System

Description and Applications

EDP Sciences – ISBN(print): 978-2-7598-2480-9 – ISBN(ebook): 978-2-7598-2536-3
DOI: 10.1051/978-2-7598-2480-9

All rights relative to translation, adaptation and reproduction by any means whatsoever are reserved, worldwide. In accordance with the terms of paragraphs 2 and 3 of Article 41 of the French Act dated March 11, 1957, "copies or reproductions reserved strictly for private use and not intended for collective use" and, on the other hand, analyses and short quotations for example or illustrative purposes, are allowed. Otherwise, "any representation or reproduction – whether in full or in part – without the consent of the author or of his successors or assigns, is unlawful" (Article 40, paragraph 1). Any representation or reproduction, by any means whatsoever, will therefore be deemed an infringement of copyright punishable under Articles 425 and following of the French Penal Code.

© Science Press, EDP Sciences, 2021

Preface

Aim of the Book

This book describes and comments on the results of research devoted to the studies of phase assemblages in the $CaO-SiO_2-Al_2O_3$-Fe oxides chemical system (mentioned 'the system' thereafter), their stability and their evolution in our environment (temperature, pressure). Its aim is to be a research support, not only for researchers and development engineers but also more generally for others interested in materials sciences. Indeed; it concerns oxygen, silicon, aluminium, calcium and iron that are by mass the most abundant elements in the earth crust and the most widely used by human beings. Thus, not only does this show the diversity of the properties of the combinations of these elements, but also reflects the abundance as a primary reason for the diversity of their uses.

Consequently, the book is divided into two parts; the first one is devoted to a description of 'the system' using phase diagrams and the second one presents the properties of some of the constituents that are in widespread industrial and commercial use.

Much of the work reported in this book is fully original, as it reports records of the research that Dr François Sorrentino undertook in the chemistry department of Aberdeen University in the early seventies and then the research he carried out as manager of *Mineral Research Processing Company* devoted to the synthesis of minerals. The laboratory synthesis of calcium silicates and calcium aluminates, and their industrial manufacture were the main interests of the development of this Company.

Summary of Part I – Description of the $CaO-SiO_2-Al_2O_3$-Fe Oxides Chemical System

The first part is an exhaustive literature survey of 'the system'. References are quoted using the first three letters of the name of the first author followed by the year of publication. For example, ARA 1959 corresponds to Aramaki S. and Roy R.

(1959) Revised equilibrium diagram for the system Al_2O_3–SiO_2, Nature **184**, 631. The major data relevant to this system and their combinations, useful for their applications in the daily lives of human being have been collected. Entirely reconstructed, revisited and redrawn phase diagrams are then described from the collected data, and experimental data from personal experience added. When data overlap or interfere, the author takes the responsibility of the choices that were made.

In each chapter, the constituents are identified by their formula in the introduction. A list of the most probable existing constituents is given and written both as chemical formula (e.g. Ca_3SiO_5) and oxides-based formula (e.g. $3CaO \cdot SiO_2$). Common/historical names are also given when known, for instance when named after a locality such as the mineral, Andalusite (named after Andalusia), or when named after a person, typically the discovering geologist or chemist, for instance Wollastonite, named after William Hyde Wollaston (1766–1828). Then, the sub-chapter, entitled mineralogy, gives the mineralogical composition, the location and the mechanism of its formation. The main crystallographic characteristics and the polymorphic forms are then described in the sub-chapter related to crystallography. Solid solutions are only treated in the context of 'the system' that is narrower than the global chemical system. The stability of the phase assemblages depending on temperature and pressure is then reported in the sub-chapter dealing with phase diagrams. The experimental phase diagrams are redrawn from sets of experimental data found in the literature. Moreover, thermodynamic models have been used to calculate the phase diagram and compare it to its experimental counterpart. Finally, the last sub-chapter provides methods to synthesize the constituents that have been described thanks to the knowledge of their conditions of formation gained from phase diagrams.

Summary of Part II – Applications of the CaO–SiO$_2$–Al$_2$O$_3$–Fe Oxides Chemical System

Part II describes the main applications of the oxides within 'the system' (except for the primary oxides where applications are reported in part I). It is hardly surprising to find that there is a large diversity of applications found with minerals and substances containing these elements. The most common combinations are oxides (silicon, aluminium, iron), or non-oxide minerals such as carbonates or sulphides. Other combinations between these elements allow us to extend the fields of applications of the phase relations known in the simpler system. The work presented in this part, is not only a bibliography synthesis, although the data come, in a large part from the literature, but they have been enriched by personal experience, in the field of hydraulic binders, refractory glass and ceramics that are part of Dr François Sorrentino's professional career.

The applications are divided into three levels of applications, the last one being the closest to the end-user or the customer (table P.1). The first level describes the application of the CaO–SiO$_2$–Al$_2$O$_3$–Fe oxides chemical system with a minimum of

TAB. P.1 – Examples of the three levels of applications of the $CaO-Al_2O_3-SiO_2-Fe$ oxides system.

Starting material	First level	Second level	Third level (end use)
CaO	Agriculture	Slag forming, cement	Civil engineering (buildings)
SiO_2	Aggregates	Glass, cement	Civil engineering (cladding material for buildings), cooking ware, container
Al_2O_3	Refractory	Aluminium (Al), gems, glass (with Ge), abrasive	Vessel, kiln lining, building, aircraft devices, wave guide
Fe oxides	Pigments	Steel, slags	Car industry, steel bar, military devices

treatment such as crushing or grinding. For example, the use of Fe_2O_3 as a pigment, the use of CaO in agriculture or silica as quartz for aggregates. The second level concerns the use of 'the system' as a precursor for the synthesis of other materials such as the manufacture of steel by reduction of Fe_2O_3 or the combination of CaO, silica, alumina and iron oxides to manufacture cement. The third level represents the end level for which the application is close to the end-user or to the customer.

The fields of applications in 'the system' reported in part II are construction, metallurgy, glass and ceramics but also specific uses such as refractory materials and fillers.

Acknowledgement

I would like to thank Prof. Denis Damidot and Dr Charles Fentiman for taking part in the process of writing and editing this book. I also would like to thank my wife Dr Danielle Sorrentino for her unconditional support.

Dr François Sorrentino

Contents

Preface III

Part I - Descriptions

CHAPTER 1

One-Component Chemical Systems: CaO, SiO_2, Al_2O_3 and Fe Oxides 3
1.1   CaO 3
    1.1.1   Introduction 3
    1.1.2   Mineralogy, Structure and Stability 3
    1.1.3   Properties and Applications 4
1.2 SiO_2 5
    1.2.1   Introduction 5
    1.2.2   Mineralogy 5
    1.2.3   Structure and Stability 6
    1.2.4   Properties and Applications 6
1.3 Al_2O_3 7
    1.3.1   Introduction 7
    1.3.2   Mineralogy 8
    1.3.3   Structure and Stability 8
    1.3.4   Properties and Applications 9
1.4   Fe Oxides 10
    1.4.1   Introduction 10
    1.4.2   Mineralogy 11
    1.4.3   Structure and Stability 11
    1.4.4   Synthesis 14
    1.4.5   Properties and Applications 14

CHAPTER 2

Binary Chemical Systems 17
2.1 CaO–SiO_2 System 17
    2.1.1   Introduction 17
    2.1.2   Mineralogy 17

		2.1.3	Structure, Polymorphism and Solid Solution	18
		2.1.4	Stability and Phase Diagram of CaO–SiO$_2$ System	25
		2.1.5	Synthesis of Calcium Silicates	27
	2.2	Al$_2$O$_3$–SiO$_2$ System		33
		2.2.1	Introduction	33
		2.2.2	Mineralogy	33
		2.2.3	Structure, Polymorphism and Solid Solutions	33
		2.2.4	Stability and Phase Diagram	37
		2.2.5	Synthesis of Alumino-Silicates	41
	2.3	Al$_2$O$_3$–Fe Oxides System		44
		2.3.1	Introduction	44
		2.3.2	Mineralogy	44
		2.3.3	Structure, Polymorphism and Solid Solution	44
		2.3.4	Stability and Phase Diagram	45
		2.3.5	Synthesis of Iron Aluminates	48
	2.4	CaO–Al$_2$O$_3$ System		48
		2.4.1	Introduction	48
		2.4.2	Mineralogy	49
		2.4.3	Structure, Polymorphism and Solid Solution	49
		2.4.4	Stability and Phase Diagram	55
		2.4.5	Synthesis of Calcium Aluminates	57
	2.5	SiO$_2$–Fe Oxides System		62
		2.5.1	Introduction	62
		2.5.2	Mineralogy	62
		2.5.3	Structure, Polymorphism and Solid Solution	63
		2.5.4	Liquid FeO–Fe$_2$O$_3$–SiO$_2$	66
		2.5.5	Stability and Phase Diagram of the FeO–Fe$_2$O$_3$–SiO$_2$ System	67
		2.5.6	Synthesis of Iron Silicates	71
	2.6	CaO–Fe Oxides System		72
		2.6.1	Introduction	72
		2.6.2	Mineralogy	72
		2.6.3	Structure, Polymorphism and Solid-Solutions	72
		2.6.4	Stability and Phase Diagram	75
		2.6.5	Synthesis of Calcium Ferrite	81

CHAPTER 3

Ternary Chemical Systems				83
3.1	General Introduction			83
3.2	CaO–Al$_2$O$_3$–SiO$_2$ System			83
		3.2.1	Ternary Constituents of the CaO–Al$_2$O$_3$–SiO$_2$ System	83
		3.2.2	Mineralogy	84
		3.2.3	Structure and Solid Solutions	85
		3.2.4	Stability and Phases Diagrams (At One Atmosphere)	90
		3.2.5	Compounds Obtained in Special Conditions	97
		3.2.6	Stability Relative to Temperature or Pressure and Both	98

	3.2.7	Thermodynamic Models	99
	3.2.8	Synthesis of Calcium Silicoaluminate	101
3.3	Al_2O_3–SiO_2–Fe Oxides		103
	3.3.1	Introduction	103
	3.3.2	Mineralogy	103
	3.3.3	Structure and Solid Solution	103
	3.3.4	Stability and Phase Diagram	104
	3.3.5	Synthesis	108
3.4	CaO–SiO_2–Fe Oxides		109
	3.4.1	Introduction	109
	3.4.2	Mineralogy	109
	3.4.3	Structure and Solid Solution	110
	3.4.4	Stability and Phase Diagrams in Air	112
	3.4.5	Stablity and Phase Diagram in the CaO–SiO_2–FeO System	117
	3.4.6	CaO–SiO_2–FeO_x System at Various Oxygen Pressures	126
	3.4.7	Model	131
	3.4.8	Preparation and Synthesis	132
3.5	CaO–Al_2O_3–Fe Oxides		134
	3.5.1	Introduction	134
	3.5.2	Mineralogy	134
	3.5.3	Structure and Solid Solutions	135
	3.5.4	Stability and Phase Diagrams	140
	3.5.5	Liquidus in Reducing Atmosphere	143
	3.5.6	Model of the System	145
	3.5.7	Formation and Synthesis	145

CHAPTER 4

Quaternary Chemical Systems			147
4.1	Introduction		147
	4.1.1	Principle of Phase Equilibrium in a Quaternary System	147
4.2	CaO–SiO_2–Al_2O_3–Fe_2O_3 System in Air		152
	4.2.1	Quaternary Constituents of the CaO–SiO_2–Al_2O_3–Fe Oxides System	152
	4.2.2	Mineralogy	153
	4.2.3	Mineralogy Structure Stability	158
	4.2.4	Formation and Synthesis	173
4.3	CaO–Al_2O_3–SiO_2–FeO System		173
	4.3.1	Introduction	173
	4.3.2	Ternary Systems Located Within the CaO, Al_2O_3, SiO_2, FeO Tetrahedron	173
	4.3.3	Binary Systems	181
	4.3.4	Crystallized Solids	182
	4.3.5	Univariant Lines and Quaternary Invariant Points	182
	4.3.6	$2CaO·SiO_2$–$CaO·SiO_2$–Gehlenite–FeO Quaternary System	183

CHAPTER 5

Quinary Chemical Systems		189
5.1	Introduction – Presentation of Quinary Data	189
5.2	$CaO–SiO_2–Al_2O_3–FeO–Fe_2O_3$ System	189
	5.2.1 Stability and Phase Diagrams	189
	5.2.2 Conclusions	192
References of Part I		193

Part II - Applications

CHAPTER 6

Applications to Hydraulic Binders		211
6.1	General Introduction	211
6.2	Portland Cements (PC)	212
	6.2.1 Characteristics	212
	6.2.2 PC Applications	220
	6.2.3 Conclusions – Prediction of the Properties	226
6.3	Calcium Aluminate Cements (CAC)	227
	6.3.1 CAC Characteristics	227
	6.3.2 CAC Applications	234
6.4	Special Cements	237
	6.4.1 Fast-Setting Cements	237
	6.4.2 Geopolymers	243
	6.4.3 Oil Well Cements	246
	6.4.4 Expansive Cement	251
	6.4.5 Dental Cements	253
	6.4.6 Glass Cements	259

CHAPTER 7

Application to Metal Refining		261
7.1	General Introduction	261
7.2	Slags from Iron and Steel Industry	261
	7.2.1 Blast Furnace Slags (BFS)	261
	7.2.2 Converter Slags (Basic Oxygen Process or LD processing)	266
	7.2.3 EAF Slags: High Carbon Steel	269
	7.2.4 Ladle Slag (Secondary Metallurgy) – Stainless – High Alloy Steel Production	269
	7.2.5 Refining Under Reducing Slag	270
7.3	Formation and Properties of Liquid Slags	271
7.4	Slags from Non Ferrous Industry	272
	7.4.1 Copper Slag	272
	7.4.2 Silico-Manganese Slag	272

CHAPTER 8

Application to Refractory Materials		275
8.1	Introduction	275
8.2	Raw Materials Based on Al_2O_3 and SiO_2	276
	8.2.1 Natural Raw Materials	276
	8.2.2 Synthetic Raw Material	277
8.3	Applying Refractory Materials	278
	8.3.1 Brick and Monolithic Refractories	278
	8.3.2 Refractory Cement and Mortar	281
8.4	Refractories Consuming Industry	286
	8.4.1 Iron and Steel Industry	286
	8.4.2 Non-Ferrous Metal Industry	287
	8.4.3 Cement Industry	287
	8.4.4 Whiteware, Traditional Ceramic Industry	288
	8.4.5 High-Tech Ceramic Industry	288
	8.4.6 Glass Industry	288

CHAPTER 9

Application of the Glassy Products		291
9.1	Introduction	291
9.2	Structure of Glass	292
9.3	Classification of Glass Products Containing CaO, Al_2O_3, SiO_2 and Fe Oxides	293
9.4	Products, Chemistry and Process	294
	9.4.1 High Silica Glass – Vycor Glass	294
	9.4.2 Soda-Lime Glass	295
	9.4.3 Sodium Borosilicate – Glass Fibres	295
	9.4.4 Aluminosilicate Glass	297
	9.4.5 Special Applications	299

CHAPTER 10

Application of Ceramic Products		303
10.1	Introduction to Ceramics	303
10.2	Structure of Ceramics	303
	10.2.1 Processing of Manufacture	303
	10.2.2 Shaping	304
	10.2.3 Physico-Chemical Changes During Firing	305
10.3	Classification of Ceramic Products Containing CaO, SiO_2, Al_2O_3 and Fe Oxides	305
	10.3.1 Introduction	305
	10.3.2 Traditional Ceramics	305
	10.3.3 Ceramics in the Construction Sector Ceramics	312
	10.3.4 High Technology Ceramics	313

CHAPTER 11

Application as Fillers 315
11.1   General Introduction 315
11.2   Mono-Component 317
      11.2.1   Calcium Oxide 317
      11.2.2 SiO_2 318
      11.2.3   Alumina 319
      11.2.4   Fe oxides 320
11.3   Multi-Components By-Products from Industrial Process 321
      11.3.1   Slags 321
      11.3.2   Fly Ashes 324
      11.3.3   Red Mud (KUR 1997) 326
      11.3.4   Cement Kiln Dust (CKD) 327
11.4   Multi-Components from Natural Origin 328
      11.4.1   Natural Pozzolans 328
      11.4.2   Metakaolin 330
      11.4.3   Rice Husk Ashes (RHA) 331
      11.4.4   Wollastonite 332
References of Part II 335

Part I

Description of the CaO–SiO$_2$–Al$_2$O$_3$–Fe Oxides Chemical System

Chapter 1

One-Component Chemical Systems: CaO, SiO$_2$, Al$_2$O$_3$ and Fe Oxides

1.1 CaO

COC 1993/SHE 2001

1.1.1 Introduction

- Calcium oxide is produced by thermal decomposition of calcium carbonate above 825 °C. The most important starting minerals are Limestone, Calcite, Aragonite, Vaterite, seashell and corral.
- The production of CaO from limestone is one of the oldest chemical transformations known for 8000 years in the Near East and it is the most abundant chemical (283 million metric tons/year).
- CaO takes different names depending on the structure and the process of manufacture; quick lime, burnt lime, unslaked lime and pebble lime.

1.1.2 Mineralogy, Structure and Stability

- Table 1.1 shows the main characteristics of calcium carbonate and lime:

  ○ Calcite, Aragonite, and Vaterite are polymorphs of calcium carbonate (CaCO$_3$).
  ○ Calcite is a stable polymorph at ambient temperature and pressure. It crystallises as rhombohedric and it is the constituent of many sedimentary rocks.
  ○ Aragonite is the polymorph stable at high temperature and pressure of CaCO$_3$. It crystallises as orthorhombic.

DOI: 10.1051/978-2-7598-2480-9.c001
© Science Press, EDP Sciences, 2020

TAB. 1.1 – Main characteristics of calcium carbonates and lime.

	CaO	Calcite	Vaterite	Aragonite
JCPDS card N°	1305-78-8	13397-26-7	13701-58-1	14791-73-2
Molar mass (g)	56.078	100.087	100.087	100.087
Density (g/cm^3)	3.34	2.6–2.8	2.54	2.93
Melting point (°C)	2613		Decompose	
Boiling point (°C)	3850			
Crystallographic lattice	Face-centred cubic	Rhombohedral	Hexagonal	Orthorhombic
Crystal lattice parameters	$a = 4.8152$ Å		$a = 4.13$ Å $c = 8.49$ Å	$a = 4.959$ Å $b = 7.968$ Å $c = 5.741$ Å

- Vaterite is a metastable polymorph of $CaCO_3$ and it crystallises as hexagonal.
- CaO is white to pale yellow or even brown and crystallises in the cubic system.

1.1.3 Properties and Applications

- The most important properties of CaO are basicity, refractory properties, and ability to form combination with silicates in solution and to precipitate insoluble salts.

1.1.3.1 First Level of Applications

- The addition of calcium carbonate to cement and concrete modifies the particle size distribution, improves the workability, the shrinkage, then density and then mechanical strengths by a physical effect.
- Calcium carbonate is used as filler for a wide range of applications including tooth pastes and in the pharmaceutical industry in the matrix composition of drugs. It is also used as food and drink additives.
- CaO is used to neutralise flue gas pollutants, to capture heavy metals, to treat organic and mineral sludges. For example, it is used in sugar industry to precipitate impurities (10/15%).
- CaO is used in chemical industry such as leather tanning or paper industry, to precipitate $CaCO_3$.
- In agriculture, CaO is used to adjust the pH of soil by chemical effect to give optimum growing conditions (30%).
- In steel and non-ferrous metal industry, CaO forms slags.
- CaO is used as fluxing agent in the glass industry (30/40%).
- CaO can be used as a refractory considering its melting point at 2613 °C.
- In civil works, CaO stabilizes soil and road basement when clay is present.

1.1.3.2 Second Level of Application

- Calcium carbonate is the main raw material used to manufacture cement. It can also be used as filler in Porland Cement (CEM II according to CEN – EN 197-1).
- CaO can be used in the production of bricks and structural blocks (15/20%).

1.1.3.3 End Level of Applications

- Buildings and works, and the road and bridge sector...

1.2 SiO_2

ILE 1979/ANG 1996/STE 1999/BRI 1989

1.2.1 Introduction

- Silicon is the second most abundant element on earth crust.
- Silicon oxide (SiO_2) exists in nature in different forms: crystallised (Quartz, Tridymite, Cristobalite), cryptocrystalline (agate, onyx, carnelian, jasper, flint), colloidal (gel), and hydrous silica (opal).
- Most synthetic silicas are amorphous.

1.2.2 Mineralogy

- The major forms of crystallised silica found in nature under standard conditions (temperature and pressure) are Quartz, Tridymite and Cristobalite. They account for more than 10% by mass of the earth's crust.
- The structure of minerals formed at typical earth's crust pressure is based on Si atom which shows tetrahedral coordination with 4 oxygen atoms surrounding a central Si atom. All oxygen atoms of the SiO_4 tetrahedron are shared with other SiO_4 tetrahedra, yielding to a 3D network having the net formula SiO_2.
- The minerals Coesite, Keatite, Stishovite, and Seifertite are found at higher temperatures and pressure but Stishovite and Seifertite are not made of SiO_4 tetrahedra and are not classified as network silicate.
- Silicates formed at higher pressure in the earth transition zone and the lower mantel, contain mainly six-coordinated silicon. Penta-coordinated silicon is probably a component of alumino-silicate melts and glasses at mantel's temperature and pressure.
- Other forms of SiO_2 have been determined (Moganite contains 3% of water and is crystallised as monoclinic), anhydrous amorphous silica (fused silica, fumed

silica), amorphous colloidal silica (gel), massive dense amorphous silica glass, minerals in various living organisms.
- Some forms contain minor components (sulphur like Melanophlogite) or soda (Faujasite).
- SiO_2 is also found as by-products of the ferroalloys production, of coal power plants (fly ashes), of rice industry (rice husk).

1.2.3 Structure and Stability

- The different polymorphs of quartz (α and β), Tridymite (α and β) and Cristobalite (α and β) are transformed as follows with the temperature:

Quartz α $\xrightleftharpoons[]{573\,°C}$ Quartz β $\xrightleftharpoons[]{870\,°C}$ Tridymite β $\xrightleftharpoons[]{1470\,°C}$ Cristobalite β $\xrightleftharpoons[]{1700\,°C}$ silica melt

trigonal hexagonal hexagonal cubic

  ○ α-Quartz converts to β-Quartz at 573 °C.
  ○ β-Quartz converts to β-Tridymite at 870 °C.
  ○ Tridymite can occur in seven polytypes and as α and β at standard atmospheric pressure.
  ○ Below 100 °C, the triclinic form α Tridymite is stable. The orthorhombic β-Tridymite is stable at an elevated temperature (>870 °C) and it converts to β-Cristobalite above 1470 °C.
  ○ At ambient temperature, Cristobalite is tetragonal and changes as cubic at higher temperatures.
  ○ The crystallographic characteristics are shown in table 1.2.

1.2.4 Properties and Applications

- SiO_2 shows a low solubility in water, acidity and viscosity at high temperatures.

1.2.4.1 First Level of Applications

- SiO_2 as fine particles is used as additive in concrete (silica fume, rice husk), as additive in pharmaceutical drugs, where silica aids powder flow when tablets are formed.
- SiO_2 is an extender in paint and a functional filler in plastic and rubber.
- SiO_2 as lumps is used as aggregate in road construction.
- Vitreous silica glass is used for high temperature application refractory.
- SiO_2 in fibre form can be used as high temperature thermal protection.

TAB. 1.2 – Crystallographic characteristics of quartz polymorphs.

	Crystallographic lattice	Crystallographic sub-lattice	Density (g/cm^3) and volume of crystal cell (Å^3)	Crystal lattice parameters
α-Quartz (low T°) JCPDS 14808-60-7	Rhombohedral	$P3_121$ $P3_221$	$d = 2.648$ $V = 113$	
β-Quartz (high T°) JCPDS 99439-28-8	Hexagonal	$P6_422$ $P6_222$	$d = 2.533$	$a = 4.91$ Å $c = 5.405$ Å $\alpha = 90°$ $\beta = 90°$ $\gamma = 120°$
α-Tridymite JCPDS 15468-32-3	Orthorhombic	Metastable	$d = 2.26$	$a = 9.88$ Å $b = 17.1$ Å $c = 16.3$ Å
β-Tridymite	Hexagonal			
α-Cristobalite (low T°)	Tetragonal		$d = 2.33$ $V = 170$	$a = 4.97$ Å $c = 6.91$ Å
β-Cristobalite (high T°)	Cubic	$P4_232$		$a = 7.12$ Å

1.2.4.2 Second Level of Application

- SiO_2 after reduction is a precursor for microelectronics and photovoltaic.
- SiO_2 is present in association with other oxides in cements, ceramics, refractories, metallurgy and glass industry (alumino-silicate glass).

1.2.4.3 End Level of Applications

- SiO_2 is used in furnace tube, melting crucibles, and equipment requiring thin film of silica.

1.3 Al_2O_3

GIT 1970/HAR 1990

1.3.1 Introduction

- Aluminium is the third most abundant element on earth crust.
- Aluminium exists mainly as oxide, hydroxide, and in different forms: crystallised, amorphous and gel.

TAB. 1.3 – Chemical composition of sources of aluminium oxide coming from different countries and suppliers.

Oxide (weight %)	CaO	SiO_2	Al_2O_3	Fe_2O_3	TiO_2	MgO	LOI (%)
Bauxite (France)	0.60	10.45	49.15	21.55		0.20	15.20
Bauxite (Greece)	4.15	6.40	48.75	22.50	2.90	0.80	13.50
Bauxite (Boké, Guinea)		0.35	63.40	1.80	3.50		31.20
Ball clay	0.20	26.40	59.80	1.00	1.40	0.50	7.90
Kaolin	0.03	47.65	36.75	0.75	0.22	0.22	12.80

1.3.2 Mineralogy

- The principal aluminium oxide is the native corundum (α-Al_2O_3). Sapphire and Ruby are gemstone varieties of corundum. Emery is a mixture of corundum and other oxides (Spinel, Hercynite).
- Aluminium hydroxide exists as trihydrate (Gibbsite, Bayerite, Nordstrandite) and mono-hydrates (Boehmite, Diaspore) found in bauxite.
- Aluminium exists as silicate in clay (Kaolinite, ball clay) but the industrial production of alumina derives from sources of aluminium oxide having variable chemical compositions (table 1.3).

1.3.3 Structure and Stability

- The crystallographic characteristics of the main aluminium oxide are shown in table 1.4:

    ○ α-Al_2O_3 crystallises in the rhombohedral system. The oxygen positions form a hexagonal compact packing (hcp) with the trivalent cations occupying 2/3 of the octahedral interstice.
    ○ Gibbsite (trihydrate) is rare in nature. It crystallises as monoclinic. Pure fine gibbsite is obtained by slow spontaneous hydrolysis of alkali aluminate at room temperature.
    ○ Bayerite (trihydrate) crystallises as orthorhombic. It is obtained by moderate slow carbonation of sodium aluminate at 30/35 °C.
    ○ Nordstrandite (trihydrate) crystallises as triclinic. It is obtained by precipitation of a gel from aluminium chloride by ammonium hydroxide at 27/49 °C.
    ○ Boehmite (mono hydrate) occurs in nature. It crystallises as orthorhombic. It is prepared by aging aluminium hydroxide gel or thermal dehydration of gibbsite.
    ○ Diaspore (monohydrate) occurs rarely in a well crystallised form but mixed with Flint. It crystallises as orthorhombic.
    ○ Polymorphs of Al_2O_3 form a group of phases called transition alumina (γ, δ, β, κ, θ) which fall between Boehmite and Corundum during thermal treatment.
    ○ The sequence of dehydration (transformation) of transition alumina from aluminium hydroxide is shown in table 1.5.

TAB. 1.4 – Crystallographic characteristics of the aluminium oxides.

Mineral JCPDS card N°	Chemical composition	Crystallographic lattice	Crystallographic sub-lattice, density (g/cm^3), formula units per crystal cell and volume of crystal cell (Å^3)	Crystal lattice parameters
Corundum 10 0173	α-Al$_2$O$_3$	Rhombohedral	$R3c$ $d = 3.98$ $Z = 2$ $V = 254$	$a = 7.58$ Å $c = 12.99$ Å $\alpha = 90°$ $\beta = 90°$ $\gamma = 120°$
Gibbsite 07 0324	α-Al$_2$O$_3$·3H$_2$O	Monoclinic	$P2_1/n$ $d = 2.4$ $Z = 4$ $V = 425.17$	$a = 8.655$ Å $b = 5.072$ Å $c = 9.716$ Å $\alpha = 90°$ $\beta = 94.6°$ $\gamma = 90°$
Bayerite 20 0011	β-Al$_2$O$_3$·3H$_2$O	Monoclinic	$P2_1/a$ $d = 2.5$ $Z = 4$ $V = 206$	$a = 5.06$ Å $b = 8.67$ Å $c = 4.71$ Å $\alpha = 90°$ $\beta = 90.27$ $\gamma = 90°$
Nordstrandite 018 0031	Al$_2$O$_3$·3H$_2$O	Triclinic	$P1$ $d = 2.21$ $Z = 8$ $V = 426$	$a = 8.89$ Å $b = 5.08$ Å $c = 10.237$ Å
Boehmite 21 1307	α-Al$_2$O$_3$·H$_2$O	Orthorhombic	$Cmcm$ $d = 3.03$ $Z = 2$ $V = 127$	$a = 3.78$ Å $b = 11.8$ Å $c = 2.85$ Å
Diaspore 05 0355	α-Al$_2$O$_3$·H$_2$O	Orthorhombic	$Pbnm$ $d = 3.4$ $Z = 4$ $V = 11.7$	$a = 4.4$ Å $b = 8.67$ Å $c = 2.84$ Å

1.3.4 Properties and Applications

- The main properties of Al$_2$O$_3$ are a good hardness, good thermal (high melting point) and insulating properties, resistance to chemical attack and the ability to sintering.

TAB. 1.5 – Occurrence of transition alumina.

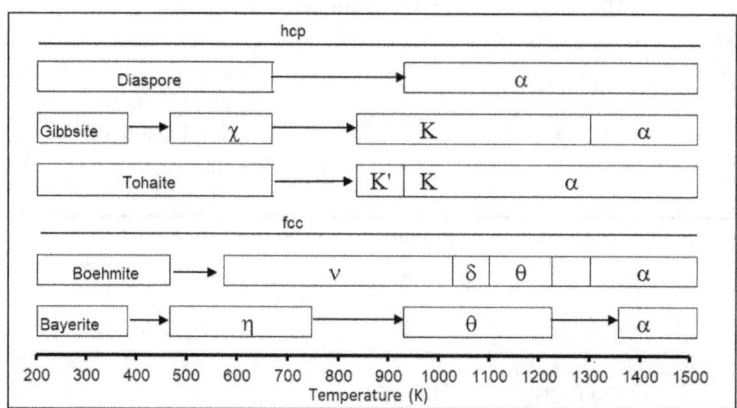

Key: hcp (hexagonal compact) and fcc (face-centred cubic) refer to the structure of the oxygen atoms sub-network.

1.3.4.1 First Level of Applications

- Al_2O_3 is used as technical ceramic devices due to its resistance to high temperatures (refractory, coating).
- Corundum is mainly used as an abrasive (mixed with other oxides: Emery).
- It is used for electrical devices (spark plug insulator) and also in jewellery (Sapphire).

1.3.4.2 Second Level of Applications

- Al_2O_3 is the precursor of the aluminium industry and is present in association with other oxides in ceramic, refractories, metallurgy, glass and ceramic industry.

1.4 Fe Oxides

MAH 2010

1.4.1 Introduction

- Iron is the fourth most abundant element on Earth's crust.
- Anhydrous iron oxides include; Feno, Fe_3O_4, five crystallized polymorphs of Fe_2O_3 (α, β, γ, ε, ζ) and amorphous Fe_2O_3 obtained at high pressure.
- Iron hydroxides include; $Fe(OH)_3$, $Fe(OH)_2$ and 5 polymorphs of FeOOH.
- Ferrihydrite is a hydrous oxyhydroxide mineral with the nominal formula Fe_5OH_8–$4H_2O$. It is poorly crystalline.

TAB. 1.6 – Typical composition of iron ore.

Oxide (weight %)	CaO	SiO$_2$	Al$_2$O$_3$	Fe$_2$O$_3$	MgO	LOI (%)
Iron ores	0.60	12.90	1.50	60.25	0.20	2.55

Key: LOI = loss on ignition.

- Iron is present at different degrees of oxidation FeO, Fe$_2$O$_3$, Fe$_3$O$_4$, Fe(OH)$_3$ and Fe(OH)$_2$.
- Typical composition of iron ore is given in table 1.6.

1.4.2 Mineralogy

- The mineral form of ferrous oxide (FeO) is known as Wüstite (named from Ewald Wüst, German metallurgist). It is a component in the mineralogy of Earth's lower mantle.
- Fe$_3$O$_4$ mineral form is Magnetite. It is found in large quantity in beach sand carried to the beach via rivers from erosion and concentrates via wave action and currents. Small grains occur in almost all igneous and metamorphic rocks.
- Fe$_2$O$_3$ (ferric oxide) is the main iron ore. Its mineral α form is Hematite that is found in magmatic, hydrothermal metamorphic and sedimentary rocks. Fe$_2$O$_3$ is paramagnetic. Maghemite is the γ form of Fe$_2$O$_3$.
- The minerals Goethite (α, β), Lepidocrocite (γ), and Akaganeite are polymorphs of FeO(OH).
- Ferrihydrate forms in several types of environment from freshwater, marine systems, aquifer to hydrothermal hot springs and soil affected by mining. Different formulas are found: Fe$_5$OH$_8$ 4H$_2$O, Fe$_5$(O$_4$H$_3$)$_3$, 5Fe$_2$O$_3$·8H$_2$O, Fe$_2$O$_3$·2FeOOH$_2$·6H$_2$O.

1.4.3 Structure and Stability

- The crystallographic characteristics of Fe oxides are summarized in table 1.7.

1.4.3.1 FeO

- Wüstite is a complex non-stoichiometric oxide of iron with the general structural form $[Fe^{2+}{}_{1-3x}\ Fe^{3+}{}_{2x-t}\ \gamma FeO_{x+t}]^{IV}Fe^{3+}{}_tO$ where $0.04 < x < 0.12$ and $0 < (x+t))/t < 4.5$.
- It crystallizes in the fcc crystal system, containing 4 formula units in the cubic unit cell. O$_2$ anions form a close packet fcc sub-lattice with small Fe^{2+} cations in the interstitial site. All irons are octahedrally coordinated to oxygen. Under thermal equilibrium this phase is stable only at low pressure and temperature 570 °C.
- The composition ranges from Fe$_{0.84}$O to Fe$_{0.95}$O. This non stoichiometric occurs because of the ease of oxidation of Fe^{II} to Fe^{III} effectively replacing a small

TAB. 1.7 – Crystallographic characteristics of Fe oxides.

Composition	Minerals	Crystallographic lattice and sub-lattice	Crystal lattice parameters	Formula units per crystal cell	Density (g/cm³)	Volume of crystal cell (Å³)
FeO	Wüstite JCPDS 06-0615	Cubic $Fm\text{-}3m$	$a = 4.302$	$Z = 4$	$d = 5.74$	$V = 79.9$
$\alpha\text{-}Fe_2O_3$ JCPDS 1309-37-1	Hematite JCPDS 33-0664	Trigonal $R\text{-}3c$	$a = 5.034$ $c = 13.75$ $\alpha = 90°$ $\beta = 90°$ $\gamma = 120°$	$Z = 6$	$d = 5.2$	$V = 301$
$\gamma\text{-}Fe_2O_3$	Maghemite JCPDS 39-1346	Tetragonal $P4_132$	$a = 8.34$ $c = 24.99$	$Z = 32$	$d = 3.85$	$V = 1734$
Fe_3O_4	Magnetite JCPDS 19 0619	Cubic $Fm\text{-}3d$	$a = 8.39$	$Z = 8$	$d = 5.17$	$V = 591$
$\alpha\text{-}FeOOH$	Goethite JCPDS 27-0713	Orthorhombic $Pbnm$	$a = 4.58$ $b = 9.93$ $c = 3.015$	$Z = 4$	$d = 4.28$	$V = 138.62$
$\gamma\text{-}FeOOH$	Lepidocrocite	Orthorhombic $Cmcm$	$a = 3.87$ $b = 12.51$ $c = 3.06$	$Z = 4$	$d = 4.09$	

proportion of Fe^{II} with 2/3 of their number which takes up tetrahedral positions in the close packed oxide lattice.

1.4.3.2 Fe_2O_3

- Six polymorphs are known:
  - α-Fe_2O_3 crystallizes with rhombohedral structure similar to that of Corundum and consists of a dense arrangement of Fe^{3+} ions in octahedral coordination with oxygen in hexagonal close packing. The structure can be described as a stacking of sheets of octahedral coordinated Fe^{3+} ions between two close-packed layers of oxygen ions.
  - β crystallizes with fcc structure and transforms into the α phase above 500 °C.
  - γ phase (Maghemite) crystallizes as cubic and converts to α at high temperatures.
  - ε phase crystallizes with rhombohedral structure. It has not been prepared in pure form. It transforms to the α phase between 500 and 750 °C.
  - Amorphous Fe_2O_3 exists at high pressure.
  - ζ is a polymorph of Fe_2O_3 formed at 30 GPa and crystallised with monoclinic structure.

1.4.3.3 Fe_3O_4

- Magnetite mineral contains Fe(II) and Fe(III). In stoichiometric magnetite Fe(II)/Fe(III) = 0.5. The crystal structure of magnetite is inverse spinel with a cell consisting of 32 oxygen atoms in an fcc structure. Fe(II) and half of Fe(III) occupy octahedral sites and the other half Fe(III) occupies tetrahedral sites.
- It exhibits permanent magnetism and is ferromagnetic. The ferrimagnetism of Fe_3O_4 arises because the electron spins of Fe^{II} and Fe^{III} in the octahedral sites are coupled. The spins of the Fe^{III} in the tetrahedral sites are also coupled but anti-parallel to the former with net effect that the magnetic contribution of each is unbalanced causing permanent magnetism.

1.4.3.4 FeOOH

- Goethite (α-FeO(OH)) exhibits an orthorhombic symmetry. The structure is described in three dimensional terms as built up with $FeO_3(OH)_3$ octahedra forming large tunnels where hydrogen atoms are located. Each octahedron is linked to eight neighbours by four edges and three vertices. Oxygen atoms are in tetrahedral surroundings either with OFe_3H or OFe_3H bond.
- Lepidocrocite (γ-FeO(OH)) has a double layer Fe octahedral structure with the hydroxyl group located on the external surface providing hydrogen bonding between these layers.

- Akaganeite (β-FeO(OH)) has a large tunnel type structure where iron atoms are strongly bound to the framework.

1.4.4 Synthesis

- The most common ways to synthesize iron oxides or hydroxides are; chemical precipitation, sol gel methods, surfactant mediated precipitation, emulsion precipitation, micro-emulsion, electro deposition and micro wave assisted hydrothermal technique.

1.4.5 Properties and Applications

- The main useful characteristics of Fe oxides are catalytic and sorbent properties and they are used as pigments, flocculants, coatings, ion exchanger and lubricants.

1.4.5.1 The First Level of Applications

- FeO is used as a pigment and used in cosmetics. Fe_3O_4 is used as a black pigment.
- The iron oxides such as Magnetite, Hematite or Maghemite are used as pigments for black, red and brown colours, respectively, in the construction industry (roof tiles, paving slabs...).
- Iron oxide based materials are efficient catalysts. They are used as catalyst in the Haber process and in the water gas shift reaction.
- For oxidation–reduction and acid-base reactions, the most applied iron oxides as catalysts are Magnetite and Hematite.
- γ-Fe_2O_3 exhibits good sensing characteristics towards hydrocarbon gases, carbon monoxide and alcohol.
- Many toxic cations (Zn, Pb...) or anions (CrO_4^{2-}) are removed using various phases containing iron oxide. The high specific surfaces and surface charge of iron oxide regulate free metal and organic matter concentration.
- γ-Fe_2O_3 as nano-ferromagnetic particles is used as a magnetic storage media in audio and video recording and magnetic optical devices.

1.4.5.2 The Second Level of Applications

- Use as raw feedstock of the steel industry.
- Analysis of iron ore is given in table 1.6.
- Iron oxides find application in magnetic recording, magnetic storage devices, toner and ink, magnetic resonance imaging, wastewater treatment, bioseparation and medicine.

- Nano particles of Fe_3O_4 are used as contrast agent in magnetic resonance imaging (MRI).
- In therapeutic applications, iron based magnetic nanoparticles are used for hyperthermia and drug delivery. In diagnostic application, they are used in nuclear magnetic resonance imaging. In *in vitro* applications, the main use is in diagnostic application (separation/selection and magneto-relaxometry).

Chapter 2

Binary Chemical Systems

2.1 CaO–SiO$_2$ system

2.1.1 Introduction

- The CaO–SiO$_2$ system includes five crystallized compounds; Ca$_3$SiO$_5$, Ca$_2$SiO$_4$, Ca$_3$Si$_2$O$_7$, CaSiO$_3$, CaSi$_2$O$_5$ (short hand cement notation written C$_3$S, C$_2$S, C$_3$S$_2$, CS, CS$_2$) – and amorphous state compositions (gel, glass, etc....).

2.1.2 Mineralogy

- Tricalcium silicate (C$_3$S) is a rare mineral in the nature. It is found as Hatrurite in a sedimentary rock sequence of cenomanian to Eocene age in Hartrurim (Israel).
- Two minerals close to dicalcium silicate (C$_2$S) are found from the limestone contact zone of Scawth Hill (Northern Ireland); Larnite is structurally close to the polymorph β-C$_2$S and Bredigite (Ca$_{1.75}$Mg$_{0.25}$O$_4$) has been recognised to be chemically and structurally different from the polymorph α'-C$_2$S.
- Calcium metasilicate (CS) naturally occurs as Wollastonite in different countries, named after the English chemist and mineralogist W.H. Wollaston. There are two polymorphs; Wollastonite is the low temperature form and Pseudowollastonite (Bourgeoisite) is the high temperature mineral. With increasing pressure Wollastonite transforms to the Walstromite structure and further dissociates into Larnite (Ca$_2$SiO$_4$) and Titanite (CaSi$_2$O$_5$). Titanite is found as inclusion in Kimberlite diamonds (JOS 1999). It is found when there is a reaction between limestone and intrusive igneous rocks.

DOI: 10.1051/978-2-7598-2480-9.c002
© Science Press, EDP Sciences, 2020

- Two polymorphs of tricalcium disilicate (C_3S_2), Rankinite and Kilkoanite, exist in the nature (Scawth Hill – Northern Ireland, Tokatoka – New-Zealand) (BLA 1969).
- The conditions for the formation of CS_2 (high pressure (10 GPa) and temperature (1500 °C) are similar to those found in specific parts of the earth's crust mantel and could possibly be found in these areas (ANG 1996).

2.1.3 Structure, Polymorphism and Solid Solution

2.1.3.1 Ca_3SiO_5 (C_3S)

2.1.3.1.1 Structure and Polymorphism

- During heating, pure C_3S exhibits 7 polymorphs forms (triclinic –T_1, T_2, T_3–, monoclinic –M_1, M_2, M_3– and rhombohedral R (table 2.1) whose range of stability as a function of temperature is described according to the following sequence:

  T1 ← 620 °C → T2 ← 920 °C → T3 ← 980 °C → M1 ← 990 °C → M2 ← 1060 °C → M3 ← 1070 °C → R.

  ○ They have been identified by thermal analysis, microscopic techniques, X-ray and neutron diffraction.
  ○ Because of the difficulty to determine the unit cell of complex structure by powder X-ray methods, a structural model has been established for single crystal of T_1, M_3 and R stabilized by Mg, Sr, but no structural model is available for T_2, T_3 and M_2.
  ○ The structure is built from isolated SiO_4 tetrahedra, calcium ions and additional oxygen atoms, the latter being coordinated around to Ca only. C_3S contains 3 non-equivalent Ca^{2+} sites. The oxygen coordination around calcium is very irregular and the structure contains voids.
  ○ C_3S undergoes numerous small amplitude transformations around a constant atomic arrangement. All the transformations are displacive: they are produced by small movement of the atoms from their position in the original structure without any alteration of the general arrangement of the chemical bonds.
  ○ The structural difference between the polymorphs affects the coordination of the Ca^{2+} ions and the oxygen atoms of the SiO_4 tetrahedra.
  ○ It is not possible to stabilise the polymorphic form uniquely by quenching. The stabilisation requires the addition of foreign elements (table 2.2).
  ○ It is worth to mention that the polymorphism of C_3S exists in the range of temperature where C_3S is not stable.

2.1.3.1.2 Solid Solution

- The presence of impurities (other elements) in C_3S structure allows the stabilization of polymorphic form, modifies the percentage of solid solution in a polyphasic mixture, as well as its reactivity.

TAB. 2.1 – Crystallographic characteristics of C_3S polymorphs.

	R (SC)	M3 (SC)	M2 (P)	M1 (P)	T3 (P)	T2 (P)	T1 (SC)
Lattice	Rhombohedral	Monoclinic	Monoclinic	Monoclinic	Triclinic	Triclinic	Triclinic
Sp. G	$R3m$	Cm	Cm	Cm	$C1$	$C1$	$P\text{-}1$
Z	9				18		18
a (Å)	12.124	12.242	12.342	12.332	11.7	24.528	11.67
b (Å)	7	7.027	7.143	7.142	14.29	14.27	12.24
c (Å)	27	9.307	25.434	25.42	13.72	25.298	13.72
V (Å^3)	2122	723	2242	2167	2190	8854	2190
d (g/cm^3)	3.168						
JCPDS	73-599	49-442			70-1846		
Ref	ILL 1985	MUM 1995	NIS 1985	NOI 2003	BIG 1967	B-G 1967	GOL 1975

Key: R for powder, SC for single crystal diffraction.

TAB. 2.2 – Effect of other minor elements on the stability of C_3S polymorphs.

R (SC) Rhombohedral 1070 °C	4.5/5% ZnO 1.8 Al_2O_3 0.5 Sr	T3 (P) Triclinic 920 °C	1% ZnO
M3 (SC) Monoclinic 1060 °C	$MgO \cdot Al_2O_3$ 0.5% ZnO 2% MgO 1.02A + 2.18M + 0.09S 0.9A + 0.5F + 0.6M	T2 (P) Triclinic 620 °C	0.8/1.8 ZnO 0.47/1.3% MgO Na_2O/K_2O 0.454/1% Al_2O_3 0.9/1.1% Fe_2O_3 0.09% BaO
M2 (P) Monoclinic 990 °C	1.13A + 1.08F + 1.06M + 0.07S 2.2/4.5% ZnO	T1 (SC) Triclinic 20 °C	0/0.8% ZnO 0/0.47% MgO Na_2O/K_2O
M1 (P) Monoclinic 980 °C	1.8/2.2% ZnO 1.3/1.80% MgO 1.05A + 1.05F + 0.99M + 1.55S 0.5A + 0.5F + 0.2M 0.96M + 4$CaSO_4$		0/0.50% Al_2O_3 0/0.85% Fe_2O_3 0/1.4% Cr_2O_3

Key: A = Al_2O_3, M = MgO, F = Fe_2O_3 and S = SiO_2.

- Many minor elements can enter the structure of C_3S and most of the studies aim at the synthesis of alite (the solid solution of C_3S found in Portland cement).
- The percentage of minor elements depends on the temperature of reaction, rate of cooling and their amount.
- Fe^{2+} substitutes Ca^{2+}, while the most plausible distribution of Al^{3+} and Fe^{3+} involves their distribution over the 3 sites. One set of sites normally occupied by Ca, one set by Si and a third set which is normally occupied in pure C_3S.
- A general formula involving Me = Mg/Zn can be written as follows:

$$(Ca_{(1-y-x/6)}(Me)_y(Fe,Al)_{x/6})_{3-z}{}^{VI}(Fe,Al)_z{}^{VI'}Si_{1-x/2}(Fe,Al)_{x/2}{}^{IV}O_5.$$

- Both Al_2O_3 and Fe_2O_3 accelerate the decomposition of C_3S solid solution at 1180 °C; if MgO is present the effect is neutralised, but the mechanism of stabilization of the polymorphic forms remains unclear.

2.1.3.2 Ca_2SiO_4 (C_2S)

2.1.3.2.1 Structure and Polymorphism

- Pure C_2S exhibits 5 polymorphs; hexagonal (α), orthorhombic (α'_L, α'_H and γ) and monoclinic (β) (table 2.3).
- The polymorphic forms of C_2S have been identified by thermal and microscopic techniques, high temperature XRD, electron diffraction, IR and Raman spectroscopy and NMR.

TAB. 2.3 – Crystallographic characteristics of C_2S polymorphs.

C_2S	γ	β	$α'_L$	$α'_H$	α
Lattice	Orthorhombic	Monoclinic	Orthorhombic	Orthorhombic	Hexagonal
Sp. G	$Pbnm$	$P2_1/n11$	$Pn2_1a$	$Pnma$	$P6_3/mmc$
Z	4	4	16	4	2
a (Å)	5.124/5.076	5.502/5.558	10.8/11.18	5.605/5.53	5.416/5.525
b (Å)	6.756/6.782	6.753/6.684	18.42/18.98	9.41/9.534	
c (Å)	11.371/11.224	9.28/9.365	6.76/6.85	6.81/6.86	6.76/7.311
V (Å^3)	385	342	1449	365	178
d (g/cm^3)	2.96	3.33	3.158	3.127	3.2
JCPDS	49-1672	29-0371	31-299	31-298	23-1042
Ref	CZA 1981	JOS 1977	REG 1979	NIE 1972	YAM 1963

- The polymorph γ is stable at temperatures below 870 °C. During heating, it transforms into $α'_L$-C_2S and the temperature stability range of the polymorphic form is described according to the following sequence:

γ-C_2S → 870 °C $α'_L$-C_2S → 1160 °C $α'_H$-C_2S → 1425 °C α-C_2S → 1538 °C-liquid.

- During cooling, the polymorph β-C_2S occurs. It is metastable and crystallises as monoclinic:

α-C_2S → 1425 °C $α'_H$-C_2S → 1160 °C $α'_L$-C_2S → 672 °C β-C_2S → 500 °C γ-C_2S.

- While for C_3S, all the forms derive from a single structure by slight variations, nevertheless, the five structures of C_2S are distinct as depicted by DTA and XRD.
- It is not possible to stabilise α and α' polymorphs without the addition of foreign elements.
- Structural determinations of the polymorphs have been carried out on single crystal solid solution of C_2S which can be considered as isostructural with high temperature modifications.
- However, the C_2S solid solution, quenched from elevated temperatures, does not necessarily give exactly the same unit cell dimensions as corresponding high temperature modifications of C_2S.
- The structure of the C_2S polymorphs has been described by single crystal measurement. Table 2.4 gives examples of foreign elements used to stabilize different polymorphic forms.
- α-C_2S forms at 1425 °C and melts at 2130 °C. It crystallizes as hexagonal and can be stabilized by V_2O_3 and $Na_4P_2O_7$. Its structure has not been fully elucidated and the proposed space groups are $P6_3/mmc$ or $6_3/mc$ that differs from the orientations of SiO_4 tetrahedra.
- α'-C_2S crystallizes between 830 and 1447 °C by transformation of α-C_2S.

TAB. 2.4 – Effect of foreign elements on the stability of C_2S polymorphs.

γ	β	α'm
Orthorhombic	Monoclinic	Orthorhombic
0.5 FeO (BEN 1979)	0.25/0.5 Cr_2O_3 (MUM 1995)	3.6% K_2O (MID 1971)
GeO_2 (EYS 1970)	B_2O_3 (MID 1974)	15/25% $CaNaPO_4$ (GUI 1968)
$α'_L$	$α'_H$	α
Orthorhombic	Orthorhombic	Hexagonal
SrO, BaO·SrO (SUZ 1968)	10% $CaMgSiO_4$ + K_2O (NIE 1972)	V_2O_5 (SAA 1967)
7.2 C_3P (SAA 1967)		$Na_4P_2O_5$ (HAH 1968)
$CaCl_2$ (YAM 1966)		2.5A + 2.5F + 6N (MID 1952)
		18% C_3P (MUL 2003)

- $α'_H$-C_2S is orthorhombic and it has a structure similar to β-K_2SO_4 with space group *Pnma*.
- $α'_L$-C_2S is orthorhombic and can be stabilised by Sr, Ba, K, Na, P. In the presence of $CaCl_2$, it shows a super structure based on $α'_H$-C_2S.
- β-C_2S is metastable below 670 °C. Stabilized by B_2O_3 (0.5%), it crystallizes with monoclinic crystal structure, with independent SiO_4 tetrahedra and 2 sorts of Ca atoms; 4 of 8 Ca atoms are placed alternatively above and below the SiO_4 tetrahedra in the direction of the *b* axis.
- BAR 1980 found no significant evidence to show that the β-angle of unit cell of β-C_2S changes gradually on heating and cooling to the step change that is the β- to $α'_L$-C_2S transition when the structure changes from monoclinic to orthorhombic symmetry.
- γ-C_2S forms at 830 °C and crystallises as orthorhombic structure identical to Olivine (Mg_2SiO_4). The positions of SiO_4 tetrahedra and half the Ca atoms are similar to those of β-C_2S. The remaining Ca atoms are surrounded by corresponding tetrahedra but displaced. The SiO_4 tetrahedra are irregular probably due to the distortion in the hexagonal arrangement of the oxygen ions. All Ca atoms are octahedrally coordinated. The low coordination of Ca explains the greater molar volume of the low temperature phase.
- The polymorphic transition of the β to γ form involves a great increase in specific volume. The β form can be stabilized either by adding small amounts of specific elements (B_2O_3 is the most frequently used) or by a control of the crystal growth during thermal transformation.
- The ways of suppressing this transition β to γ is achieved by the control of the growth of β crystals (as follows):

    (1) Reheating γ-C_2S.
    (2) Addition of excess of lime.

Binary Chemical Systems

2.1.3.2.2 Solid Solutions

- The effect of iron on α- to β-C_2S synthesis was investigated by Mössbauer and XRD. It was confirmed that they include only Fe^{3+} at octahedral and tetrahedral sites with the ratio 30/70. In β-C_2S, Fe^{3+} was mostly situated at tetrahedral sites.
- In coexistence of Na or K with Fe, β-C_2S is stabilized:

$$\alpha\text{-}(Ca_{1.88}\ Fe_{0.05}\ Na_{0.24})\ (Si_{0.88}\ Fe_{0.11})\ O_4.$$
$$\alpha\text{-}(Ca_{1.94}\ Fe_{0.09}\ K_{0.18})\ (Si_{0.88}\ Fe_{0.05})\ O_4.$$
$$\beta\text{-}(Ca_{1.93}\ Fe_{0.03}\ Na_{0.04})\ (Si_{0.99}\ Fe_{0.05})\ O_4.$$
$$\beta\text{-}(Ca_{1.94}\ Fe_{0.01}\ K_{0.04})\ (Si_{0.92}\ Fe_{0.13})\ O_4.$$

- The study by ^{29}Si NMR spectroscopy of the substitution of Al_2O_3 shows that the structural disorder brought in the C_2S lattice by partial replacement of Si^{4+} by Al^{3+} can be described as a continuous distribution of the SiO_4 tetrahedra geometry with the mean geometry being identical to that which occurs in pure β-C_2S.
- C_2S and C_3P form a wide range of solid solutions with a modification of the polymorphism form of C_2S.

2.1.3.3 $Ca_3Si_2O_5$ (C_3S_2)

2.1.3.3.1 Structure and Polymorphism

- Pure C_3S_2 exhibits two polymorphs, Rankinite and Kilchoanite, and both sometimes occur together with Rankinite retrograde replacement by Kilchoanite.
- Rankinite crystallises as monoclinic and Kilchoanite as orthorhombic. It contains isolated SiO_4^{4-} tetrahedra and isolated $Si_3O_{10}^{8-}$ groups (TAY 1971) (table 2.5).
- Rankinite formed when Afwillite is heated at 1000 °C is monoclinic (AGR 1961, KUS 1975, MOO 1952).
- Synthetic Kilchoanite is obtained under hydrothermal conditions (1100 bars, 700 °C). A single crystal of Kilchoanite inverted to a polycrystal of Rankinite at 1000 °C.
- Kilchoanite can be considered as the lower temperature form of Rankinite. Their equilibrium boundary lies on 810 °C at 2000 psi and 825 °C at 21 000 psi.

2.1.3.4 $CaSiO_3$ (CS)

2.1.3.4.1 Structure and Polymorphism

- Pure CS exhibits two polymorphs at ordinary pressure (table 2.5):

  (1) Low temperature Wollastonite (named β in cement technology) occurs at least as three polytypes related by a simple packing modification: ordered triclinic (Wollastonite 1 T_C) that is the most common natural mineral, ordered monoclinic (Parawollastonite or Wollastonite 2 M) and disordered Wollastonite.

TAB. 2.5 – Crystallographic characteristics of CS and C_3S_2 polymorphs.

Composition	β-CaSiO$_3$	β-CaSiO$_3$	α-CaSiO$_3$	Ca$_3$Si$_2$O$_7$	Ca$_3$Si$_2$O$_7$
Oxide formula	CaO·SiO$_2$	CaO·SiO$_2$	CaO·SiO$_2$	3CaO·2SiO$_2$	3CaO·2SiO$_2$
Cement formula	β-CS	β-CS	α-CS	C_3S_2	C_3S_2
Mineral	Parawollastonite b	Native	Pseudowollastonite (Bourgeoisite)	Rankinite	Kilchoanite
Lattice	Monoclinic	Triclinic	Triclinic	Monoclinic	Orthorhombic
SP. G	$P2_1/a$	C-1	C-1	$P2_1/a$	$Imam$ or $Ima2$
Z	12	12	8	4	4
a (Å)	15.409/15.429	7.9258/7.894	6.82/6.9	10.55	11.42
b (Å)	7.27/7.327	7.27/7.371	6.82/11.95	8.88	5.09
c (Å)	7.063/7.07	7.037/7.07	19.65/19.67	7.85	21.95
A		90°02/90°06	90°40/90°12		
B		95°22/95°32	90°40/90°55	120°1	
γ	95.4	102°93/103°43	119°30/90		
d (g/cm^3)	2.910	2.87/3.09	2.905	2.887	2.980
V (Å^3)	795	394	801	641	1668
JCPDS N°	43-1460	42-547	74-874	22-539	29-368
Ref	KEN 2001	SMI 1965	YAM 1957	AGR 1961	TAY 1971

(2) High temperature Wollastonite, commonly known as Pseudowollastonite (Bourgeoisite or α in cement notation), is the most common synthetic calcium metasilicate. Transition occurs at 1125 °C. The structure of low temperature Wollastonite has an infinite chain of SiO_4 tetrahedra parallel to the b-axis (as it is expected from its fibrous habit). The repeating tetrahedra in the chain consists of a pair of tetrahedra joined apex to apex as in the Si_2O_7 group alternating with a single tetrahedron with one edge parallel to the chain direction. They are linked by calcium in irregular octahedral coordination with 6 oxygen atoms.

- The structure of the Pseudowollastonite is composed of Si_3O_9 rings.
- With increasing pressure, Wollastonite transforms to have a structure similar to that of barium calcium silicate, Walstromite and then dissociates to $Ca_2SiO_4 + CaSi_2O_5$ (Titanite), and finally a cubic Perovskite modification of $CaSiO_3$ can be formed above 150 kbars.
- The crystal structure of pure calcium silicate version of Walstromite is almost identical to the Walstromite ($Ca_2BaSi_3O_9$) formed by Si_3O_9 ring) (JOS 2003).

2.1.3.4.2 Solid Solutions

- Wollastonite is able to accept a limited percentage of iron as solid solution. Fe^{2+} cannot substitute more than about 10% of Ca^{2+}.

2.1.4 Stability and Phase Diagram of CaO–SiO₂ System

2.1.4.1 Experimental Phase Diagram

- The principal experimentally or by modelling works dealing with the system $CaO-SiO_2$ are given by table 2.6.
- Experimental data of the liquidus temperatures are represented in the diagram redrawn from literature data given in references (figure 2.1):

  ○ CaO, C_3S, C_2S, C_3S_2, CS and silica occur as primary phases at ordinary pressure.
  ○ The melting temperature of CaO is given between 2617 and 2897 °C.
  ○ Pure C_3S is a stable compound between 1250 and 2150 °C. Below 1250 °C, it decomposes into C_2S and CaO. Above 2150 °C, it decomposes into liquid and CaO.
  ○ C_2S melts congruently at 2130 °C.
  ○ C_3S_2 decomposes incongruently at 1464/1474 °C.
  ○ CS melts congruently at 1544 °C (ADA 1980).
  ○ Silica rich melts exhibit a broad liquid immiscibility region. The miscibility gap over the SiO_2 rich composition liquidus curves shown as dashed and dot dash lines in figure 2.1 is from GRE 1927–1, HAG 1986, OLS 1951, TEW 1979.

TAB. 2.6 – Principal works dealing experimentally or by modelling with the system CaO–SiO$_2$.

Experimental works	Modelling works
Rankin et al. (RAN 1915)	Laptev (LAP 1970)
Greig (GRE 1927–1)	Kaestle (KAE 1976)
Carlson (CAR 1931)	Kaufman (KAU 1979)
Olshanski (OLS 1951)	Byker et al. (BYK 1981)
Roy (ROY 1958)	Lee (LEE 1982)
Welch et al. (WEL 1959)	Berman et al. (BER 1984)
Phillips et al. (PHI 1959)	Pelton et al. (PEL 1986)
Trömel (TRO 1949)	Hillert et al. (HIL 1990–2, HIL 1991)
Obst et al. (OBS 1970)	Taylor et al. (TAY 1993)
Hanic et al. (HAN 1987)	Erikson et al. (ERI 1993)
Tewhey et al. (TEW 1979)	Huang et al. (HUA 1995)
	Zaitsev et al. (ZAI 1997)
	De Capitani et al. (DEC 1998)
	Pytel et al. (PYT 1999)

2.1.4.2 Thermodynamic Approach of the Phase Diagram

- Heat capacities and standard state properties (enthalpy, entropy and Gibbs energy) are summarized in table 2.7.
- Calculation of the phase diagram is based on a system of semi-empirical equations that describe the excess free energy of the melt (subregular solution, margules...). These equations were coupled with an 'optimization computer program' to analyse available thermodynamic data (Gibbs energy, heat of formation of compounds, activities, enthalpies of mixing, entropy of fusion).
- The most striking feature is the break in properties at the orthosilicate composition of C$_2$S, which divides the region of acidic and basic liquids (slag).
- Six-term polynomial regressions are incapable of fitting both the large region of small curvature and the break near C$_2$S. This forced us to use separate sets of Redlich–Kister coefficient for the acidic and basic regions.
- The calculated phase diagram of the binary CaO–SiO$_2$ join is shown in figure 2.2. The diagram includes C$_3$S, C$_2$S and CS compounds in agreement with experimental diagrams.
- In the diagram pressure/temperature, two triple points occur:

  (1) Liquid + CS Perovskite + C$_2$S + CS$_2$ (13.5 GPa, 2150 °C).
  (2) Liquid + CS Walstromite + C$_2$S + CS$_2$ (10.4 GPa, 1800 °C).

- The decomposition of CS Walstromite to Ca$_2$SiO$_4$ and CaSi$_2$O$_5$ is described by the equation P = 9 + 0.0021 × T°C.
- The boundary between C$_2$S + Titanite and CS Perovskite has been calculated (slope of 1.3/1.8 MPa/K).

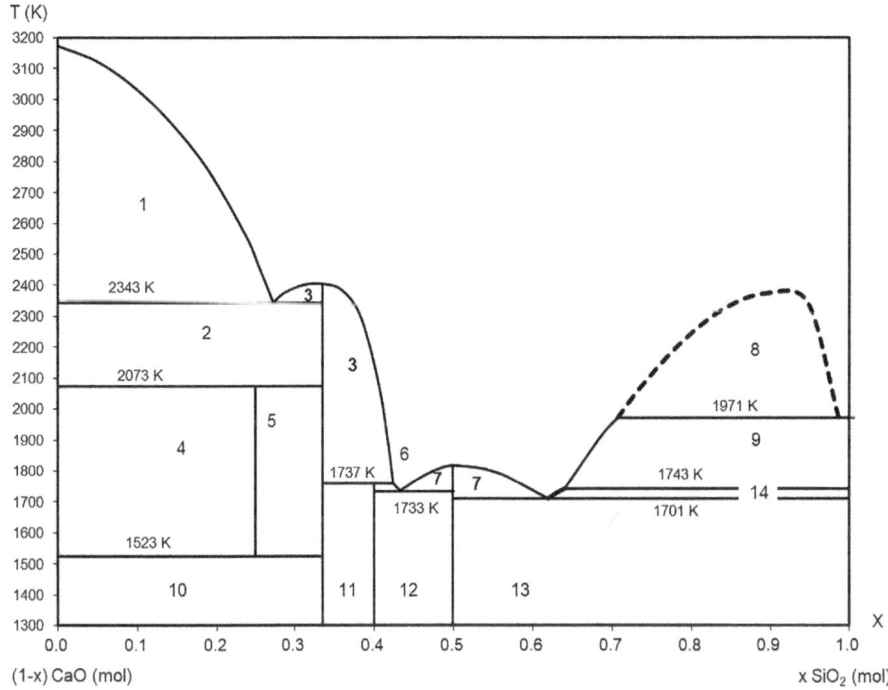

FIG. 2.1 – Experimental phase diagram of CaO–SiO$_2$ system. Key: (1) CaO + liquid, (2) CaO + Ca$_3$SiO$_5$, (3) Liquid + Ca$_2$SiO$_4$, (4) CaO + Ca$_3$SiO$_5$, (5) Ca$_3$SiO$_5$ + Ca$_2$SiO$_4$, (6) Liquid + Ca$_3$Si$_2$O$_7$, (7) Liquid + CaSiO$_3$, (8) two liquids, (9) Liquid + Tridymite, (10) CaO + Ca$_2$SiO$_4$, (11) Ca$_2$SiO$_4$, Ca$_3$Si$_2$O$_7$, (12) Ca$_3$Si$_2$O$_7$ + CaSiO$_3$, (13) SiO$_2$ + CaSiO$_3$, (14) Tridymite + Silica.

2.1.5 Synthesis of Calcium Silicates

2.1.5.1 General Features

- C$_3$S and C$_2$S (usually with other phases formed at the same time, such as; tricalcium aluminate and calcium alumina-ferrite) are produced industrially as hydraulic binders.
- Wollastonite, both natural (after beneficiation) and synthetic, is used in many industries.
- Different types of research and applications depend on the synthesis and production of pure calcium silicates or their solid solutions, for:
  (1) Identification and characterization of the pure products, such as; C$_3$S, C$_2$S, CS and their polymorphs for analysis.
  (2) Crystal chemistry and derivation of the mechanism of substitution of minor elements and their impact on reactivity.

TAB. 2.7 – Thermodynamic data of the CaO–SiO$_2$ system.

	H 298	S 298	ΔG 298	a	b
CaO·SiO$_2$	−81 928	2	−91 282	44.085	0.1797
	−95 001	3	−106 345		
C$_3$S$_2$	−206 000	9	−237 081	140.9	0.3395
	−246 524	15	−268 614		
β-C$_2$S	−125 166	6	−128 075	91.04	0.1763
	−136 686	11	−170 468		
α-C$_2$S			−138 244		
α'-C$_2$S			−126 685		
			−134 573		
γ-C$_2$S	−136 690				
	−144 113				
C$_3$S	−112 860	8	−115 034	103.17	0.293
	−140 480	17	−135 881		

Key: enthalpy, entropy and Gibbs' free energy are given in J mol^{-1}.
The molar specific heat at constant pressure Cp can be calculated at a given temperature (T in K) with the following equation; $Cp = a + bT + 10^{-3}\, T^2 + 10^{-8}\, T^3$ (J K^{-1} mol^{-1}).

- (3) Measurement of their thermodynamics characteristics (*i.e.* heat of formation) in order to calculate the equilibrium with other phases.
- (4) Evaluation of the rate of formation (activation energy).
- (5) Calibration of equipment (XRD, SEM...) for further analysis in mixtures.

- The sources of SiO$_2$ are quartz, flint, amorphous SiO$_2$, silica gel, colloidal silica (ludox) and ethyl silicate (TEOS).
- The sources of calcium oxide are calcium carbonate, hydroxide, oxalate, nitrate and alcoholate.
- Minor modifier elements such as Mg, Zn, Na, K, Al, Fe, Cr, Mn and Zr (stabilizers) are introduced to form the desired polymorphs or to facilitate the reaction of combinations (mineralisers).
- Parameters of the process are: the mixing time and energy, heating and cooling conditions (T°, rate of heating and cooling, residence time, atmosphere), intermediate grinding, simultaneous grinding and heating, the quality of the vessel (platinum and rhodium, refractory).
- The final product is usually quality assured by XRD, SEM and the measurement of un-combined calcium oxide.
- The methods of synthesis can be divided into two main groups:
  (1) Direct sintering from oxides or salts.
  (2) Indirect synthesis from precursors (salts, hydrates).

2.1.5.2 Direct or Solid-State Sintering

- Calcium salts (carbonate, oxalate, sulphate, nitrate...) and silica are mixed and heated at the appropriate high temperature (1450–1650 °C). If the total reaction

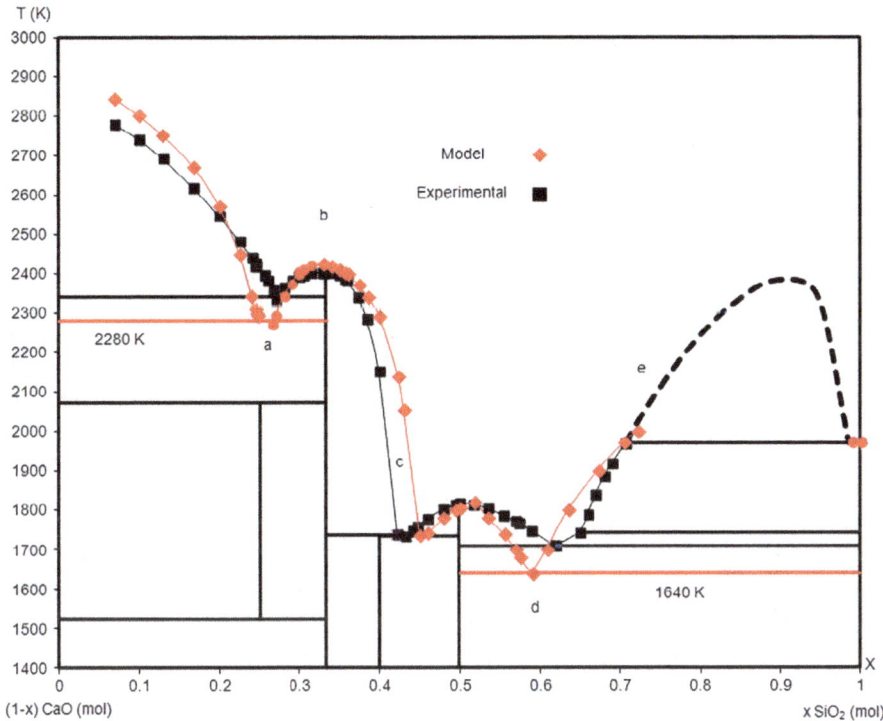

FIG. 2.2 – Experimental (black curve) and calculated (red curve) phase diagram of CaO–SiO$_2$ system.

is not achieved (indicated by the measurement of the un-combined free lime (CaO)), the product is ground and reheated at high temperatures. The operation is repeated until the amount of free lime is reduced to the desired low level.
- The solid-state reactions take place at high temperatures and are controlled by the diffusion rate of the chemical species moving to the phase boundaries. Two techniques are used to study the kinetics of the reaction between CaO and SiO$_2$; the phase boundaries can be created between 2 pellets of CaO/SiO$_2$, CaO/C$_2$S, C$_3$S C$_3$S$_2$, C$_2$S/CS, C$_3$S$_2$/SiO$_2$ and CS/SiO$_2$ or between fine particles of CaO and SiO$_2$. In the first technique, it is difficult to define the boundary between the pellets. In the second case, it is difficult to identify precisely the reacted compounds (fineness). The rate of calcium salt decomposition, rate and temperature of the occurrence of calcium silicate, and time to reach a given percentage of reaction are the main determinations (CHE 1989, GHO 1979).
- C$_2$S forms directly from lime and silica without intermediate compounds and its formation kinetics between 950 and 1150 °C is described by the Ginstling-Brounhstein 'diffusion controlled' equation.

- Ca_3SiO_5 forms according to the reaction: $Ca_2SiO_4 + CaO \rightarrow Ca_3SiO_5$. The reaction is favoured by the presence of a liquid phase. C_3S is metastable at temperatures below 1100 °C and decomposes into dicalcium silicate and free lime according to the reaction: $Ca_3SiO_5 \rightarrow Ca_2SiO_4 + CaO$. A fast cooling prevents its decomposition.
- The direct synthesis using calcium salt and silica requires a high temperature of heating and a long residence time to achieve the complete reaction. However, once liquid phase occurs in the system, the reaction is facilitated and is achieved at a lower temperature and a shorter time. In the absence, or when only a small quantity of liquid phase is present, the temperature required for the reaction is higher (1600 °C). In this case, it is possible to add mineralisers, to decrease the working temperature. Mineralisers can act in different ways by accelerating the decomposition of calcium salts by increasing the rate of occurrence of the phases and by decreasing the temperature of combinations.
- Many mineralisers have been tried, for instance fluorine (CaF_2), $CaCl_2$, $CaBr_2$, Cr_2O_3, TiO_2 and MnO_2. Some mineralisers such as B_2O_3 act also as stabilizer of the β-C_2S.
- Water vapour also accelerates the combination of the components into silicate in the temperature below 1000 °C. In a CO_2 atmosphere, the silicates are produced by direct reaction of calcium carbonate with quartz. The reaction mechanism is based on surface diffusion of Ca^{2+} ions through the polycrystalline layer of the products namely C_2S and CS (SVE 1982).

2.1.5.3 Indirect Synthesis

- The high temperature of synthesis and the consequences of this on the energy consumption, the wear of the refractories which increases with the working temperature and the fast rate of cooling have led to the following 'soft methods'.

2.1.5.3.1 Coprecipitation of Calcium Salt

- Solution of calcium salt (nitrate for example) is mixed with colloidal silica or TEOS

$$3Ca(NO_3)_2 + SiO_2 + H_2O \rightarrow Ca_3Si(OH)_5 + HNO_3.$$

Nitric acid is evaporated at low temperature (300 °C) and an amorphous mass is obtained. Finally, the product is heated at 1450 °C.

- The method using calcium nitrate produces a homogeneous mixture that can be heated at lower temperatures. Nevertheless, it does not prevent from the use of platinum crucible.
- The necessity to evaporate large quantities of nitric acid reduces the quantity that can be produced. Handling and recycling nitric acid require special vessel and equipment.

2.1.5.3.2 Dehydration of a Synthetic Hydrate

- Preparation of the suitable hydrate is the first step. Calcium salt and silica are mixed in water medium. A calcium silica hydrate is obtained depending on the condition of the synthesis (temperature and pressure 1/20 bars).
- β-C_2S can be produced by dehydration of a calcium silicate hydrate obtained by autoclave at 205 °C and 17–19 bars of pressure.

2.1.5.3.3 Polyacrylamide Route

- A solution containing the cations is stabilized by citric acid. This solution is gelified by acrylamide and N, N' methylene diacrylamide in acid media.
- The monomer is hydrosoluble and polymerises as bidimensional. The second monomer is bifunctional and reticules the polyacrilamide chain that becomes N, N, N', N'. Tetramethyl ethylendiamide (TEMED) is activated by ammonium peroxydisulfate.
- The mixture is heated at 80 °C to obtain copolymerisation. The gel is then calcined at 500 °C in a glass container and then at higher temperatures in platinum vessel.

2.1.5.4 Sol–gel Processing Method

- The method is mainly based on polymerization reaction, *i.e.* hydrolysis of and poly-condensation of metal alkoxides. For the majority of them, the simultaneous reaction of hydrolysis and polycondensation leads to the formation of particles which undergo agglomeration. These alkoxides cannot be used for further polymerization and formation of a continuous polymeric network.
- Silicon alkoxides such as $Si(OC_2H_5)_4$ (also called TEOS) show a particular behaviour as they hydrolyse more slowly and incompletely to produce soluble species. By polycondensation reaction, these species give rise to the formation of an extended polymeric network. For single component oxide materials, the alkoxides are utilized by partially hydrolysing the alkoxide:

$$M(OR)x + yH_2O \rightarrow M(OR)x\text{-}y\ (OH)\ y + y\ ROH.$$

- The partially hydrolysed species are then allowed to link forming M–O–M' bond by polymerisation or condensation reaction. If an OH group was to react with an OR group, then another alcohol ROH would be produced by the formation of the M–O–M' bond.
- When more than a single species is involved, two types of reaction are possible: Firstly, reactions where all of the cationic constituents are supplied as alkoxide (TEOS and calcium alkoxide). Secondly, reactions where, for different reasons (costs, availability…), some components are added as soluble salts (TEOS and calcium nitrate).
- The sol–gel methods also produce a homogeneous raw material easy to transform into C_3S at rather low temperatures. This process requires the evaporation of a large quantity of alcohol.

2.1.5.5 The Pechini Method

- This method, originally applied to fabricate Perovskite powders is widely used by the ceramic industry. In this process a chelate is formed between mixed cations (dissolved as salt in a water solution) with a hydroxyl carboxylic acid (citric acid).
- The nitrate and citric acid solution are mixed with a polyhydroxyl alcohol and stirred while heating at around 80 °C. Heating to moderated temperature (150 and 250 °C) causes a condensation with the formation of water molecules.
- During heating, poly-esterification occurs and most of the excess water is removed, resulting in a polymeric resin. Polyacrylics can substitute citric acid or can be used as ion exchange resin.
- The metal cations are homogeneously distributed in the polymeric resin which is then sintered to yield the desired oxides.

2.1.5.6 Modified Pechini Process

- The modification consists of using C_3S hydrate instead of calcium salts during the phase of chelation. This improves the combination during the heating as calcium silicate hydrate gel is acting as a 'seed'. The reaction schematically gives an organo-mineral C_3S hydrate, $(HO)x–C_3S–OR$ (R organic radical).
- In a first step, an amorphous mass of hydrated C_3S–O R is obtained. A cross linking agent (triethanolamine titanate) can be used as chelating agent. Finally, the product is heated at about 1450 °C for a short time (less than 60 min depending on the quantity).

2.1.5.7 Self-Propagating Combustion Synthesis

- Metal nitrate and fuel (urea, glycine...) used as reducer are mixed in water. The stoichiometry is calculated and the thermochemical concepts from propellants chemistry are used; the total oxidising and reducing valences in the mixture should be balanced by the total valency of the fuel. The reactants are heated at 350 °C to evaporate water.
- The mixture is then heated at 500 °C to initiate the reaction. A dry foamy powder is obtained which is then heated at higher temperatures to obtain the desired calcium silicate.

2.1.5.8 Mastering the Crystallisation

- Synthesis of glass is useful to obtain a variable ratio CaO/SiO_2 in a homogeneous product. The raw material (limestone and quartz) is molten at 1560 °C. The liquid is cooled rapidly. The raw material could be obtained by gelification.
- From the glass, it is also possible to obtain crystallisation by total or partial annealing.

- Single crystals of C_3S are grown by firing a mixture of C_2S, CaO and calcium chloride.
- The synthesis of the polymorphs of C_2S and C_3S is obtained by addition of selected foreign elements and using the methods previously described (direct sintering, sol gel…).
- CS Perovskite is synthesised at 1500 °C under 15 GPa.
- CS Walstromite is synthesised at 1200–1500 °C and 5–10 GPa.

2.2 Al_2O_3–SiO_2 System

2.2.1 Introduction

- The Al_2O_3–SiO_2 system includes two crystallised compounds, Al_2SiO_5 and $Al_6Si_2O_{13}$ (short hand written AS and A_3S_2) and amorphous state compositions (Metakaolin…).

2.2.2 Mineralogy

- Mullite ($Al_6Si_2O_{13}$ or $3Al_2O_3 \cdot 2SiO_2$) is the only stable crystalline phase under normal atmospheric pressure. The orthorhombic form is a rare mineral, discovered in tertiary volcanoes in the Isle of Müll (Scotland). Cubic or tetragonal Mullite has been identified during heating of clay and aluminosilicates gels.
- Al_2SiO_5 exhibits three polymorphic modifications, Andalusite, Sillimanite and Kyanite, which are concurrently present in the earth crust.
- Andalusite forms at a high temperature and low stress and is typically of thermal aureoles of rocks produced by regional metamorphism ($d = 3.1$–3.2 g/cm^3).
- Sillimanite is formed at a high temperature and pressure and crystallises in gneiss and hornfels rock types, characteristic of the innermost zones of thermal metamorphism: it may also form with high grade regional metamorphism ($d = 3.25$ g/cm^3).
- Coexisting Sillimanite and Mullite were discovered in a metamorphosed lithomarge.
- Kyanite is characterized of regionally metamorphosed rocks of intermediate grade ($d = 3.5$–3.6 g/cm^3).
- Metakaolin is produced by heating kaolinite, a clay mineral with the composition $Al_2Si_2O_5(OH)_4$. Kaolinite occurs in soils that have formed from the chemical weathering of rocks (*e.g.* Feldspars), typically in hot moist climate.

2.2.3 Structure, Polymorphism and Solid Solutions

2.2.3.1 Structure and Polymorphism

2.2.3.1.1 General Characteristics

- The structures of the aluminosilicates have been studied by single crystal X ray diffraction, single crystals X ray refinement, high resolution electron microscopy,

TAB. 2.8 – Crystal characteristics of $Al_6Si_2O_{13}$ and of the polymorphs of Al_2SiO_5.

Composition	Al_2SiO_5	Al_2SiO_5	Al_2SiO_5	$Al_6Si_2O_{13}$
Oxide formula	$Al_2O_3 \cdot SiO_2$	$Al_2O_3 \cdot SiO_2$	$Al_2O_3 \cdot SiO_2$	$3Al_2O_3 \cdot 2SiO_2$
Cement formula	AS	AS	AS	A_3S_2
Mineral	Andalusite	Kyanite	Sillimanite	Mullite
Lattice	Orthorhombic	Triclinic	Orthorhombic	Orthorhombic
Space group	$Pnnm$	P-1	$Pbnm$	$Pbam$
Z	4	4	4	1
a (Å)	7.79	7.12	7.48	7.5785/7.584
b (Å)	7.90	7.85	7.67	7.6817/7.693
c (Å)	5.56	5.57	5.77	2.8864/2.89
A		89°98		
B		101°12		
γ		106°01		
V (Å^3)	324	4	331	168
d (g/cm^3)	3.13–3.16	3.53–3.65	3.23–3.27	3.16/3.22
JCPDS N°	13-122/39-376	11-46	38-471	15-776
Ref	BUR 1961	BUR 1963-2	BUR 1963-1	AGR 1960

IR and Raman spectrometry. Table 2.8 shows the crystal properties of the aluminosilicate $Al_6Si_2O_{13}$ and of the polymorphs of Al_2SiO_5.

- All the polymorphs share a common crystal structural feature, *i.e.* chains of AlO_6 octahedral parallel to the c crystallographic axis. The chains are cross linked by Si in tetrahedral coordination and by Al with different coordination in each polymorph, Al VI in Kyanite, Al V in Andalusite and Al IV in Sillimanite. These chains account for the prismatic crystal habit parallel to 001 and the well-developed hk0 cleavage.

2.2.3.1.2 *Kyanite*

- Kyanite can be considered as having a distorted close packed arrangement of oxygen atoms that yield a density greater than the other polymer. Kyanite melts incongruently at 1500 °C at 25 kbars to Al_2O_3 (Corundum) and liquid.

2.2.3.1.3 *Andalusite*

- Andalusite consists of edge shared chains of AlO_6 octahedra that are cross linked through corner sharing with double chains consisting of SiO_4 tetrahedra and Al in five coordinated trigonal bipyramids.
- At high temperatures (between 1300 and 1530 °C), it transforms into Mullite and vitreous silica.

2.2.3.1.4 *Sillimanite*

- Sillimanite is distinguished by two contrasting habits: coarse prismatic Sillimanite and acicular fibrolite. It consists of edge shared chains of AlO_6 octahedra that are cross linked by double chains of tetrahedra containing Si and Al.

- Calorimetric studies, heat treating experiments and theoretical consideration indicate that some degree of Al/Si disorder occurs in Sillimanite.
- The Al/Si disorder in Sillimanite facilitates the transformation Andalusite to Sillimanite.
- In each structure of Andalusite and Sillimanite, the stiffness measured parallel to c axis is greater than that measured normal to c axis. The shear modulus can be directly correlated with the relative rigidity of the cross-linking structures. At high temperature, Sillimanite transforms into Mullite.
- The Mullite structure is derived from the Sillimanite one by the replacement of one SiO_4 tetrahedron by one AlO_4 tetrahedron per unit cell of Sillimanite. The Mullite structure has a c-axis repeat of about 2.9 Å half of that of Sillimanite.
- X-ray pattern can hardly be used to differentiate Mullite from Sillimanite. The strong reflexions of both patterns are almost identical but the weak reflexions differ, neither, the variation of lattice constant of the Mullite solid solution is an unequivocal guide to composition since Mullite structure at any one composition can be ordered or disordered depending on the heat treatment IR spectra analyses allow a distinction between Sillimanite and Mullite. The main difference lies in the region from 9.897 to 10.57 μm because there is no band in this region of the Mullite spectrum whereas the Sillimanite spectrum has two distinct bands.

2.2.3.1.5 Metakaolin

- Metakaolin is a dehydroxylated form of Kaolinite (BRI 1959). Kaolinite is a phyllosilicate formed with one tetrahedral sheet of silica (SiO_4) linked through oxygen atoms to one octahedral sheet of alumina (AlO_6) octahedral. Above 530 °C, Kaolinite dehydroxylates and transforms into Metakaolin, having a complex amorphous structure which retains some long-range order due to layer stacking. Much of the aluminium of the octahedral layer becomes tetrahedrally and pentahedrally coordinated. Overheating causes the formation of Mullite and a defect in Al–Si spinel.

2.2.3.1.6 Mullite/Sillimanite Solid-Solutions

- Given the similarity of the structure, a large range of solid solution can be expected between Sillimanite, Mullite and Alumina nevertheless, the presence of a miscibility gap between Sillimanite and Mullite is confirmed experimentally.
- The most aluminous Sillimanite contains 51.4 mol % Al_2O_3 and the most siliceous natural Mullite contains 57 mol % Al_2O_3.
- Natural Mullites have a restricted field of solid solution 57–60 mol %.
- When held in the stability field of Sillimanite, Mullite (59 mol % Al_2O_3) exsolved non-stoichiometric Sillimanite leaving another Mullite with a higher aluminium content (64 mol %).
- The composition of $Al_{4+2x}Si_{2-2x}O_{10-x}$ ranges with x lying between 0.17 and 0.59, but concentrates from 3:2 and 2:1.
- Table 2.9 summarizes the range of solid solution of Mullite.
- The range of solid solution of Mullite depends on the experimental conditions of formation (sintering, devitrification, solidification from a melt, transformation of Sillimanite...).

TAB. 2.9 – Solid solution of Mullite and Sillimanite.

	Sillimanite		Natural Mullite		Alumina rich Mullite			Alumina
	AS		A_3S_2	A_2S	A_3S			A
mole%	50	51.4	57	60	66.6	75	76	100
weight %	62.9	64.3	69.3	71.8	77.2	83.6	84.4	100
x	0	0.04	0.17	0.25	0.4	0.57	0.59	1

- Formation of successive phases, or phase associations, passes from thermodynamically unstable to a stable through some metastable intermediary steps. The ordering can occur during continuous or discontinuous change tacking place in the framework, or it may be accompanied by the formation of a new structure. From an energetic point of view, it is advantageous for the Si and Al atoms to be distributed regularly. Alumino-silicates of framework type may be considered as solid solution in which the Si–Al ordering phenomena come when temperature falls.
- Mullite exsolved from the liquid phase appears to contain more aluminium (2:1 type) with a disordered structure. With continued heating Al_2O_3 exsolved and Mullite close to that of 3:2 type (A_3S_2) grows slowly by solid state reaction.
- Mullite cell dimensions vary almost linearly with composition and when extrapolated beyond those being the most aluminous found in this study $x = 0.59$ i.e. 76 mol %, that falls very close to a metastable polymorph of alumina κ-Al_2O_3.
- Mullite forms a limited series of solid solutions when fired in air.
- The studies of the solid solution in Mullite are orientated to the influence of the elements contained by the Kaolinite that form Mullite by heating: Fe_2O_3, TiO_2, MgO and Na_2O.
- Fe_2O_3 content of natural Mullite is about 1.5 mol % whereas synthetic sintered Mullites contain up to 18 mol % at 1300 °C. The iron content of fused Mullite is very low. Fe_2O_3 increases the refractive index and lattice spacings of the Mullite crystal. Fe_2O_3 content decreases when the temperature increases.
- Iron is essentially located in octahedral sites of the Mullite structure (Mossbauer).
- No solid solution exists with ferrous iron. $FeAl_2O_4$ and Fe_2SiO_4 are formed at 1192 °C.

2.2.3.2 *Structure of Liquid and Glass – Transition from Liquid to Solid*

- ^{27}Al NMR experiments and the molecular dynamic simulations indicate that Al is present in IV, V and VI coordinations to oxygen in the glass and liquid.

- At high silica contents (mole % $Al_2O_3 < 10\%$) both Si and Al occupy IV coordinated sites. In addition, the presence of some III coordinated Al is indicated by molecular dynamic simulation. As the Al_2O_3 content increases, the proportion of Al V and Al VI increases (average coordination number for Al_2O_3 liquid near V).
- The short-range structure of Al_2O_3 rich glass is similar to that of Mullite. The average SiOAl bond distance is 0.181 nm and the coordination number is 4.7.
- In the IR absorption spectra of the glasses, the band in the regions 700/900, 600/700 and 500/600 cm^{-1} are associated to the presence of AlO_4, AlO_5 and AlO_6 polyhedra.
- The cooling rate has a large effect on the NMR and IR spectra of the glass. Al V disappears if the glass is prepared at a slow cooling rate, due to a structural relaxation time determined from viscosity measurements.
- From oxygen diffusion and ^{29}Si NMR studies of high silica content liquid, it appears that the major structural relaxation mechanism associated with viscous flow is related to oxygen ion transfer between silicate species in the liquid.
- Glasses with added Al_2O_3 were found to devitrify to Cristobalite more rapidly than ultra-pure silica glasses at a given temperature and yet their fluidities (*i.e.* reciprocal to viscosities) were found to be less.

2.2.4 Stability and Phase Diagram

2.2.4.1 At Atmospheric Pressure

2.2.4.1.1 Stability in Presence of Liquid

- The melting point of Al_2O_3 and SiO_2 is, respectively, 2047 and 1453 °C.
- Table 2.10 summarizes the literature regarding the phase diagram at 1 atmosphere, carried out experimentally or calculated from thermodynamic data.
- Three types of phase diagrams under atmospheric pressure are found experimentally (figure 2.3) and explained thermodynamically.
- In diagram I figure 2.3, semi-infinite diffusion-couple experiments of α alumina and fused silica yield to an incongruent melting behaviour of Mullite at about 1830 °C. Its solid solution ranges from 58 to 69 mol%.
- Quenching in air of sample previously held at a given temperature yields to a congruent behaviour of Mullite at 1869 °C and solid solution extends to 74 mol% (diagram II figure 2.3).
- The replacement of polycrystalline alumina and fused silica by Corundum or Sapphire (Al_2O_3) and Cristobalite yields diagram III in figure 2.3 with an eutectic point at 1282 °C without the formation of Mullite. The differences of the results come from the conditions in which the experiments are carried out, the starting materials and the presence of minor elements in the lattice of the Mullite.
- The various results for phase equilibria can be explained by the slowness of diffusion process of Si and Al in silicate leading to difficulties associated with nucleation phases and to disordered Al/Si structure. The presence or absence of α-Al_2O_3 and the intermediate formation of liquid phase are a major factor of the stability and metastability of the phase equilibria.

TAB. 2.10 – Main references describing the stability behaviour in the Al_2O_3–SiO_2 system at atmospheric pressure.

Experimental works	Modelling works
Bowen et al. (BOW 1924)	Horibe et al. (HOR 1967)
Morris et al. (MOR 1935)	Risbud et al. (RIS 1977)
Toropov et al. (TOR 1951)	Howald et al. (HOW 1978)
Filonenko et al. (FIL 1953)	Dörner et al. (DOR 1979)
Budnikov et al. (BUD 1953)	Kaufman et al. (KAU 1979)
Konopicky (KON 1956)	Berman et al. (BER 1984)
Trömel et al. (TRO 1957)	Hillers et al. (HIL 1989)
Arakami et al. (ARA 1959)	Eriksson et al. (ERI 1993)
Welch (WEL 1960)	Zaitsev et al. (ZAI 1995)
Staronka et al. (STA 1968)	Li et al. (LI 1999)
Mac Dowel et al. (MAC 1969)	Kirschen et al. (KIR 1999)
Davis et al. (DAV 1972)	Pytel (PYT 1999)
Aksay et al. (1975) (AKS 1975)	
Risbud et al. (RIS 1978)	
Weissweiller et al. (WEI 1981)	
Klug et al. (KLU 1987)	

2.2.4.1.2 Immiscibility Gap

- A metastable region of liquid immiscibility exists at subsolidus temperature (7 and 55 mol % Al_2O_3) (diagram IV – figure 2.3).
- Between 35 and 55 mol % Al_2O_3 the decomposition of rapidly quenched SiO_2–Al_2O_3 glasses occurs by a spinodal decomposition mode (demonstrated by optical and electron microscopy and small angle neutron scattering experiments).
- In the spinodal region, the host phase is unstable with respect to continuous separation. Outside this area, a nucleus barrier exists and a critical nucleus different in composition from the host phase is necessary to allow the separation to proceed.
- The position of the immiscibility gap has been calculated from thermodynamic data. A structural interpretation of the immiscibility is a process involving arrangement of unstable high energy triclusters – *i.e.* one alumina and two silica tetrahedra are quenched in a cluster, forcing alumina into a corner-shared tetrahedral arrangement with silica with densification, a higher energy state and therefore promotes glass separation.

2.2.4.1.3 Thermodynamic Modelling

- Thermodynamic calculations are based on the substitutional model on the regular or subregular solution model. The data required for this calculation are summarized in table 2.11.
- Figure 2.4 shows an example of the thermodynamic calculation of the liquidus of the experimental phase diagram II. It is based on the basic Redlich Kister model of the excess free energy.

Binary Chemical Systems

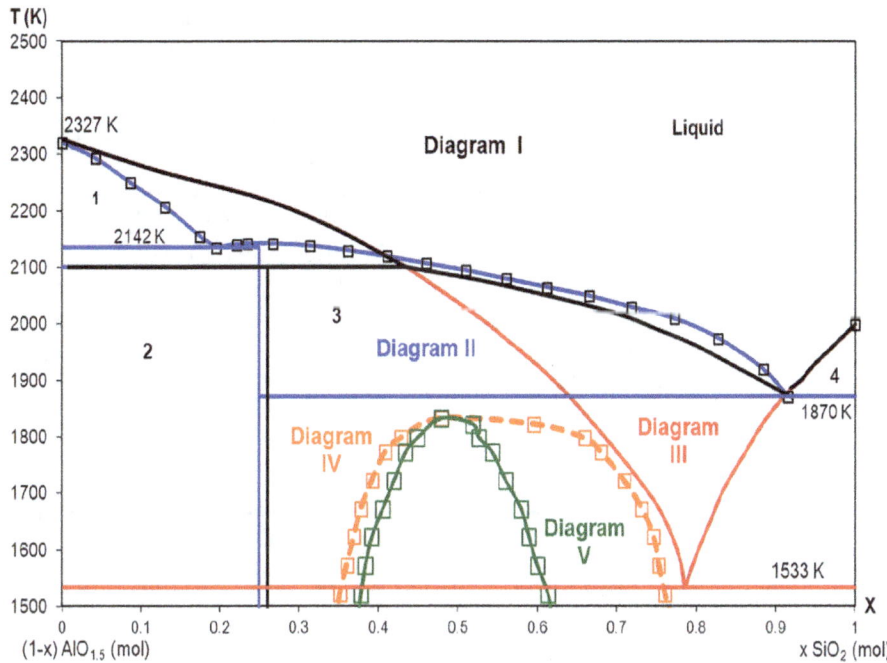

FIG. 2.3 – Different descriptions of the stability in the Al_2O_3–SiO_2 system. Key: Diagram I (black) incongruence of Mullite, diagram II (blue) congruence of Mullite, diagram III (red) no occurrence of Mullite, diagram IV (pink) metastable boundary of immiscibility, diagram V (Green) spinodal region area: (1) liquid + Alumina area, (2) Alumina + Mullite area, (3) liquid + Mullite area, (4) SiO_2 + liquid area.

2.2.4.2 At High Pressure

- The phase equilibrium P–T diagram has been studied experimentally (hydrothermal conditions, solubilities at atmospheric pressure...) determined from lattice vibration or calculated from thermodynamic data.
- Table 2.11 summarizes the main published literature.
- There are considerable experimental discrepancies regarding this phase equilibrium diagram, particularly the position of the univariant lines (Kyanite–Sillimanite, Andalusite–Sillimanite, Kyanite–Andalusite) and the location of the triple point Sillimanite–Kyanite–Andalusite (from 2, 4 to 8 kbars and 300 to 750 °C). The calculation based on available thermochemical data does not solve totally the problem because the small values of ΔG for transformations involving the Al_2SiO_5 polymorphs yield to considerable error in the P–T location of the Al_2SiO_5 univariant equilibria. On the other hand, important variables such as non-stoichiometric (solid solution), lattice defect and disorder significantly perturb the chemical potentials of the aluminium silicates.

TAB. 2.11 – Main references describing the stability behaviour in the system Al_2O_3–SiO_2 experimentally and thermodynamic model at high pressure.

Experimental works	Modelling works
Evans (EVA 1965)	Weill (WEI 1966)
Newton (NEW 1966–1 NEW 1966–2)	Day et al. (DAY 1980)
Richardson et al. (RIC 1967 RIC 1968)	Salje et al. (SAL 1982)
Holdaway (HOL 1971)	R.A. Robie et al. (ROB 1984)
Brown et al. (BRO 1971)	Holland et al. (HOL 1985)
Bowman (BOW 1975)	
Heninger (HEN 1984)	
Bohlen et al. (BOH 1991)	

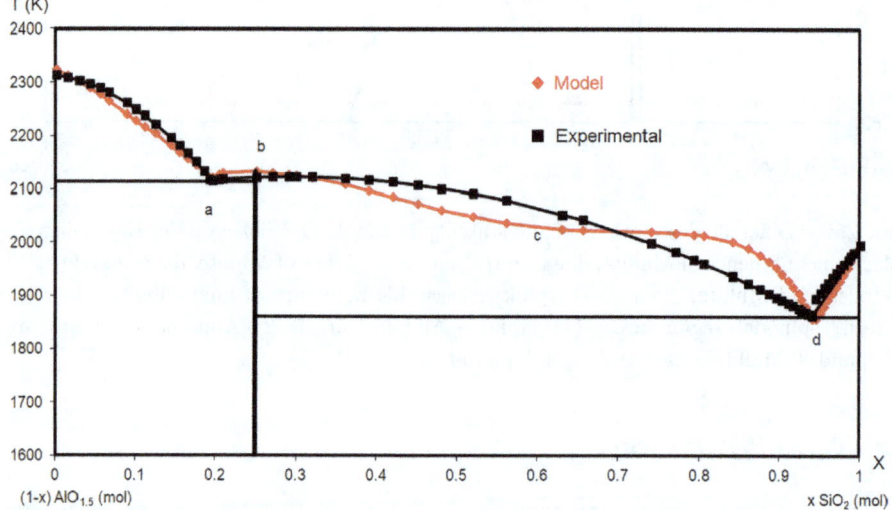

FIG. 2.4 – Comparison of experimental and model of diagram II (liquidus). Key: area 1 – liquid + alumina, area 2 – Alumina + Mullite, area 3 – liquid and Mullite, area 4 – Silica + liquid.

- With the help of petrologic analyses, the triple point position has been defined from 3.8 to 4.5 kbars and 505–540 °C (figure 2.5).
- The uncertainties in experimental results can be explained by the structure of the aluminosilicate compounds and their evolution.

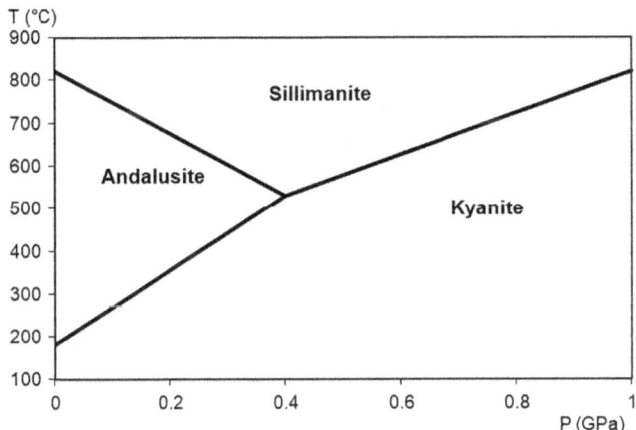

FIG. 2.5 – Triple point Andalusite + Kyanite + Sillimanite (BOH 1991).

2.2.5 Synthesis of Alumino–Silicates

LIU 1991, CHO 2002

2.2.5.1 Synthesis Depending Mainly on the Starting Materials

- Mullite is obtained by (1) heating natural alumino silicates such as Kaolinite, Kyanite, Sillimanite or synthetic ones such as mixture of oxides, salts, sol or metal alkoxide, (2) devitrification of alumino silicate glass during annealing or by cooling of an alumino silicate liquid.
- Aluminium bearing compounds are solid mineral (Bauxite, Kaolinite), synthetic (αAl_2O_3), colloidal, Boehmite, Gibbsite, Diaspore, salts (chloride, nitrates, sulphates...) or alkoxides (Al ethyl ester, isopropoxides, sec butoxide...).
- Silica bearing compounds are crystallised (Quartz, Cristobalite, Kaolinite), glassy (fused silica), colloidal (ludox, silicic acid), salt (Si chloride), alkoxides (Si ethyl ester or (tetraethyl orthosilicate (TEOS)).
- Starting materials, process of mixing and drying result in different sequences of formation.
- Two main routes of formation are followed during heating; a direct formation of Mullite or a formation through intermediated compounds (spinel phase) and mullitization at temperatures higher than 1200 °C.
- Table 2.12 summarizes the different possibilities to obtain Mullite.
- During sintering reaction of a mixture of solids (α-Al_2O_3 (0.3/0.5 μm) and quartz particle, (<2 μm) mullitization begins at 980 °C with an exotherm at 980 °C and is complete between 1200 and 1400 °C. The exotherm at 980 °C is attributed to the crystallisation of γ-alumina, the occurrence of an intermediate compound and the formation of Mullite. The degree of mixing of Al_2O_3 and SiO_2 in the grains determines the temperature range of mullitization.

TAB. 2.12 – Different possibilities to obtain Mullite.

Si bearing ⟹ Al bearing ⟱	Salt (SiCl4)	Sol (ludox)	Alkoxides (i.e. TEOS)	Oxides (SiO$_2$)	Clay (kaolinite)
Oxides (Al$_2$O$_3$)	Si salt Al oxides	Si sol Al oxides	Si alkoxide Al oxides	*Si oxides* *Al oxides*	*Clay* *Al oxides*
Salt (i.e. ANN)	*Si salt* *Al salt*	*Si sol* *Al salt*	*Si alkoxide* *Al salt*	Si oxides Al salt	*Clay* *Al salt*
Sol (i.e. Boehmite)	Si salt Al sol	*Si sol* *Al sol*	*Si alkoxide* *Al sol*	Si oxides Al sol	*Clay* *Al sol*
Alkoxide	*Si salt* *Al alkoxide*	Si Sol Al alkoxide	*Si alkoxide* *Al alkoxide*	Si oxides Al alkoxide	Clay Al alkoxide

Note: The most commonly used syntheses are shown in italics.

- Mullite is obtained by thermal decomposition reaction of Kaolinite. During heating, Kaolinite forms Metakaolinite by dehydroxylation. Metakaolinite shows very little powder X-ray diffraction pattern and contains 11% of the hydroxyl groups. At 970 °C, amorphous silica, (poorly crystalline), alumina rich Mullite and a spinel phase occur with a loss of the residual hydroxyl groups associated with an exothermic peak. Further heating results in a change in the composition of Mullite (reaching the 3:2 Mullite) and in the conversion of the spinel into Mullite by reaction with silica. Kinetics of formation of Mullite and its grain growth is modified by the presence of minor elements and mechanical activation of the starting materials (Kaolinite, Andalusite or Kyanite). The addition of alumina bearing compound to Kaolinite leads to the formation of a primary Mullite and Cristobalite between 970 and 1250 °C, and a secondary Mullite by reaction of with aluminium compound (oxide, hydroxides and salts). Mullite can also be obtained by mixture of sols, coprecipitation of salts, mixing of alkoxides and salt or a mixture of alkoxides. A Single phase xerogel is formed by mixing TEOS and Al salts (aluminium nitrate nanohydrate (ANN)) dissolved in alcohol form at 60 °C. During heating of a single phase xerogel, if the solution is slowly hydrolysed by aqueous ammonia, a sharp thermal peak occurs at 960 °C attributed to a direct crystallisation of Mullite. If the solution is rapidly hydrolysed by ammonia, a spinel phase crystallises at 960 °C from the amorphous state and transforms into Mullite by further heating (1200 °C).
- This exotherm is similar to the one reported for the Metakaolinite to Mullite reaction and the sintering reaction Al$_2$O$_3$–SiO$_2$.
- The composition, the stability field and the condition of formation of the intermediate compound (called spinel or cubic Mullite) are not totally clear. It is obtained by exothermal reaction between 950/1050 °C of Kaolinite or alumino-silicate gel. It contains around 8% of SiO$_2$ and is very similar to γ-Al$_2$O$_3$.

- Mixing aluminium and silicon alkoxide as a precursor yields Mullite by either spinel phase formation before mullitization (basic catalyst), either reaction similar as the mixture aluminium salt, Si alkoxide.
- A diphasic xerogel is obtained by mixing alumina sol (Boehmite sol) and aqueous silica sol dispersed in ethanol. During heating of a diphasic Xerogel, a very broad exotherm associated with Mullite formation occurs from 760 to 1200 °C.
- The phase development of Al_2O_3–SiO_2 gel at 980 °C is a function of both the starting materials and the process of gelation.
- A single phase xerogel is prepared by mixing TEOS and ANN after exchange in methanol and dried at 60 °C. A sharp exotherm occurs at 980 °C. The same sample produced by methanol or ethanol after drying at 260 °C under 8.27 MPa in an autoclave (leading to the formation of aerogel) shows no exotherm.
- With the same starting material (TEOS/ANN), Mullite is obtained at low temperatures (<1000 °C) when the precursor solution was aged at 60 °C, whereas it is the spinel phase when they were aged under 60 °C.
- Starting from aluminium salt and Si sol, Mullite is obtained through spinel phase as intermediate compound, whereas tetragonal like Mullite is formed by mixing monosilicic solution and aluminium hydroxide (OKA 1991).

2.2.5.2 Synthesis Dependent Mainly on the Process

- Crystallization of glass, crystallisation from Al_2O_3–SiO_2 melts, rapid quenching of glass, splat cooling or flame spraying).
- Devitrification: when a glass is heated to temperatures between its transformation range and the liquidus temperature, a process known as devitrification will occur. Crystalline nuclei form and grow. However, sometimes this never occurs (polystyrene, polymethyl metacrylate).
- Flame fusion.
- Chemical vapour deposits (CVD) process mixed vapour of $SiCl_4$ and $AlCl_4$ that are fed into H_2/O flame and oxidised at 1600–1800 °C. The produced powder is quickly quenched to suppress grain growth or methanolic solution of aluminium nitrate.
- Metal powder combustion or mixed vapour of Al and Si produced by rapid heating at high temperatures reacts with oxygen to form metal oxide vapour. During cooling, the vapour condenses into Mullite.
- Oxidation of Al–Si alloy. Hydrothermal processing using alkoxide solution at 600 °C under 20 MPa.
- Ultrasonic spray pyrolysis of metal alkoxide solution using nanometric particles (400/500 nm).
- Radio frequency sputtering.
- Microwave sintering.
- Freeze drying process; silica and alumina sulphate into solution sprayed into liquid N_2 and frozen dry at −10 °C under vacuum. The solid heated at 1000 °C reveals traces of cubic Mullite and γ alumina (WHE 1979).

- Zeolite type reaction; $Al_2(SO_4)_3 16H_2O$ and sodium silicate followed by electrodialysis to removed Na^+. At 986 °C, Mullite occurs as orthorhombic and cubic which transforms at 1250 °C into orthorhombic (INS 1935).
- Microwave treatments have been reported to decrease the temperature of sintering (acceleration of diffusion kinetics) and for densification without chemical change.

2.3 Al_2O_3–Fe Oxides System

2.3.1 Introduction

- Two main compounds occur in the Al_2O_3–Fe oxides system:
  (1) $AlFe^{3+}O_3$ ($Al_2O_3 \cdot Fe_2O_3$ or in cement chemist short hand notation AF).
  (2) Hercynite, $Fe^{2+}Al_2O_4$ ($FeO \cdot Al_2O_3$ or in short hand notation Af).
- The existence of the compounds $Al_3Fe^{3+}{}_5O_{12}$ ($3Al_2O_3 \cdot 5Fe_2O_3$ or A_3F_5) and $Al_2Fe^{2+}{}_3O_6$ ($2Al_2O_3 \cdot 6FeO$ or A_2f_6) has to be confirmed.

2.3.2 Mineralogy

- Hercynite ($Fe^{2+}Al_2O_4$) takes its name after a Bohemian forest (Latin name). It was first found in Czech Republic (Pobezovice). It belongs to the spinel group. It occurs in high grade metamorphosed iron rich argillaceous sediment as well as in mafic and ultra-mafic igneous rock.

2.3.3 Structure, Polymorphism and Solid Solution

- $AlFeO_3$ crystallises in the orthorhombic system and is isostructural with $GaFeO_3$.
- Mössbauer analysis shows that Fe^{3+} occupies both tetrahedral and octahedral sites. All but 10% of the tetrahedral sites are occupied by Fe^{3+} and the remainder is being occupied by Al (Al^{3+} having a strong octahedral difference by contrast with almost no site preference for Fe^{3+}).
- Table 2.13 gives the crystal data of $AlFeO_3$.
- The solid solution on the join Al_2O_3–Fe_2O_3 does not show a complete range of solid solution but exhibits a large immiscibility gap. The limit can be placed at 24 wt % Al_2O_3 in Fe_2O_3 and 12 wt % Fe_2O_3 in Al_2O_3.
- In compound $AlFeO_3$, the balance of the composition ranges from 4% excess Al_2O_3 to 4% excess Fe_2O_3.
- $FeAl_2O_4$ (Hercynite) crystallises in the cubic system and belongs to the spinel group. It is made up of approximately cubic close packed anion sublattices with divalent cations occupying the tetrahedral and trivalent cations occupying octahedral interstitial site. 1/8 of the tetrahedral holes are filled with M^{2+} cation

TAB. 2.13 – Crystal characteristics of $FeAl_2O_4$ and $AlFeO_3$.

	Mineral	Lattice	Z	a (Å)	b (Å)	c (Å)	V (Å^3)	d (g/cm^3)	JCPDS N°
$FeAl_2O_4$	Hercynite	Cubic	8	8.15			542	4.26	34-192
$AlFeO_3$		Orthorhombic	8	8.56	9.24	4.98	395	4.39	30-24

and ½ of the octahedral holes are filled with X^{3+} cations. The cation to anion ratio is ¾.
- The unit cell contains 8 molecules. Each oxygen ion has four nearest neighbour cations, 3 in octahedral position and one in tetrahedral position.
- Complete solid solution exists between Fe_3O_4 and $FeAl_2O_4$ except at a lower temperature (858 °C) where miscibility gap occurs. $FeAl_2O_4$ is a normal spinel $Fe^{2+}(Fe^{3+}Al^{3+})O_4$ whereas Magnetite is an inverse spinel $Fe^{3+}(Fe^{2+}Fe^{3+})O_4$.
- The entropy of structural transformation in spinel has been evaluated by a semi empirical correlation involving the entropy of cation mixing on the tetrahedral and octahedral sites of the spinel and a statistical factor of distortion. It gives information on the relative stabilities of oxides in different crystal structures.
- At a given temperature, the composition of the spinel phase depends on PO_2. The spinels of higher aluminium contents are stable at the lower oxygen pressure.
- Spinel phase may contain up to 64.2% Al_2O_3 at 1750 °C.

2.3.4 Stability and Phase Diagram

- The Fe–Al–O phase diagram may be divided into two parts (figures 2.6 and 2.7). The first area is bounded by Fe–Spinel–Al_2O_3–Al where oxygen pressure is low and the second area bounded by spinel–Al_2O_3–O contains compositions which show a marked change in oxygen content with increasing temperature even below the solidus. Oxygen pressure control is an important variable in this area.
- It is possible to represent the Fe–Al–O system in simplified form Al_2O_3/Fe_2O_3 under oxidizing conditions or Al_2O_3/FeO in the presence of metallic iron. Table 2.14 lists the main references.
- In the liquidus of the system Al_2O_3–Fe_2O_3, only the solid solutions of Spinel and Corundum occur as primary phase solutions (Spinel in the iron rich melts and Corundum in alumina rich melts). Spinel is a solid solution of Hercynite and Magnetite. The other crystalline compounds are Hematite with Corundum as solid solution and $AlFeO_3$.
- The system is binary of the Hematite ss, Corundum ss and $AlFeO_3$ are the only phases present. As the temperature increases, ferric oxide decomposes into FeO and a Spinel phase form. For pure Fe_2O_3, the decomposition takes place at 1390 °C in air and 1445 °C at $PO_2 = 1$ atm. The phase $AlFeO_3$ (AF) exists between 1318–1410 °C in air and 1318–1495 °C at 1 atm PO_2. It decomposes to Spinel and Corundum at 1410 °C in air and 1490 °C at $PO_2 = 1$ atm).

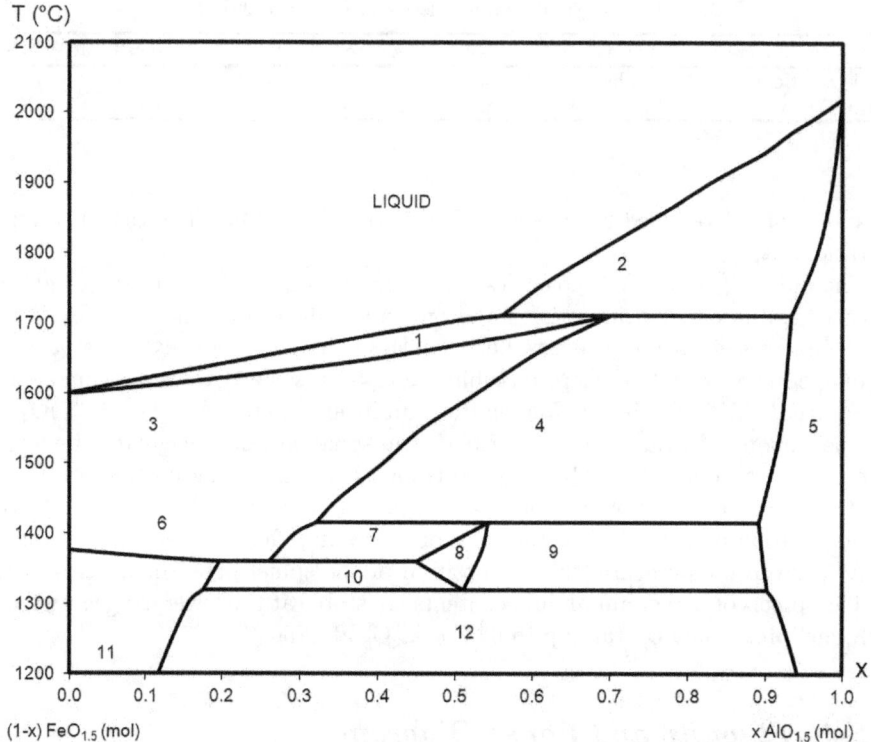

FIG. 2.6 – Phase relations in Al_2O_3–Fe_2O_3 the system in air. Key: (1) Spinel ss + Liq, (2) Corundum ss + Liq, (3) Spinel ss, (4) Spinel ss + Corundum ss, (5) Corundum ss, (6) Hematite ss + Spinel ss, (7) Spinel ss + $AlFeO_3$ ss, (8) $AlFeO_3$, (9) $AlFeO_3$ + Corundum ss, (10) $AlFeO_3$ + Hematite ss, (11) Hematite ss and (12) Hematite ss + Corundum ss.

- When PO_2 decreases, the stability field of $AlFeO_3$ reduces and this phase disappears. Figure 2.7 represents the diagram with $PO_2 = 1$ atm ('a' in black). The diagram in air (similar to figure 2.6) is added as 'b' in red whereas 'c' in green represents the diagram in reducing condition $PO_2 \leq 0.03$ atm.
- The Al_2O_3–Fe_2O_3 system exhibits a large miscibility gap, 24% Al_2O_3 in Fe_2O_3 and 12% Fe_2O_3 in Al_2O_3, in air between 1250 and 1350 °C.
- In equilibrium with iron, the phase diagrams can be represented in the FeO–Al_2O_3 system (figure 2.8). $FeAl_2O_4$ is the only binary compound. Different versions of this system exist depending on the congruency or not of $FeAl_2O_4$.
- Hercynite ($FeAl_2O_4$) melts congruently at 1780/1820 °C or incongruently at 1750 °C.
- Activity and free energy of mixing of Fe_3O_4 and Hercynite have been determined experimentally or from thermodynamic data derived from calculated cation distribution.
- From the thermodynamic data, it is possible to model the experimental results of the Al_2O_3–Fe_2O_3 system with a good agreement (figure 2.9)

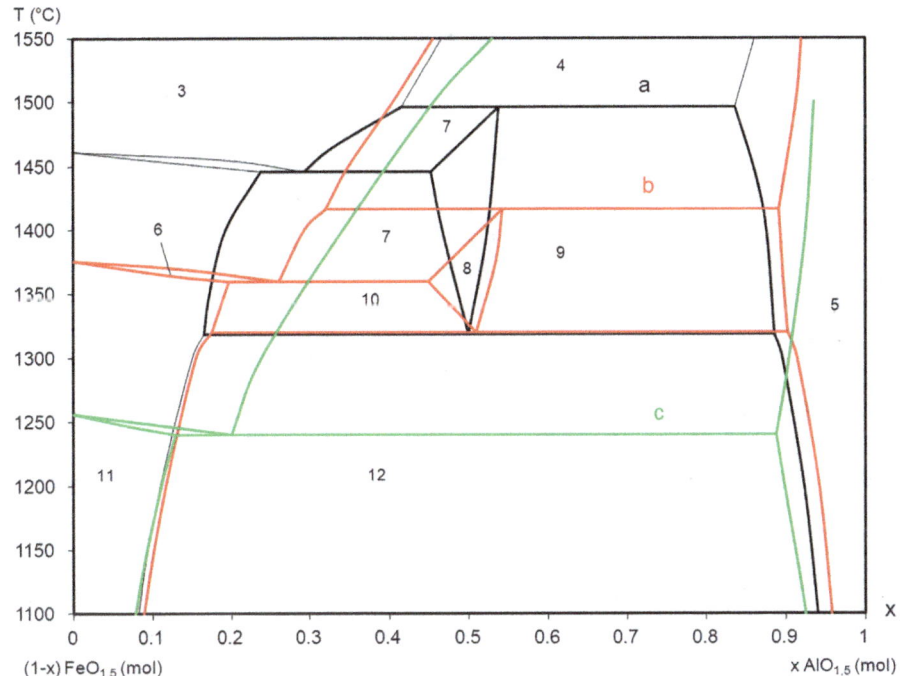

FIG. 2.7 – Effect of PO$_2$ on subsolidus equilibria of the Al$_2$O$_3$–Fe$_2$O$_3$ system. Diagram 'a' for PO$_2$ = 1 atm, diagram 'b' for air and diagram 'c' for PO$_2$ ≤ 0.03 atm. Key: (3) Spinel ss, (4) Spinel ss + Corundum ss, (5) Corundum ss, (6) Hematite ss + Spinel ss, (7) Spinel ss + AlFeO$_3$ ss, (8) AlFeO$_3$, (9) AlFeO$_3$ + Corundum ss, (10) AlFeO$_3$ + Hematite ss, (11) Hematite ss, (12) Hematite ss + corundum ss.

TAB. 2.14 – Main references describing the Fe–Al–O system.

Experimental works	Experimental works
McIntosh et al. (MCI 1937)	Turnock et al. (TUR 1962)
Richards et al. (RIC 1954)	Roiter (ROI 1964)
Oelsen et al. (OEL 1955)	Novokatskii et al. (NOV 1965)
Fisher et al. (FIS 1956)	Willshee et al. (WIL 1968)
Hoffman et al. (HOF 1956)	Rosenbach et al. (ROS 1974)
Muan (MUA 1958)	Meyers et al. (MEY 1980)
Atlas et al. (ATL 1958)	
	Modelling works
	Barry et al. (BAR 1993)

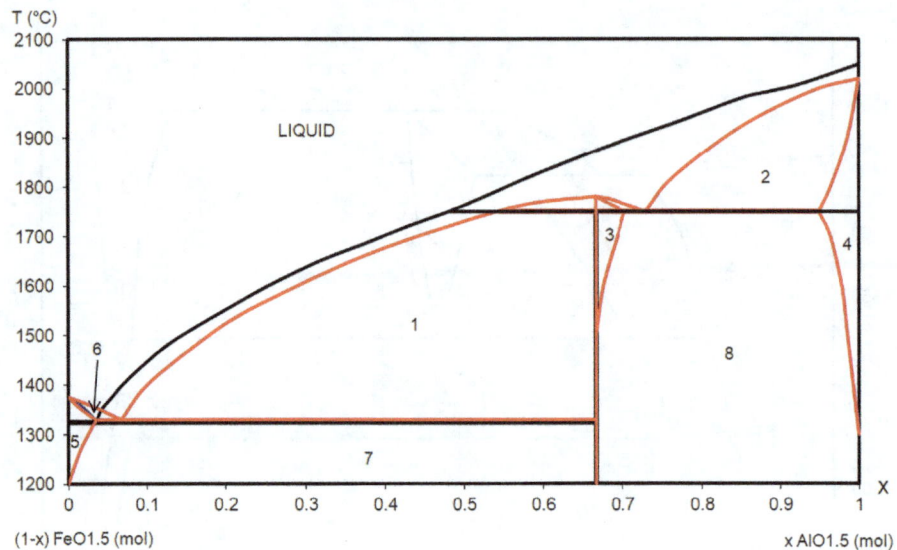

FIG. 2.8 – Two versions of the equilibrium diagram of the FeO–Al_2O_3 system. Key: (1) Spinel ss + liquid, (2) Corundum ss + liquid, (3) Spinel ss, (4) Corundum ss, (5) FeO, (6) FeO + liquid, (7) Spinel ss + FeO, (8) Spinel ss + Corundum ss.

2.3.5 Synthesis of Iron Aluminates

- $FeAlO_3$ is obtained by heating a mixture of Al_2O_3–Fe_2O_3 in oxygen at 1400 °C or by coprecipitation of hydrous oxides at 1370 °C during 4 days in air.
- $FeAl_2O_4$ (Hercynite) is prepared by heating the calculated preparation of alumina and ferrous oxalate in an iron crucible in a muffle furnace at 1200 °C.
- Aluminium and ferric nitrate are added to ammonium hydroxide solution. After filtration, the precipitate is fired at 100 °C in oxygen for 12 h. Five gram disks are pressed and fired at 1300 °C in air. The pellets are then equilibrated in a mixture of CO_2/CO to obtain the required reduction.
- $Al_3Fe_5O_{12}$ is obtained by firing oxide at 1000 and 1400 °C under oxygen (SAL 1994).

2.4 CaO–Al_2O_3 System

2.4.1 Introduction

- The CaO–Al_2O_3 system includes five compounds: $Ca_3Al_2O_6$, $Ca_{12}Al_{14}O_{33}$, $CaAl_2O_4$, $CaAl_4O_7$ and $CaAl_{12}O_{19}$ (short hand written C_3A, $C_{12}A_7$, CA, CA_2, CA_6).

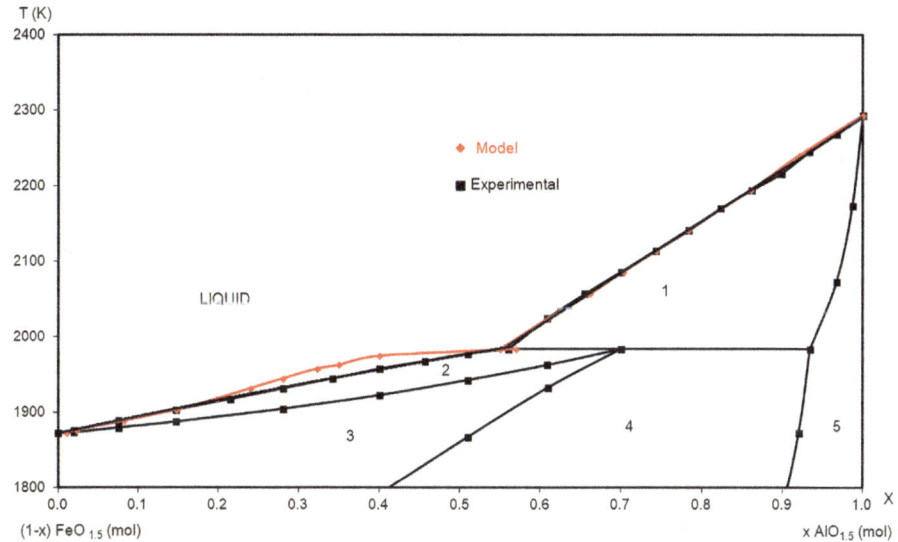

FIG. 2.9 – Experimental and modelled diagram of the Al_2O_3–Fe_2O_3 system. Key: (1) Corundum ss, (2) Spinel ss + Liquid, (3) Spinel ss, (4) Spinel ss + Corundum ss, (5) Corundum ss.

- Other compounds have been found in special conditions; $Ca_2Al_2O_5$ (C_2A), $Ca_5Al_6O_{13}$ (C_5A_3), $CaAl_{20}O_{31}$ (CA_{10}).

2.4.2 Mineralogy

- $Ca_3Al_2O_6$ (C_3A). The mineral Katoite (named for the Japanese mineralogist Akira Kato) has the same ratio $C/A = 3$ and crystallises as cubic but it contains silica and hydroxyl; $Ca_3Al_2(SiO_4)_{1.5}(OH)_6$.
- $Ca_{12}Al_{14}O_{33}$ ($C_{12}A_7$). The mineral Mayenite found in metamorphosed clayey limestone inclusions in volcanic lava and is named after the locality (near Mayen, Germany).
- $CaAl_2O_4$ (CA). It is not found as mineral in the nature but was found in a meteorite in 2011 and given the name Krotite.
- $Ca\,Al_4O_7$ (CA_2). The mineral Grossite is found in haturim formation at Negev desert (Israel). It is named after Shulamit Gross.
- $CaAl_{12}O_{19}$ (CA_6). The mineral Hibonite is named for Paul Hibon who discovered the mineral near Ambindandrakemba (Madagascar).

2.4.3 Structure, Polymorphism and Solid Solution

- Table 2.15 summarizes the main crystal properties of these compounds:

TAB. 2.15 – Crystal characteristics of the principal calcium aluminates.

Composition	$Ca_3Al_2O_6$	$Ca_{12}Al_{14}O_{33}$	$CaAl_2O_4$	$CaAl_4O_7$	$CAl_{12}O_{19}$
Oxide formula	$3CaO \cdot Al_2O_3$	$12CaO \cdot 7Al_2O_3$	$CaO \cdot Al_2O_3$	$CaO \cdot 2Al_2O_3$	$CaO \cdot 6Al_2O_3$
Cement formula	C_3A	$C_{12}A_7$	CA	CA_2	CA_6
Mineral	Katoite	Mayenite	Krotite	Grossite	Hibonite
Lattice	Cubic	Cubic	Monoclinic	Monoclinic	Hexagonal
Space group	$Ia3d$	I-$43d$	$P2_1/n$	$C2/c$	$P6_3/mmc$
Z	4	2	12	4	2
a (Å)	15.262	11.982	8.697	12.897	5.536
b (Å)			8.092	8.879	21.825
c (Å)			15.208	5.454	
B			90°14	106°93	
V (Å^3)	3555	1720	1070	596	586
d (g/cm^3)	3.03	3.06		2.86	3.69
JCPDS N°	35 1429	9 413	23-1036	23-1037	38 470
Ref	MON 1975	NUR 1965	BAL 1970	BAL1970	BUI 1968

2.4.3.1 Structure and Polymorphism

2.4.3.1.1 $Ca_3Al_2O_6$

- C_3A crystallized as cubic.
- The structure of this compound has rings of six AlO_4 tetrahedra of formula $(Al_6O_{18}^{18-})$ surrounding holes of radius 1.47 Å and interconnected by calcium atoms. These rings are formed by a corner sharing of two oxygen atoms per tetrahedron to give a structure with two non-bridged oxygen atoms per AlO_4 tetrahedron. Aluminium ion is present in four-fold coordination.

2.4.3.1.2 $Ca_{12}Al_{14}O_{33}$

- The structure of $C_{12}A_7$ has a cubic unit cell containing $Ca_{24}Al_{24}O_{66}$.
- It is based on tetrahedral AlO_4 units. Twelve tetrahedra within the unit cell are fully polymerised by corner sharing to other group while the remaining $16AlO_4$ tetrahedra of each have one non-bridged oxygen $(Ca_{24}[Al_{16}O_{40}][Al_{12}\ O_{24}])^{4+}$.
- The oxygen atoms O_1 and O_2 account for $40 + 24$ of the oxygen atoms in the unit cell. Two O_2 per unit cell are not bound to aluminium but are coordinated by Ca giving a total of 66 oxygen atoms (BAR 1972). Two polymorphic forms, called alpha and gamma with a transition temperature 1200 °C in air have been detected by SEM, optical microscopy and DTA.

2.4.3.1.3 $Ca_2Al_2O_7$

- C_2A is produced at 1250 °C and high pressure (25 kbars).
- The structure of C_2A is based on body centred orthorhombic unit cell. It is isostructural with C_2F whose crystal structure was shown to be based on interconnected FeO_6 octahedral and FeO_4 tetrahedra (AGG 1972).
- Partial solid solution exists along the join C_2A–C_2F.

2.4.3.1.4 $Ca_5Al_6O_{13}$

- C_5A_3 crystallises in the orthorhombic system. Its structure is related to that of Gehlenite (C_2AS).
- All aluminium atoms in C_5A_3 are in tetrahedral coordination with oxygen. The structure is based on alternating twisted sheets of distorted AlO_4 tetrahedra and layers of Ca atoms which lie perpendicular to 001 plane. The tetrahedra are linked through corners to form a network of five membered rings. The arrangement of the Ca atoms in a layer presents some unusual features in that some of the Ca–Ca distances are very short 3.200–3.267 Å resulting in face sharing of certain of the Ca tetrahedra (ARU 1957).

2.4.3.1.5 $CaAl_2O_4$

- The structure of CA is based on a monoclinic unit cell.
- Its structure, studied from a sample of phosphorus furnace slag is related structurally to β-Tridymite and based on a fully polymerized three-dimensional arrangement of corner shared AlO_4 tetrahedra. The calcium atoms are accommodated in cavities caused by the distortion of the Tridymite like structure (BAL 1970). A metastable orthorhombic form of CA is obtained by crystallisation of a gel at 950 °C. The gel is prepared by coprecipitation of calcium and aluminium nitrate in alcohol. The orthorhombic form changes to the stable monoclinic form at a higher temperature (1000 °C).
- Infrared spectrometry allows the distinction between the 2 polymorphs (absence of bands at 342, 415, 529, 717, 763, 832 and 851 cm^{-1} for the orthorhombic form and different intensity of the band at 720 cm^{-1}.

2.4.3.1.6 $CaAl_4O_7$

- The structure is based on a monoclinic unit cell.
- Its structure is similar to the Feldspar structure and is based on a fully polymerized network of AlO_4 tetrahedra sharing one oxygen atom per molecule unit between $3AlO_4$ tetrahedra. The Ca atoms are irregularly coordinated with 5 bonds Ca–O shorter than the other 4 (2.6 against 3.5 Å). The compact situation causes a different length of the Al–O bonds (average 1.76 Å). There are 4 formula units $CaO \cdot 2Al_2O_3$ in a unit cell and the volume of the unit cell is 591.54 $Å^3$ and the calculated density is 2.915 g/cm^3. CA_2 does not exhibit polymorphism.

2.4.3.1.7 $CaAl_{12}O_{19}$

- The crystal symmetry of CA_6 is hexagonal. Its structure is isomorphous with Magnetoplumbite ($PbO \cdot 6Fe_2O_3$) and similar to that of β-aluminate ($Na_2O \cdot 11Al_2O_3$). The unit cell contains two formula units of CA_6 with 20 aluminium atoms in six-fold coordination and two tetrahedra coordinated to form a spinel like unit. These units are linked by the remaining two aluminium atoms per unit cell which are in irregular five-fold coordination to oxygen (BUI 1968). Analysis by TEM of CA_6 containing syntactic intergrowths phases shows that CA_6 can accommodate Ca/Al non stoichiometry in a relatively wide range (SCH 1983).

2.4.3.2 Solid Solutions

2.4.3.2.1 $Ca_3Al_2O_6$

- There is no polymorphic transformation of pure C_3A but alkalis substituting for calcium produce changes in symmetry of pure C_3A crystal lattices. The replacement of Ca^{2+} by Na^+ is accompanied by the occupation of the second Na^+ in the centres of the rings. Five crystalline forms from cubic to orthorhombic and monoclinic are found by increasing the Na content of the solid solution (GRZ 1986).
- A chemical formula, $Na_{2x} Ca_{3-x-y} (Al_{1-y})Si_y)_2O_6$, has been proposed for C_3A doped with Na and Si.
- The isomorphous substitution of Sr^{2+} for Ca^{2+} and Ga^{3+} for Al^{3+} is possible.
- A comparison between the absorption frequency for a SiO_4 group and an AlO_4 group is used to identify the IR absorption bands in C_3A.
- Mg, Cd, Pb and Ba could all partially replace Ca to form an oxide of the type $Ca_{3-x}M_xAl_2O_6$ similar to that described recently for the Sr substituted series $Ca_{3-x}M_{Sr}Al_2O_6$.
- Fe^{3+} replaces Al in tetrahedral site (IR spectrometry).
- Table 2.16 represents the maximum values currently found as solid solution in C_3A.

2.4.3.2.2 $Ca_{12}Al_{14}O_{33}$

- The presence of large voids in calcium aluminate structure causes the formation, in the T° range of 900–1400 °C, of solid solution $C_{11}A_{7-x}Fe_xCaX_2$ with variable content of CaO and CaX_2 as well as part substitution of Al_2O_3 (up to 0.7 mol) for Fe_2O_3.
- With a structure similar to $C_{12}A_7$, single crystal of $Ca_{12}Al_{10.6}$–$Si_{3.4}O_{32}Cl_{5.4}$ has been prepared and shown to be structurally similar to Garnet $Ca_3Al_2(SiO_4)_3$.
- IR spectrum of $C_{12}A_7$ with 1.8% Na_2O does not reveal any significant difference compared with the spectrum of pure $C_{12}A_7$.
- Structural changes generated by paramagnetic species Mn/Fe are found by EPR techniques (STO 1989).
- Iron incorporation in $C_{12}A_7$ occurs, but only at very limited extent.

2.4.3.2.3 $CaAl_2O_4$

- The Al^{3+} in AlO_4 tetrahedra can be partially replaced by iron (11 wt %). At 1200 °C, 16.5% wt of CF substitutes CA. The formula $Ca_xAl_yFe_zSi_tO_{x+3/(x+y)+2t}$ with $4.4 < x < 4.8 < z < 1.3$ and $4.7 < y < 5.0 < t < 0.1$ has been proposed (GUL 1994).
- A little quantity of silica can enter the structure of CA. In a laboratory clinker, the formula $Ca_{1.04}Al_{1.896}Fe_{0.094}Si_{0.012}$ has been proposed (SCH 1983).

2.4.3.2.4 $CaAl_4O_7$

- The structure of CA_2 appears to be a suitable host lattice capable of being doped with quantities of minor elements (Nd^{3+}...) (GOO 1970).

TAB. 2.16 – Extent of solid solutions of C_3A (weight %).

SiO_2	Fe_2O_3	MgO	Na_2O	K_2O	P_2O_5	SO_3	Mn_2O_3	TiO_2
5	7	2	4	0.8	0.1	0.4	0.1	0.4

- The solid solution series with iron extends from CA_2 to 12 mol % CF_2 in the range 1000/1320 °C.

2.4.3.2.5 $CaAl_{12}O_{19}$

- CA_6 can accept a wide range and quantity of elements in solid solution.
- Addition of Mn, Fe, Co, Ni, Cu, Zn, as nitrate or acetate and Ti or Si as alkoxide to the citrate-based sol, yields the modification of the sequence and the temperature of formation of solid solution of CA_6. Direct formation of Ca $Al_{11.5}M_{0.5}O_{19}$ is promoted by the incorporation of MnO, Fe_2O_3, CoO, NiO, CuO, ZnO, while TiO_2 and V_2O_5 enhance the γ to α Al_2O_3 phase transformation (CIN 1998).
- It is proved by XRD and ^{27}Al NMR that by treating samples of Na β alumina in a $CaCl_2$ melt at 900 °C, the sodium ions completely exchange for bivalent calcium ions to form $CaO \cdot 10Al_2O_3$ with a β alumina structure. Between 900 and 1100 °C, $CaO \cdot 10Al_2O_3$ decomposes to $CaO_6 \cdot Al_2O_3$ and Al_2O_3 (VAN 1991).

2.4.3.3 Structure of Liquid, Glass and Transition from Liquid to Glass and Crystals

- ^{27}Al NMR MAS, Raman spectroscopy, ion dynamic simulation are the main techniques to study the structure of the liquid phase and glass of calcium aluminates.
- Calcium aluminate does not vitrify easily. A small quantity of glass can be obtained in a narrow range of compositions, by solar furnace melting, followed by a splat-quench technique, or by addition of traces of silica. A necessary but insufficient condition for easy glass formation is that the ratio of oxygen ions to network forming cations should be 2.5 (assumed to be Si^{4+} and Al^{3+}) (ONO 1970).
- The linear progression of the ^{27}Al NMR isotropic shift and the results for ionic dynamic simulation of the calcium aluminate liquid indicate that the average Al coordination increases as a function of increasing Al_2O_3 content (five and six coordination being more abundant).
- The comparison of variation *versus* Al_2O_3 content, of the correlation time for ^{27}Al spin lattice relaxation estimated from NMR and the shear relaxation time determined from viscosity measurements indicate that oxygen exchange between neighbour aluminate polyhedra plays an important role in the mechanism for the viscous flow. These two phenomena have the same microscopic origin.

- The chemical shift, measured by ^{27}Al NMR during the cooling of a liquid droplet of $CaAl_2O_4$, exhibit a negative temperature dependence consistent with the isotropic chemical shift measured on the glass and reported at Tg (glass transition) measured on the same sample (MAS 1995).
- Variations in the isotropic field and quadrupolar parameters ^{27}Al MAS-NMR *versus* composition and rate of cooling are explained by the structural arrangement in the next nearest neighbour (NNN) environment of Al.
- ^{27}Al NMR isotropic chemical shifts and quadrupolar coupling constants obtained for glasses along the $CaO-Al_2O_3$ join prepared by fast (splash) cooling quench technique shows Al in IV, V and VI coordination to oxygen at high alumina content ($CaO:Al_2O_3 < 1$).
- The averaged isotropic chemical shift of the glass increases with increasing CaO content which may reflect a decrease in the polymerisation state of the aluminate units although a similar trend cannot be detected in the corresponding crystal.
- ^{27}Al NMR studies on polycrystalline calcium aluminate show that only a slight variation in the chemical shift values is observed between the various AlO_4 tetrahedra with different numbers of bridging oxygen atom. The structural effects are reflected more sensitively in the quadrupole coupling data than in the chemical shifts (MUL 1986).
- The observed quadrupole coupling constant (C_Q) for the polycrystalline calcium aluminate may be related to the distortion of the AlO_4 tetrahedron which are quantified using the mean bond angle deviation from perfect tetrahedral symmetry and a calculated estimate of the geometrical dependence of the electric field gradient tensor. Correlation between these parameters and the C_Q values are observed even though only effects from the first coordination sphere are considered. The correlations allow assignment of the quadrupole coupling constant for the calcium aluminates CA_2, $C_{12}A_7$, and C_3A each containing two non-equivalent AlO_4 tetrahedra in the asymmetric unit.
- In situ high temperature Raman spectroscopy demonstrated that liquid is composed of molecular aluminate species similar to those found in the composition of glass.
- Glasses of CaO/Al_2O_3 (50/50 wgt %) prepared by sol gel technique (as in conventional sintering synthesis) contain Al atoms in tetrahedral and octahedral coordination. The hydroxyl groups are responsible for the larger coordination of Al atoms and no essential differences of structure exist between the sol–gel and melt quenched glasses.
- The Raman spectrum of vitreous $CaAl_2O_4$ is consistent with a polymerized network of tetrahedral aluminate. Addition of CaO ($C/A > 1$) results in the appearance of a progressive depolymerisation of tetrahedral network. At higher alumina contents than $CaAl_2O_4$, highly condensed alumina tetrahedra begin to appear in the glass structure.
- The radial distribution function method associated with the intensity comparison method shows that the coordination numbers of Ca and Al in $C_{12}A_7$ glass are, respectively, 4.2 and 5.6.

2.4.4 Stability and Phase Diagram

- Table 2.17 reports the main works dealing experimentally or by modelling with the $CaO-Al_2O_3$ system.
- Figure 2.10 shows the most common representation of the $CaO-Al_2O_3$ phase diagram in air from experimental data (redrawn from the published data). It is a part of the most complete diagram Ca, Al, Si, O, H_2 involving the partial pressure of oxygen and water.
- Figure 2.10 illustrates the following phase assemblages:

  (1) C + Liq.
  (2) C_3A + Liq.
  (3) $C_{12}A_7$ + Liq.
  (4) CA + Liq.
  (5) CA_2 + Liq.
  (6) $CA6$ + Liq.
  (7) A + Liq.
  (8) C + C_3A.
  (9) C_3A + $C_{12}A_7$.
  (10) $C_{12}A_7$ + CA.
  (11) CA + CA_2.
  (12) CA_2 + CA_6.
  (13) CA_6 + A.

- CaO, C_3A, $C_{12}A_7$, CA, CA_2, CA_6 and Al_2O_3 crystallise as primary phases in the $CaO-Al_2O_3$ phase diagram. The melting behaviour and the stability field of the calcium aluminates are not totally defined.
- C_3A melts incongruently at temperatures between 1541 and 1544 °C.
- CA is found to melt congruently or incongruently between 1602 and 1606 °C (ZAI 1990, ZAI 1991).
- CA_2 is found to melt congruently or incongruently between 1759 and 1775 °C (GOR 1951).
- CA_6 melts incongruently between 1783 and 1833 °C. Its stability field is not well defined because of its structural similarity with β-alumina and its affinity to form non-stoichiometry compounds.
- The absence of C_5A_3 and the presence of $C_{12}A_7$ only in moisture atmosphere require explanation. $C_{12}A_7$ is not strictly anhydrous and required hydroxyl group for the stability of its structure.
- The presence of oxidising atmosphere seems also to stabilize the structure of $C_{12}A_7$ but the real limit related to oxygen partial pressure is not known. $C_{12}A_7$ quenched in ambient air from temperatures higher than 900 °C, always contains small amounts of excess oxygen.
- By heating $C_{12}A_7$ in low H_2O and O_2 (<5 pm) atmosphere it transforms to C_5A_3 and C_3A).
- The heating of amorphous of calcium and aluminium oxide mixture at 900 °C in air leads to the formation of C_5A_3.

TAB. 2.17 – Principal works dealing experimentally or by modelling with the system CaO–Al$_2$O$_3$.

Experimental works	Experimental works
Sheperd *et al.* (SHE1909)	Kaufman (KAU 1979)
Rankin *et al.* (RAN 1915)	Eliezer *et al.* (ELI 1981)
Filonenko *et al.* (FIL 1949)	Chang *et al.* (CHA 1982)
Wisnyi (WIS 1955)	Berman *et al.* (BER 1984)
Langenberg *et al.* (LAN1956)	Koch (KOC 1988)
Gentile (GEN 1963)	Hallstedt (HAL 1991)
Nurse *et al.* (NUR 1965)	Zaitsev *et al.* (ZAI 1990 ZAI 1991)
Rolin *et al.* (ROL 1965)	Erickson *et al.* (ERI 1993)
Imlach *et al.* (IML 1971–1)	Pytel *et al.* (PYT 1999)
Udalov *et al.* (UDA 1969)	Mao *et al.* (MAO 2004)
Chatterjee *et al.* (CHA 1972)	
Nityanand *et al.* (NIT 1983)	
Jerebtsov *et al.* (JER 2001)	

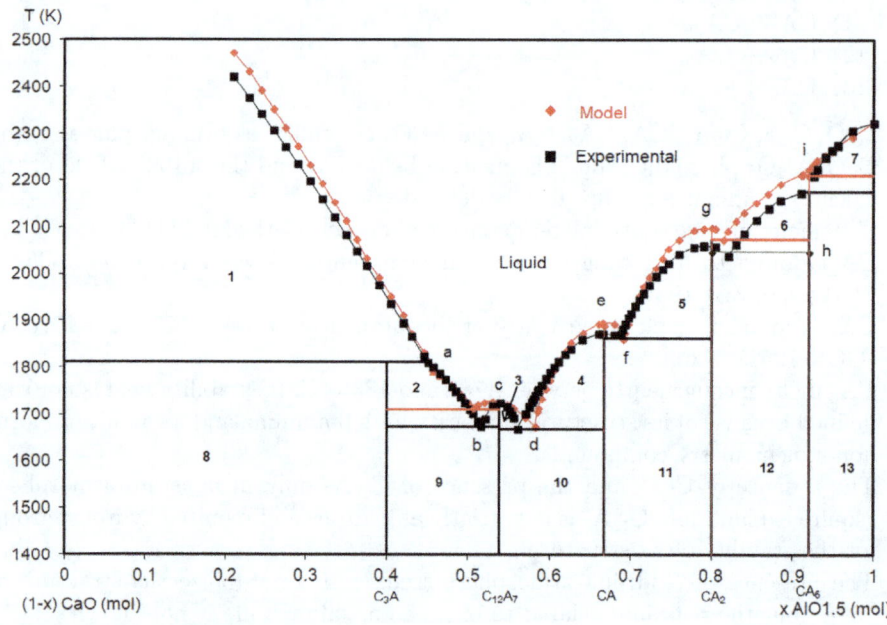

FIG. 2.10 – Experimental (black curve) and calculated (red curve) corresponding to the phase diagram of CaO–Al$_2$O$_3$ system. Key: Ta = 1539 °C, Tb = 1392 °C, Tc = 1421 °C, Td = 1373 °C, Te = 1603 °C, Tf = 11 592 °C, Th = 1757 °C, Tg = 1762 °C, Ti = 1902 °C.

Binary Chemical Systems

TAB. 2.18 – Probable phase found according a defined atmosphere.

Oxidising	Oxidising	Reducing	Reducing
Low H_2O	High H_2O	Low H_2O	High H_2O
$Ca_{12}Al_{14}$ CaO_2	$Ca_{12}Al_{14}$ $Ca(OH)_2$	$Ca_5Al_6O_{14}$	$Ca_{12}Al_{14}$ $Ca(OH)_2$

- By heating in air C_5A_3, it transforms in $C_{12}A_7$ slowly at 900 °C and faster at 1200 °C. The conditions of synthesis are important (sintering, coprecipitation, crystallisation from liquid, annealing from glass...).
- $C_{12}A_7$ and C_5A_3 were grown in the high temperature microscope, from melts in molar ratio ranging from 12:7 to 5:3 depending on the heat treatment.
- By heating at 900 °C in dry ($H_2O + O_2 < 5$ ppm) and reducing atmosphere (H_2) of an amorphous mixture of dried calcium and aluminium nitrate (ratio Ca/Al = 5/6), C_5A_3 is identified.
- Table 2.18 shows qualitatively the probable phase found according a defined atmosphere.
- $C_{12}A_7$ melts congruently between 1415 and 1477 °C in the oxidising or moisture atmosphere.
- C_5A_3 melts congruently in the reducing atmosphere at about $T = 1477$ °C.
- Phase assemblages can be determined from thermodynamic data (free energy or enthalpy of formation and entropy).
- Thermodynamic data have been measured using the Knudsen diffusion mass spectrometry method, calorimetry (dissolution or drop technique), electromotive force method, slag metal equilibria, but also calculated according to thermochemical cycles from the Janaf database, Thermodata or from an estimation calculated from existing phase diagrams.
- Table 2.19 shows the free energy of formation of the main compounds found in the binary system.
- Despite the uncertainty of the thermodynamic data, calculation of the diagram $CaO-Al_2O_3$ has been assessed. Solid phases are treated as stoichiometric compound and the liquid phase is described by a regular or sub-regular solution model. Figure 2.10 shows the result of a calculation based on an evaluation of the free energy of liquid from Riedler Kister coefficient. It is in agreement with the experimental results.

2.4.5 Synthesis of Calcium Aluminates

2.4.5.1 Introduction

- Calcium aluminates associated with other phases (calcium silicates, calcium alumino-ferrite) are produced industrially as hydraulic binder and refractory cements.
- Special applications use the optical or mechanical properties (IR transmission, high strengths optical fibre...) and require the synthesis and production of pure calcium aluminates or single crystals.

TAB. 2.19 – Gibbs free energy of formation depending on temperature is given by equation $A + (B \times T)$ in $J\,mol^{-1}$ (ALL 1981), T is the temperature in K.

	A	B
CA_6	−17 430	−37.27
CA_2	−16 400	−26.8
CA	−18 120	−18.62
$C_{12}A_7$	−12 300	−29.3
C_3A	−17 000	−32

- The sources of calcium oxide are calcium carbonate, $Ca(OH)_2$, oxalate, nitrate or alcoholate.
- The sources of Al are oxides or hydroxides (α-Al_2O_3, Boehmite, Gibbsite, Diaspore), salts (chloride, nitrates, sulphates...) or alkoxides (Aluminium; ethyl ester, isopropoxides and sec-butoxide...).
- Minor modifier elements, Mg, Zn, Na, K, Al, Fe, Cr, Mn and Zr, are introduced to form the desired polymorphs (stabilizers) or to facilitate the reaction of combination (mineralisers).
- The main parameters of the process are; the mixing time and energy, heating and cooling conditions (temperature, rate of heating and cooling, residence time, atmosphere), intermediate grinding, the quality of the vessel (platinum and rhodium or refractory).
- The final product is usually controlled by XRD, SEM and uncombined lime measurement.
- Usually, in most of the applications, calcium aluminate powders are produced by solid-state reactions between lime (CaO) or calcium carbonate and alumina powders at temperature in excess 1400 °C. Powders produced by this method typically have a very low specific surface area ($<1\ m^2/g$).
- It can be stated that wet chemical synthesis produces ceramics powder with high sinterability, high surface area, well defined chemical reaction, and homogeneous distribution of elements.
- Powder synthesis can be achieved at low temperatures. Many solution processes can be applied at small scale but expensive starting materials (metal alkoxide...) can be a serious limitation. Hydrolysis of organometallic compound, coprecipitation and hydrothermal synthesis are complicated procedures that are obstacles to reproducibility, cost and reliability of desired powders.
- The methods of synthesis can be divided into two main groups:
  - Direct sintering from oxides or salts.
  - Indirect synthesis from precursors (salts, hydrates...).

2.4.5.2 Direct or Solid-State Synthesis

- Calcium oxide, hydroxide or salts (carbonate, oxalates, sulphate nitrates...) and alumina (Bayerite, α-alumina) are mixed and heated at high temperatures (1450–1650 °C). If the total reaction is not achieved (indicated by the measurement of the uncombined lime), the product is ground and reheated at high temperatures. The operation is repeated until the desired percentage of free lime is reached (ALI 1989). High energetic attrition allows a sintering of pure calcium aluminate at lower temperatures than those required for the traditional sintering process, provided to prevent contamination during the attrition process.
- The techniques used to study the rate of formation of calcium aluminate *versus* temperature are:

    (1) Mixing different proportions of calcium oxide or carbonate and alumina (Gibbsite as example), heating the mixture at a given condition (temperature, time of sintering) and analysing the results (XRD, optical microscopy, SEM, uncombined lime...).
    (2) Studying the interface (layer thickness) between pellets of calcium oxide/alumina, or couple of calcium aluminate C_3A/CA_2, $C_{12}A_7/CA_2$, $C_{12}A_7/Al_2O_3$ at different temperatures.

- The reactions occur via diffusion of calcium, and the formation of calcium aluminates appears to fit the diffusion-controlled models of Jander or Ginsling–Brounstein. Intermediate or minor phases are formed depending on the operating conditions which increase the velocity of the reaction. Experimental and calculated phase diagrams show that calcium fluoride forms a ternary compound $3CaO \cdot 3Al_2O_3 \cdot CaF_2$. Its behaviour depends on the composition of the contact gaseous atmosphere.
- By means of EPR spectroscopy, it was possible to characterize the state of mechanically and thermally activated starting material using the signals of the paramagnetic species Mn^{2+}, Mn^{4+}, Fe^{3+} and O. Structural changes which were caused by the solid-state reaction were clearly indicated by the EPR fine structure of Fe^{3+}. In addition, Fe^{3+} ions favour the solid-state reaction which is characterized by the diffusion of Ca^{2+} ions. The conclusion about the function of the paramagnetic centres in the starting materials and intermediates was confirmed by external pressure treatment of the samples in the 1 GPa range (STO 1995).
- The high temperature of synthesis and its consequences on the energy consumption, the wear of the refractories which increases with the working temperature and the fast rate of cooling have led to the following soft methods.

2.4.5.3 Indirect Synthesis

2.4.5.3.1 Coprecipitation of Calcium and Aluminium Salt (Nitrate or Oxalate)

- A mixture of nitrate produces a homogeneous mixture that can be evaporated directly or treated with an ammonium salt and heated between 300 and 600 °C:

$$xCa(NO_3)_2 + yAl(NO_3)_3 + (2x + 3y)H_2O \rightarrow Ca_xAl_y(OH)_{(2x+3y)} + (2x + 3y) HNO_3.$$

- Nitric acid or ammonium nitrate is evaporated at a low temperature (300 °C) and an amorphous mass is obtained. Finally, the product is heated at 1450 °C. Aluminium sulphate or chloride can be used. The method using a precipitation of nitrate produces a homogeneous mixture that can be heated at lower temperatures. Nevertheless, it does not prevent from the use of platinum crucible. The necessity to evaporate a large quantity of nitric acid reduces the quantity that can be produced. Handling and recycling nitric acid require special vessel and equipment.
- Metastable orthorhombic CA is obtained by firing at 950 °C a gel of calcium and aluminium nitrate.

2.4.5.3.2 Dehydration of Synthetic Hydrates

- Preparation of the suitable hydrate is the first step. It can be prepared by coprecipitation and the method is similar to the preceding process. A mixture of Al–Ca hydroxide is treated by adding ammonium hydroxide and then fired after pre-treatment as washing, pelletizing to the desired calcium aluminates. C_3A, $C_{12}A_7$, CA, CA_2, CA_6 were prepared by this method at temperatures of 300–500 °C lower than that of conventional synthesis using oxide or carbonates as raw materials. For example, it is possible to obtain C_3A by dehydration of C_3AH_6.

2.4.5.3.3 Sol–gel Method

- Aluminium di(sec-butoxide)acetoacetic chelate is mixed with $Ca(NO_3)_2$ in isopropyl alcohol without precipitation. Addition of water induces a precipitate; it is dissolved by 1N HCl. It is dried by 14 à 18 days; heating at 900 °C results in the formation of amorphous Ca aluminate powder (UBE 1990).
- Aluminium chelate and calcium nitrate precursors were used to synthesize amorphous calcium aluminate powders by sol–gel processing at <900 °C. The method of preparation and results of characterization of the gels by thermogravimetric analysis, X-ray diffraction, scanning Auger microscopy, and single-point BET analysis are presented. An optimum heat treatment schedule consisting of heating the gel to 900 °C at 5 °C/min and holding for 16 h was developed to produce highly reactive, X-ray amorphous calcium aluminate powders.
- Aluminium sec butylate $(Al(OC_4H_9)_3$ was dissolved (added dropwise) in H_2O/HCl solution. In another beaker $Ca(NO_3)_2 \cdot 4H_2O$ was dissolved in ethanol. This solution was added into the first one. The solution was poured into a plastic beaker and aged 1–2 months (GOK 1991).

2.4.5.3.4 The Pechini Method

- Calcium and aluminium nitrates are dissolved in water, citric acid (C_3H_7OH) and ethylene glycol are added at 900 °C at 3 h. Citric acid forms polybasic acid chelates with metal cations. These chelates undergo polymerisation when heated with a polyhydroxyl alcohol such as ethylene glycol at 150 °C to form a polymeric resin. Further heating was performed to remove the resin at higher temperatures. Surface with high purity and high specificity was chemically synthesised at T° or below 900 °C. Specific surface areas of 10 m²/g were routinely obtained at 900 °C after 3 h of calcination.

2.4.5.3.5 Self-Propagating Combustion Synthesis

- Calcium aluminate can be synthesised using exothermic redox reactions between an oxidizer (metal nitrates) and a fuel (amides, hydrazides...). This method exploits an exothermic very rapid and self sustaining chemical reaction. The heat required to drive the chemical reaction is provided by the reaction itself (and not by an external source). The process involves the combustion of the corresponding metal nitrate plus either urea or carbohydrazide mixture at a temperature of 500 or 250 °C, respectively, under atmospheric pressure. The process yields foamy voluminous and fine oxide powders in 5 min.
- Nitrates decompose at lower temperatures than 700 °C as NO_2, NO and N_2O_5. Urea decomposes in Biuret (H_2N–CO NH–CO–NH_2), cyanuric acid, HCNO and NH_3 at 200 °C. Biuret decomposes at temperatures higher than 300 °C. CA, CA_2 and $C_{12}A_7$ are formed at 850 °C. C_3A is formed at 1050 °C and CA_6 at 1200 °C in a dry air atmosphere (CUN 1998, FUM 1966).

2.4.5.4 Special Process

2.4.5.4.1 Crystallisation from the Melt, Devitrification of Glass, Single Crystal Synthesis

- Uncertainty in the melting behaviour (congruency or not) and difficulties in obtaining glass by quenching pure calcium aluminate render difficult the synthesis of several aluminates by crystallisation of liquid. Nevertheless, it is possible to crystallize $C_{12}A_7$ by cooling. Single crystal is obtained by crystallisation of industrial products (phosphorus furnace slag) or cooling from the melts of synthetic mixture, or zone melting process. Rate of cooling, oxygen partial pressure and quality of the crucibles are the main parameters to control the size of the crystal.
- Single crystals of CA_2 and $C_{12}A_7$ are grown from the melt at the end of an iridium road at 1500–1700 °C by the Czochralski technique.

2.4.5.4.2 Ion Exchange

- Exchange of Na^+ by Ca^{2+} is performed by immersing commercial sodium beta alumina (Na β-alumina) in molten $CaCl_2$ at 1000 °C or eutectic composition $CaCl_2/Ca(NO_3)_2$ at 500 °C. The XRD peaks of the CA isomorph resemble the pattern of its parent Na β-alumina and are different from CA_6.

2.4.5.4.3 Synthesis of Special Aluminates

- C_2A is produced by mixing $C_{12}A_7$ and C_3A at 25 kbars and 1250 °C.
- C_5A_3 can also be obtained within the solid state by heating at 1200 °C C_3A and CA at 1200 °C in a flow of very dry N_2 or O_2.

- Orthorhombic CA is synthesised by heating at temperatures between 600 and 900 °C. An amorphous compound is obtained by coprecipitation of Al and Ca nitrate and 10% tartaric acid or organic polyol and citric acid (similar to Pechini process).
- CA_{10} is produced by ion exchange with samples of Na β-alumina treated in a $CaCl_2$ melt at 900 °C. Because the sodium ions in beta alumina are extremely mobile, it exchanges completely for bivalent calcium ions. By XRD and 27 Al NMR, it was proved that CA_{10} has the β alumina structure. Heat treatment at 900 and 1100 °C showed that the compound decomposes to CA_6 and Al_2O_3 and therefore is metastable at these temperatures.

2.5 SiO$_2$–Fe Oxides System

2.5.1 Introduction

- Five compounds are found in the SiO_2–Fe oxides system: $2FeO\text{–}SiO_2$ (Fayalite), $FeO\text{–}SiO_2$ (Ferrosilite), $FeO\text{–}Fe_2O_3\text{–}2SiO_2$ (Laihunite), $5FeO\text{–}Fe_2O_3\text{–}3SiO_2$ (Iscorite), $3FeO\text{–}Fe_2O_3\text{–}3SiO_2$ (Skyagite).
- Other compounds are stabilized by OH and Mg:
  - $Fe_7^{2+}Si_8O_{22}(OH)_2$ (Grünerite).
  - $Fe_{2.3}^{2+}Fe_{0.5}^{3+}Si_{2.2}O_5(OH)_{3.3}$ (Greenalite).
  - $Fe_2^{2+}Fe_2^{3+}SiO_5(OH)_4$ (Cronstedite).
  - $(Fe^{2+}Mg^{2+})Si_4O_{10}(OH)_2$ (Minesotaite).
  - $Fe_4^{2+}Mn_2^{2+}Fe_{2.25}^{3+}Al_{0.75}Si_6O_{20}(OH)_5$ (Deerite) (LAN 1977).

2.5.2 Mineralogy

2.5.2.1 Fayalite ($Fe_2^{2+}SiO_4$)

- Fayalite is the iron rich end member of the mineral Olivine $(Fe,Mg)SiO_4$ orthosilicate mineral solid solution. There is complete diadochy between the end members and Fayalite is the iron silicate end member. It is found as a mineral in ultra-mafic silica-poor igneous rocks and is named from the type locality, the Azores volcanic island, Fayal. The term Olivine Group also has a wider meaning for other minerals with similar crystal structures.

2.5.2.2 Ferrosilite ($Fe^{2+}SiO_3$)

- Ferrosilite is a rare mineral. Crystallographic and optical properties of minutes needles found in the lithophysae of an obsidian from Lake Naivasha (Kenya) correspond to those of pure FeSiO$_3$ as determined by extrapolation of series of monoclinic pyroxene) (BOW 1935).

2.5.2.3 Laihunite ($Fe^{2+}Fe^{3+}_2Si_2O_8$)

- Laihunite occurs in a Precambrian metamorphic magnetite deposit in Liaoning Province in Northeast China. It is associated with quartz, Fayalite and Ferrosilite.

2.5.2.4 Iscorite ($Fe_5^{2+}Fe_2^{3+}SiO_{10}$)

- Iscorite was named after a location where it was found, the floor of a reheating furnace at the Pretoria works of the South Africa Iron and Steel Industrial Corporation (ISCOR).

2.5.2.5 Skyagite ($Fe_3^{2+}Fe_2^{3+}Si_3O_{12}$)

- Skyagite is found in deeper-seated mafic granulite from kimberlite pipe in Siberia.

2.5.3 Structure, Polymorphism and Solid Solution

Table 2.20 shows the crystallographic characteristics of the principal iron silicates.

2.5.3.1 Fe_2SiO_4 (Fayalite)

- Fayalite is the iron-end member of the olivine solid-solution and as such crystallises with orthorhombic symmetry of the olivine structure (Fe,Mg)SiO$_4$. The structures consist of independent SiO$_4$ tetrahedra linked by Fe^{2+} in six-fold coordination. The oxygen atoms lie in sheets nearly parallel to the (100) plane and are arranged in approximate hexagonal close-packing. In accordance with full orthorhombic symmetry, the silicon oxygen tetrahedral points alternatively either way along both x and y directions. Half of the available octahedral voids occupied by Fe^{2+} and one eighth of the available tetrahedral voids by Si atom silicates based on stoichiometric Fayalite show a symmetric solitary quadrupole doublet. In deviation from stoichiometry, the doublets become asymmetric which is caused by the presence of Fe^{2+} and Fe^{3+} atoms.
- Orthorhombic Fayalite transforms to Spinel at high pressure and temperature (40–70 kbars and 700–1500 °C). The transition is reversible. The Spinel has a lattice constant of 8.235 Å and is 12% denser than Fayalite.
- A spineloid structure is found in a wide intermediate compositional range $x = 0.37$–0.73 in Fe$_{3-x}$Si$_x$O$_4$ at 3–9 GPa.

TAB. 2.20 – Crystallographic characteristics of the principal iron silicates.

Composition	$Fe^{2+}_2SiO_4$	$Fe^{2+}SiO_3$	$Fe^{2+}Fe^{3+}_2Si_2O_8$	$Fe^{2+}_5Fe^{3+}_2SiO_{10}$	$Fe^{2+}_3Fe^{3+}_2Si_3O_{12}$
Oxide formula	$2FeO \cdot SiO_2$	$FeO \cdot SiO_2$	$FeO \cdot Fe_2O_3 \cdot 2SiO_2$	$5FeO \cdot Fe_2O_3 \cdot SiO_2$	$3FeO \cdot Fe_2O_3 \cdot 3SiO_2$
Cement formula	f2S	fS	fFS2	f5FS	f3FS3
Mineral	Fayalite	Ferrosilite	Laihunite	Iscorite	Skyagite
Lattice	Orthorhombic	Orthorhombic	Monoclinic	Monoclinic	Cubic
Space group	$Pbnm$	$Pbca$	$P2_1/b$	$P2_1/m$	$Ia3d$
Z	4	16	4	2	
a (Å)	4.83	18.41	4.81	21.4	11.75
b (Å)	10.49	9.08	10.21	3.06	
c (Å)	6.1	5.24	5.81	5.88	
			$\alpha = 90.87°$	$\beta = 98°$	
V (Å^3)	309	876	285.6	381	1622
d (g/cm^3)	4.38	3.88	4.09	5.04	4.49
JCPDS N°	76-512	29-721	78-1435	74-535	
Ref	DIN 1990	LIN 1964	QIT 1988	SMU 1968	DEE 1962

- A complete Spinel solid solution between Fe_3O_4 and γ-Fe_2SiO_4 was found above 10 GPa. γ-Fe_2SiO_4 with a normal Spinel structure is stable at pressures above 7 GPa.
- The structure of Spinel consists of the cubic close packed oxygen atoms in which the Si^{4+} ions are coordinated with four oxygen atoms in a three tetrahedral form and the Fe^{2+} ions are coordinated with six oxygen atoms arranged in the octahedral form. Each oxygen atom coordinated to one Si^{4+} and three Fe^{2+} atoms. The site occupancy analysis indicates that the γ-Fe_2SiO_4 is of a mixed normal inverse Spinel structure since a large amount of Fe^{2+} ions has been displaced in the octahedral site.
- Electron density distribution in crystals of γ-Fe_2SiO_4, determined from single crystal X-ray diffraction data shows evidence for trigonally deformed distribution of 3d electrons around Fe^{2+}. Fe^{2+} is known to be in the high spin state in γ-Fe_2SiO_4 with six 3d electrons. The cation is octahedrally surrounded by six O^{2-} anions, but the exact symmetry of the coordination polyhedron is not cubic but trigonal, because of the flattening of the polyhedron along the direction of the threefold rotation axis. Five out of the six d electrons occupy each of the five 3d orbitals and the remaining one occupies the singlet level in the ground state. The observed peaks seem to originate in this trigonally deformed distribution of 3d electrons.

2.5.3.2 $Fe^{2+}SiO_3$ (Ferrosilite)

- It is the iron end-member of the Pyroxene family of chain silicate minerals. It crystallises as orthorhombic when it is produced at temperatures below the incongruent melting point. This inverts in a few hours at room temperature and pressure to a polysynthetic twinned monoclinic form.

2.5.3.3 $Fe^{2+}Fe^{3+}_2Si_2O_8$ (Laihunite)

- The structure of Laihunite has been derived from the frame of Olivine structure although about a quarter of the octahedral sites of Olivine are not occupied by Fe atoms in Laihunite. The unit cell of Laihunite is smaller than that of Fayalite because of the replacement of larger Fe^{2+} by smaller Fe^{3+} ions and of generation of vacancies at the V sites. The chemical formula of Laihunite is V0.4Fe^{2+}0.8-Fe^{3+}0.8SiO_4 indicating a non-stoichiometry.

2.5.3.4 $Fe^{2+}_5Fe^{3+}_2SiO_{10}$ (Iscorite)

- It is based on a cubic close packed arrangement of oxygen atoms with ferrous ions in octahedral interstices. It can also be described in terms of alternating layers of FeO and Fe_3O_4 perpendicular to 100 with chains of SiO_2 running along 010 in the Fe_3O_4 layers (SMU 1969).

2.5.3.5 $Fe_3^{2+}Fe_2^{3+}Si_3O_{12}$ (Skyagite)

- It represents a mixed valence Garnet forming a solid solution with Almandine and Andradite.

$Fe_3^{2+}Fe_2^{3+}Si_3O_{12}$	$Fe_3^{2+}Al_2^{3+}Si_3O_{12}$	$Ca_3^{2+}Al_2^{3+}Si_3O_{12}$
Skyagite	Almandine	Andradite

- The crystal structure has been carried out on samples synthesized at high T° and pressure. All Fe^{3+} ions are octahedrally coordinated and Fe^{2+} ions are in dodecahedral coordination. The stability range is found from 9 to 13 GPa. The Garnet of solid solution Andradite/Skyagite are isotropic and cover the complete range Fe^{3+}/total Fe from 0 to 1.

2.5.4 Liquid FeO–Fe_2O_3–SiO_2

- High T° X-ray diffraction technique indicates that SiO_4 tetrahedral units exist individually in the more basic region of the FeO–Fe_2O_3–SiO_2 melt and some may polymerize to form silicate anions with increasing SiO_2 content. A constant coordination number of about four is obtained for Si–O pairs in this composition range. The SiO_4 tetrahedron is thus the fundamental local ordering units in FeO–Fe_2O_3–SiO_2 melts.
- The Si–Si distance which corresponds to the inter SiO_4 distance gradually increases as the composition increases from 27.5 wt % SiO_2 and then remains nearly constant at the composition beyond 30% SiO_2. This constant value is comparable with the value of the Si–O–Si distance with a bond angle of less than 180°. This indicates that some of the SiO_4 tetrahedra polymerize to form larger silicate anions.
- The distance of the Fe–O varies in a similar way with the Si–Si correlation and the variations correspond to a change in the position of Fe from octahedral to tetrahedral sites. The small local ordering units of the SiO_4 tetrahedra are insensitive to temperature whereas the distribution of these local ordering units depends on temperature.
- With respect to the near neighbour correlations the structure of molten FeO–Fe_2O_3–SiO_2 is insensitive to changes in the Fe_2O_3 content due to the change in oxygen partial pressure. Thus, the SiO_2 content primarily determines the fundamental feature of the structure of molten FeO–Fe_2O_3–SiO_2 rather than the Fe_2O_3 content.
- The viscosity maximum near Fayalite composition can be qualitatively interpreted by a combination of silicate anion polymerization effect and cation effect that can be accounted for from the structural information.

TAB. 2.21 – Main references describing the FeO–Fe$_2$O$_3$–SiO$_2$ system.

Experimental works	Experimental works
Bowen et al. (BOW 1932)	Goel et al. (GOE 1980)
Darken (DAR 1948)	Fei et al. (FEI 1986)
Gurry et al. (GUR 1950)	Bjorman et al. (BJO 1985)
Schuhman et al. (SCH 1953)	Wu et al. (WU 1993)
Muan (MUA 1955, MUA 1956)	Sellesby (SEL1997)
Guo et al. (GUO 1988)	

2.5.5 Stability and Phase Diagram of the FeO–Fe$_2$O$_3$–SiO$_2$ System

- In the FeO–Fe$_2$O$_3$–SiO$_2$ system, the composition of the condensed phases is dependent on the partial pressure of oxygen of the gas phase.
- The principal works on the ternary FeO–Fe$_2$O$_3$–SiO$_2$ system are referenced in table 2.21.
- In the FeO–Fe$_2$O$_3$–SiO$_2$ ternary diagram, Wüstite, Hematite, Magnetite, Hematite, Fayalite and silica are the oxides occurring as primary phase (figure 2.11).
- The temperature of fusion of iron oxide FeO in contact of silica is sensitive to gas composition; hence that fusion may be accomplished in some cases by control of the gas atmosphere without changing the temperature (DAR 1948).
- In the FeO–Fe$_2$O$_3$–SiO$_2$ system, the composition of the condensed phases is dependent on the partial pressure of condensed of oxygen of the gas phase.
- The experimental method to determine liquidus surface in the ternary FeO–Fe$_2$O$_3$–SiO$_2$ system from 1250 to 1450 °C consisted of equilibrating small samples in Pt crucibles or in iron crucible or saturated with silica in bubbling CO–CO$_2$ mixture, quenching and microscopic examination.
- The pure Fayalite melts incongruently with separation of iron at 1205 ± 2 °C. The presence of Grünerite in the natural Fayalite extends the range of melting temperatures. The liquid in contact with Fayalite is 2.25% Fe$_2$O$_3$.
- There are two ternary invariant points that are found at PO$_2$ pressures ranging from $10^{-10.9}$ to 1 atm:

    ○ At one invariant point (1140 °C) the coexisting phases are; Tridymite (SiO$_2$), Fayalite (2FeO·SiO$_2$) and Magnetite (FeO·Fe$_2$O$_3$). The liquid has the composition, 35% SiO$_2$, 11% Fe$_2$O$_3$, 54% FeO and the gas phase has a partial pressure of oxygen equal to 10^{-9} atm.

    ○ At the other invariant point (1140 °C), the coexisting phases are; Fayalite, Magnetite and Wüstite with a composition 35% Fe$_2$O$_3$. The liquid has a composition; 22% SiO$_2$, 14% Fe$_2$O$_3$, 64% FeO and the gas phase has a partial pressure of oxygen of $10^{-9.9}$ atm.

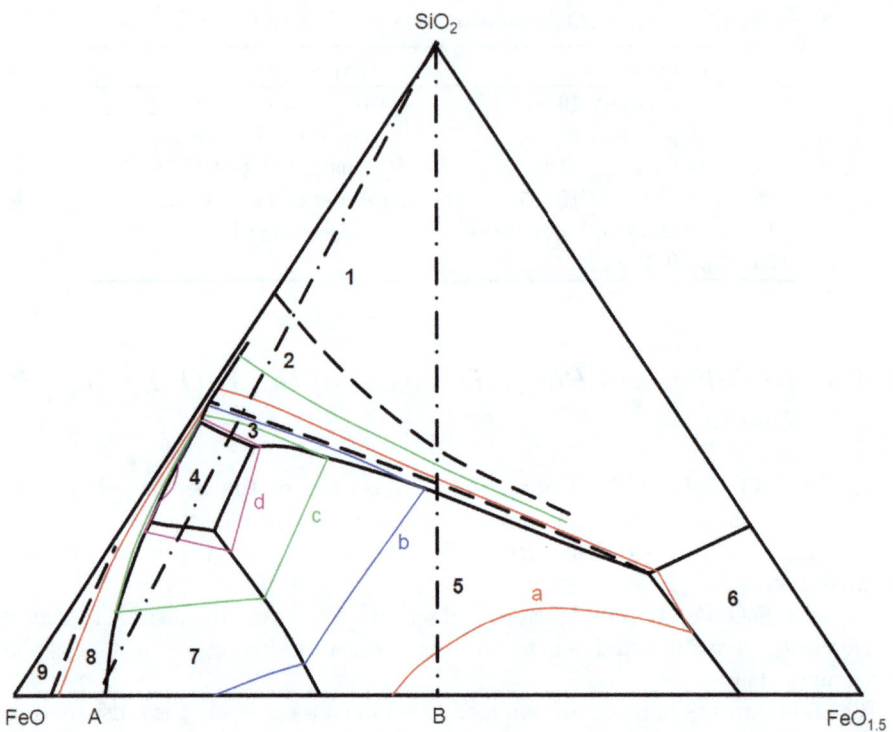

FIG. 2.11 – FeO–Fe$_2$O$_3$–SiO$_2$ ternary phase diagram. Isotherms; 1200 °C in magenta (d), 1300 °C in green (c), 1400 °C in blue (b) and 1500 °C in red (a). The line SiO$_2$-A and SiO$_2$-B refer to the pseudo-binary diagrams presented in figures 2.12 and 2.13. Key: (1) Two liquids, (2) Cristobalite, (3) Tridymite, (4) Fayalite, (5) Magnetite, (6) Hematite, (7) Wüstite, (8) Iron, (9) two liquids.

- The ternary invariant point where Tridymite, Magnetite and Hematite, liquid and gas are in equilibrium was not determined experimentally but is inferred to be located at a temperature of approximately 1455 °C with liquid composition of; 15% SiO$_2$, 69% FeO, 16% Fe$_2$O$_3$ and a PO$_2$ slightly higher than 1 atm.
- The lowest T° at which the liquid can exist under equilibrium conditions in an iron oxide silica mixture is approximately 1455 °C at PO$_2$ of 1 atm. With PO$_2$ decreasing to 10^{-4} atm the lowest temperature of liquid existence is essentially the same.
- With further decrease in PO$_2$, liquid is present at rapidly decreasing temperatures reaching a minimum of 1140 °C for PO$_2$ equal to 10^{-9} atm for the mixture above the FeO–SiO$_2$ iso-silica mole fraction line and a minimum of 1150 °C for PO$_2$ equal to $10^{-9.9}$ atm for the mixture below 2FeO·SiO$_2$ iso-silica line.
- With further decreases in PO$_2$ pressure, the lowest temperature of liquid existence increases somewhat until metallic iron appears as a phase at PO$_2$ pressure between 10^{-12} and 10^{-13} atm).

- In the FeO–Fe$_2$O$_3$ rich part of the system (FeO/Fe$_2$O$_3$ = 8/2 and 6/4), a two liquid field area is proposed based on the assumed position of a metastable eutectic involving a dehydroxylated compound Fe$_3$SiO$_2$O$_7$ analogous to the phyllosilicate Greenalite.
- Ferrosilite melts incongruently in the range 20–40 Kb to a Fe$_2$SiO$_4$ rich liquid plus quartz.
- No crystalline compound of the composition FeSiO$_3$ forms at any temperature where liquid occurs in the system.
- Iscorite does not occur in the phase diagram probably because of the unusual (industrial rather than natural) condition of formation.
- With respect to Wüstite, Magnetite and Fayalite, X-ray results remains constant after a heating 30 min at 1100 °C).
- Laihunite is a phase that is stable, not only at high pressures but also at ordinary pressures. Its formation and stabilization are closely related to oxygen fugacity. However, its mode of formation is different at different pressures. At relatively low pressure, its formation is closely related to the oxidation of Fayalite. At higher pressure, the oxidation of Fayalite together with Ferrosilite is responsible for the formation of Laihunite. At even higher pressures, the formation of Laihunite is only pertinent to the oxidation of Ferrosilite. Therefore, the association of natural Laihunite with Fayalite and Ferrosilite provides a clue to estimate the pressure conditions of its formation.
- The composition of the liquids cannot be accurately expressed in a binary diagram and full relations are expressed only with the aid of a ternary diagram in which FeO/Fe$_2$O$_3$/SiO$_2$ are taken as component. It is possible to draw pseudo-binary diagrams *i.e.* SiO$_2$-A or SiO$_2$-B (figures 2.12 and 2.13).
- The pseudo-binary diagram of the SiO$_2$–Fe oxide system in air is shown in figure 2.12. A part of this diagram has been calculated using a Redlich Kister model for the excess free energy of the liquid.
- In the pseudo diagram SiO$_2$–FeO (figure 2.13), the quantity of Fe$_2$O$_3$ varies up to 12%.
- Gibbs Duhem equations are used to calculate activities of SiO$_2$, FeO and O in ternary silicate melt. A quantitative relation is derived between tangent intercept of isoactivity curves for the various components and this relation is the basis of a graphical procedure of constructing isoactivity curves.
- Calculated from experimental data on the activity of oxygen in Fe–Si–O melts at 1550 °C it appears that the temperature has little or no noticeable effects on the oxide activities in SiO$_2$–FeO–Fe$_2$O$_3$ melts.
- Thermochemical data set which may be used to compute experimentally determined phase equilibrium, as closely as possible, are established.
- The models assessed to describe the system FeO–Fe$_2$O$_3$–SiO$_2$ differ by the description of the liquid phase (as a solution of the components Fe, FeO, FeO$_{1.5}$, SiO$_2$) using binary interaction by a form of the 'Margules activity model' or using an ionic sublattice model while Fayalite is treated as a stoichiometric phases.

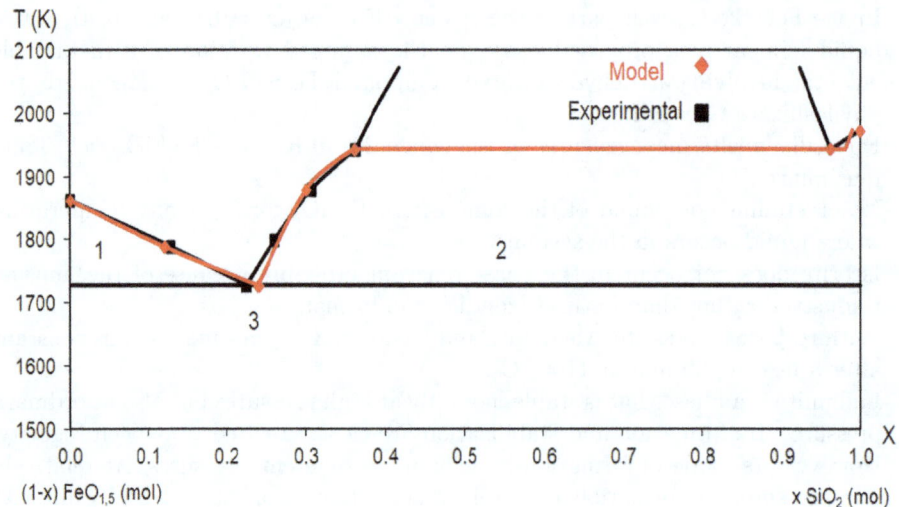

FIG. 2.12 – Pseudo-binary diagram of the SiO_2–Fe oxide system in air. Key: (1) Magnetite + liquid, (2) Cristobalite + liquid, (3) Hematite + Tridymite. The area Tridymite + liquid is not shown.

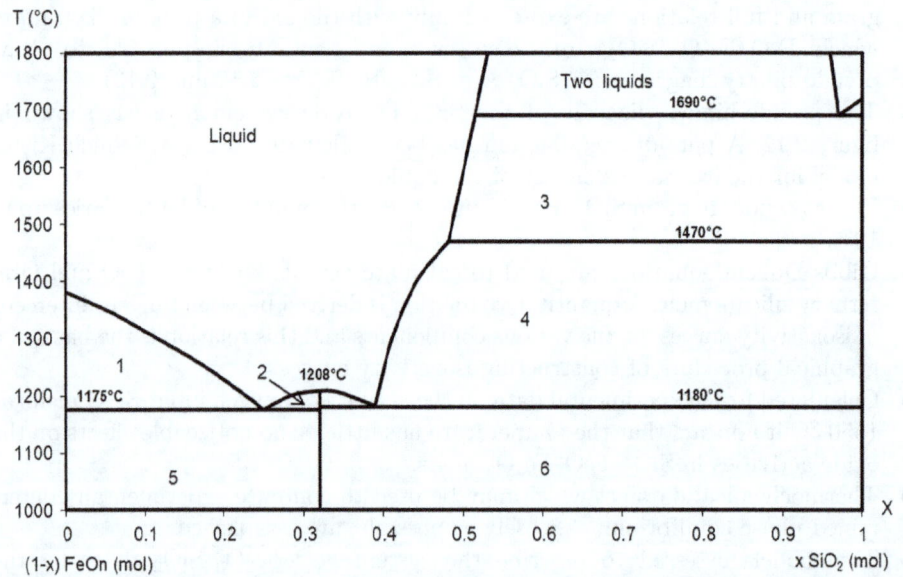

FIG. 2.13 – Pseudo binary diagram SiO_2–FeO. Key: (1) Wüstite, (2) Fayalite + liquid, (3) Cristobalite + liquid, (4) Tridymite + liquid, (5) Fayalite + Wüstite, (6) Fayalite + SiO_2.

2.5.6 Synthesis of Iron Silicates

2.5.6.1 Fayalite

- Fayalite ($2FeO-SiO_2$) is prepared by heating the calculated preparation of silica and ferrous oxalate in an iron crucible in the furnace at 1100 °C.
- Pure Fayalite (Fe_2SiO_4) is obtained by sintering SiO_2 + Wüstite at 1150 °C during 24 h under a controlled partial pressure of oxygen. A stoichiometric mixture of FeO and SiO_2 was intimately mixed and reacted at 1000 °C and 30 kbars to obtain α-Fayalite with Olivine structure.
- α-Fayalite was reacted at 900 °C and 70 kbars to obtain the Spinel structure.
- Spinel Fe_2SiO_4 was synthesised from Fayalite at a pressure of 65 kb at 800 °C for 40 min). Ferrous oxalate ($FeC_2O_4 2 \cdot H_2O$) and silica in a flow of argon. At 1250 °C in an iron crucible.

2.5.6.2 Ferrosilite

- In the $FeO-SiO_2$ system, $FeSiO_3$ does not occur at liquidus T° (but, it could be obtained at lower T° by devitrification of glass at 660 °C).
- Presence of Orthoferrosilite ($FeSiO_3$) which was recently found to be stable at pressures greater than 20 kbars and high temperature was detected in the run at T° above 850 °C. It suggests that Fe_2SiO_4 decomposes into $FeSiO_3$ + FeO.

2.5.6.3 Glass Formation

- Gels in the $FeO-SiO_2$ Fe_2O_3 system have been prepared from $Si(OC_2H_5)_4$ and Fe $(OC_2H_5)_3$ (ethanol and $FeCl_3$) and Fe_2O_3 in the following ratios 5/10/20/40.
- The transformation gel to glass can be been studied by Mössbauer spectroscopy up to 1000 °C.
- It was possible to prepare amorphous iron silicate with all the iron atoms present as Fe^{3+} by the gel route: this is not possible by melting the oxides.
- A progressive reduction of iron was observed in the samples with less iron oxide in spite of the oxidising conditions of the thermal treatment. The crystallization of SiO_2 was coincident with the presence of ferrous iron. No discontinuity was observed during the passage from gel to glass.
- It is the end member of the Pyroxene. It crystallises as orthorhombic when it is produced at temperatures below the incongruent melting point. This inverts in a few hours at room temperature and pressure to a polysynthetic twinned monoclinic form.
- Gels in the $FeO-SiO_2$ Fe_2O_3 system have been prepared from $Si(OC_2H_5)_4$ and Fe $(OC_2H_5)_3$ (ethanol et $FeCl_3$) and Fe_2O_3 in the following ratios 5/10/20/40.
- It was possible to prepare amorphous iron silicate with all the iron atoms present as Fe^{3+} by the gel route: this is not possible by melting the oxides.

- A progressive reduction of iron was observed in the samples with less iron oxide in spite of the oxidising conditions of the thermal treatment. The crystallization of SiO_2 was coincident with the presence of ferrous iron. No discontinuity was observed during the passage from gel to glass (GUG 1982).

2.6 CaO–Fe Oxides System

2.6.1 Introduction

- The following compounds occur in the system:
  (1) CaO–Fe_2O_3 (F) in the presence of a little amount of FeO (f); $Ca_2Fe_2O_5$ (in short hand notation C_2F), Ca Fe_2O_4 (CF), $CaFe_4O_7$ (CF_2) and $Ca_4Fe_{14}O_{25}$ (C_4fF_7).
  (2) CaO–FeO–Fe_2O_3 (table 2.22).

2.6.2 Mineralogy

- $Ca_2Fe_2O_5$ is found in the nature as Srebrodolskite (so named after the mineralogist Boris Srebrodolsky). The type location for this mineral is Kopeisk in the Chelyabinsk coal basin (Chelyabinsk Oblast, Russia).

2.6.3 Structure, Polymorphism and Solid-Solutions

2.6.3.1 $Ca_2Fe_2O_5$ (C_2F)

- $CA_2Fe_2O_5$ crystallises with orthorhombic symmetry (table 2.23).
- The structure can be derived from the parent cubic Perovskite by inserting chains of vacancies along (110) axes in the alternate layers. In $Ca_2Fe_2O_5$, iron is both octahedrally and tetrahedrally coordinated by oxygen and the crystal structure is built up from chains of alternating iron-oxygen octahedra and tetrahedra, running parallel to the b-axis with calcium atoms in the holes between the polyhedra. The octahedra and tetrahedra are however considerably distorted atoms around each calcium (BER 1971, BER 1959, COL 1970, GRE 1973, POU 1977).
- $Ca_2Fe_2O_5$ has been found to take up additional oxygen under high pressure and elevated temperatures; $Ca_2Fe_2O_{5.12}$ is obtained.
- Mössbauer spectra in the range 4.2/290 K establish that all the original material has been transformed into a new phase which is still structurally related to the original lattice. The additional oxygen enters the tetrahedral site layers. The disorder in the material results in at least six distinct lattice sites for iron of which five contain Fe^{3+} cation in both octahedral and tetrahedral coordination and the sixth contains Fe^{4+} cations (GIB 1994, GRA 1969).

TAB. 2.22 – Main stoichiometric compounds in the CaO–FeO–Fe$_2$O$_3$ system.

Chemical composition	Simplified formula	Authors
CaFe^{2+}Fe$^{3+}{}_2$O$_5$	CfF	Cirilli et al. (CIR 1952–1)
CaFe$^{2+}{}_2$Fe$^{3+}{}_2$O$_6$	Cf2F	Alota et al. (ALO 1972)
CaFe$^{2+}{}_3$Fe$^{3+}{}_2$O$_7$	Cf$_3$F	Cirilli (CIR 1952–1)
Ca$_2$Fe$^{2+}{}_2$Fe$^{3+}{}_{14}$O$_{25}$	C$_2$f$_2$F$_7$	Ara et al. (ARA 1988)
Ca$_2$Fe$^{2+}{}_3$Fe$^{3+}{}_6$O$_{14}$	C$_2$f$_3$F$_3$	Malaman et al. (MAL 1988)
Ca$_2$Fe$^{2+}{}_3$Fe$^{3+}{}_4$O$_{11}$	C$_2$f$_3$F$_2$	Martin et al. (MAR 1935)
Ca$_2$Fe$^{2+}{}_4$Fe$^{3+}{}_{18}$O$_{33}$	C$_2$f$_4$F$_9$	Ara et al. (ARA 1990)
Ca$_2$Fe$^{2+}{}_5$Fe$^{3+}{}_4$O$_{13}$	C$_2$f$_5$F$_2$	Schenk et al. (SCH 1937)
Ca$_{2.99}$Fe$^{2+}{}_{0.55}$Fe$_{14.3}$O$_{25}$	C$_3$f$_{0.55}$F$_7$	Kar et al. (KAR 1987)
Ca$_3$Fe^{2+}Fe$^{3+}{}_{14}$O$_{25}$	C$_3$fF$_7$	Holmquist et al. (HOL 1960)
Ca$_{3.56}$Fe$^{2+}{}_{0.06}$Fe$^{3+}{}_{14.25}$O$_{25}$	C$_{3.6}$f$_{0.06}$F$_7$	Kar et al. (KAR 1985)
Ca$_4$Fe^{2+}Fe$^{3+}{}_8$O$_{17}$	C$_4$fF$_4$	Philip et al. (PHI 1960)
Ca$_4$Fe^{2+}Fe$^{3+}{}_{14}$O$_{25}$	C$_4$fF$_7$	Schurmann et al. (SCH 1973)
Ca$_4$Fe^{2+}Fe$^{3+}{}_{16}$O$_{29}$	C$_4$fF$_8$	Philips et al. (PHI 1960)
Ca$_4$Fe^{2+}Fe$^{3+}{}_{18}$O$_{32}$	C$_4$fF$_9$	Burdese et al. (BUR 1952)
Ca$_4$Fe$^{2+}{}_2$Fe$^{3+}{}_{18}$O$_{33}$	C$_4$f$_2$F$_9$	Braun et al. (BRA 1960)
Ca$_{7.2}$Fe$^{2+}{}_{0.8}$Fe$^{3+}{}_{30}$O$_{53}$	C$_{7.2}$f$_{0.8}$F$_{15}$	Hughes et al. (HUG 1967)

2.6.3.2 CaFe$_2$O$_4$ (CF)

- CF crystallises with orthorhombic symmetry (table 2.23).
- The structure is made up of layers of atoms at ¼ and ¾ in z axis. There are 2 non-equivalent Fe ions, each octahedrally surrounded by oxygen atoms (Fe–O distances varying from 1.98 to 2.09 Å). Each calcium atom has eight oxygen atoms surrounding it. It is at the centre of a triangular prism of six oxygen atoms with Ca–O distances of 2.36–2.51 Å: one rectangular face of the prism is narrower (than the other two) and oxygen atoms coming out of the centres of the two wider faces are at a distance of 2.53–2.58 Å: a ninth oxygen atom coming out of the centre of the narrow face might be considered as a neighbour of the calcium, but the distance to it is 3.41 Å (DEC 1957, HIL 1956).

2.6.3.3 CaFe$_4$O$_7$ (CF$_2$)

- CF$_2$ crystallizes in the monoclinic system.
- Its structure is related to that of the hexagonal ferrite. It is constituted by an alternating stacking process along c axis of two structural blocks. They are characterized by a plane of trigonal based FeO$_5$ bipyramids surrounded by two mixed Fe–Ca layers.
- A triple layer of iron atoms is formed by a plane of mixed tetrahedral and octahedral polyhedra surrounded by two octahedral polyhedral. This ordering is

TAB. 2.23 – Crystal characteristics of $Ca_2Fe_2O_5$, $CaFe_2O_4$, $CaFe_4O_7$ and $(CaFe)_4Fe_{14}O_{25}$.

Chemical formula	$Ca_2Fe_2O_5$	$CaFe_2O_4$	$CaFe_4O_7$	$(CaFe)_4Fe_{14}O_{25}$
Lattice	Orthorhombic	Orthorhombic	Monoclinic	Rhombohedral
Space group	$Pnma$	$Pnam$	$C2$. Cm or $C2/m$	$R3c$
Z	4	4		
a (Å)	5.4253	9.23/9.16	10.409	6
b (Å)	14.7687	10.705/10.67	6.005	95
c (Å)	5.598	3.024/3.012	31.640	
β			96°30	
V (Å^3)	446.3	298.8	1964	2961.81
d (g/cm^3)	4.00	4.53/4.81	4.52/4.44	4.6
JCPDS N°	38 408	32 168	39 1033	13 343
Ref	BER 1959	DEC 1957	PHI 1958	BRA 1960

also encountered in the spinel structure (mixed system). It is ferromagnetic with a curie point of 135 °C (MIL 1986, CHE 1962).

2.6.3.4 Compounds of the CaO–FeO–Fe_2O_3 System

- These compounds are located either along the CF/FeO line or close to CF_2 (circle in figure 2.14).
- The compounds along the line CF/FeO are; $Ca_4Fe_9O_{17}$ ($Cf_{0.25}F$), $CaFe_3O_5$ (CfF), $Ca_2Fe_{11}O_{11}$ ($Cf_{1.5}F$), $CaFe_4O_{11}$ (Cf_2F), $Ca_2Fe_9O_{13}$ ($Cf_{2.5}F$) and $CaFe_5O_7$ (Cf_3F).
- They can be represented by the general formula $CaFe_{2+n}O_{4+n}$ equating to one of the following; $n = 0.25, 1, 1.5, 3, 2.5$ and 3.
- They crystallise in the orthorhombic system and the structure derive from each other by a stacking process of blocks with the $CaFe_2O_4$ composition and blocks with the FeO (Wüstite) composition like for the other terms of the series with the integer n used to distinguish them.
- CF* and f* are the block deduced from CF and f by rotation of circa 180° around axis c: the following stacking sequences is obtained: $CaFe_{2+n}O_{4+n}$ (EVR 1977, HOL 1960, HUG 1967, JAS 1975):
    - $n = 1$ $CaFe_3O_5$ CF–f–CF*–f*.
    - $n = 1.5$ $Ca_2Fe_7O_{11}$ f*–CF–f–CF*–f*.
    - $n = 2$ $CaFe_4O_6$ CF–f–f–CF*–f*–f*.
    - $n = 2.5$ $Ca_2Fe_9O_{13}$ f*–CF–f–f–f–CF*–*.
    - $n = 3$ $CaFe_5O_7$ CF–f–f–f–CF*–f*–f*–f*.
- Figure 2.14 represents the position of the phases Fe_2O_3, Fe_3O_4, FeO, C_2F, CF, C_4f F_8, C_4fF_4, Cf_3F and CfF at the temperature of the solid state.

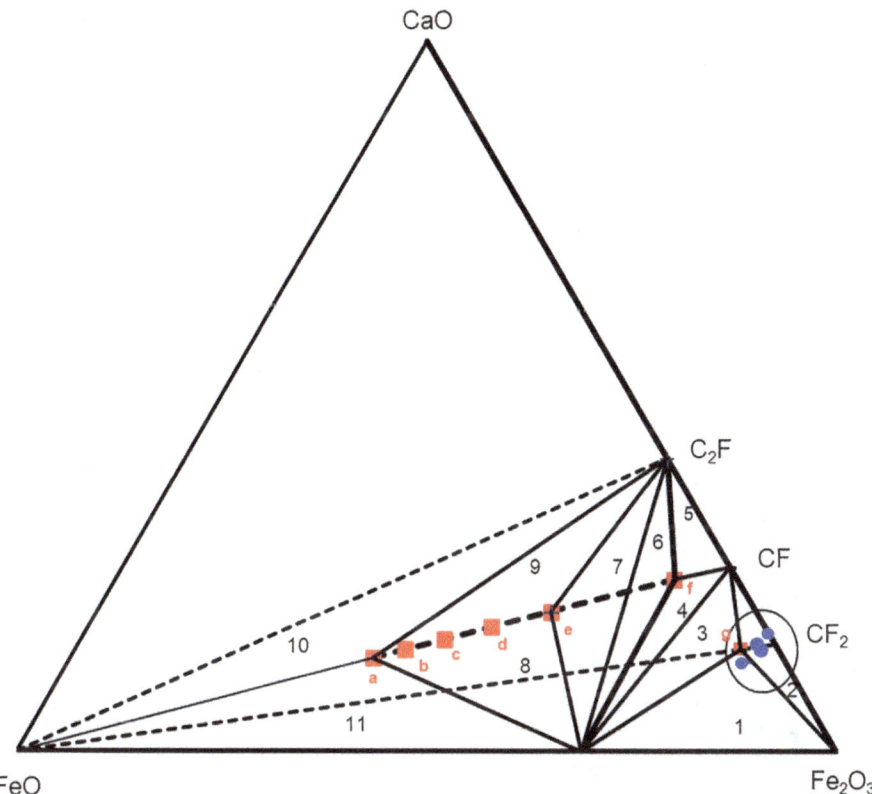

FIG. 2.14 – Position of the compounds in the CaO–Fe oxide system. Key: a – (Cf_3F), b – ($Cf_{2.5}F$), c – (Cf_2F), d – ($Cf_{1.5}F$), e – (CfF), f – ($Cf_{0.25}F$) or (C_4fF_4). Circle = compounds close to CF_2 i.e. g – (C_4fF_8). Numbers 1–11 represent the phase assemblages at the solid state.

- Compounds close to CF_2 are: $Ca_4Fe_{17}O_{29}$, (C_4fF_8), $Ca_4Fe_{19}O_{32}$ (C_4fF_9), $Ca_3Fe_{15}O_{25}$ (C_3fF_7), $Ca_2Fe_{22}O_{33}$ ($C_2f_4F_9$), $Ca_4Fe_{20}O_{33}$ ($C_4f_2F_9$), $Ca_{7.2}Fe^{2+}_{0.8}Fe^{3+}_{30}O_{53}$ ($C_{7.2}f_{0.8}F_{15}$) and $(CaFe)_4Fe_{14}O_{25}$ ($(Cf)_4F_7$).
- Calcium iron oxide compounds $Ca_4Fe_{14}O_{25}$ are stabilized by small amounts of a third component that can be Y^{3+}.
- $Ca_4Fe_{14}O_{25}$ and $Ca_4Fe^{2+}_2Fe^{3+}_{18}O_{33}$ are stabilized by Mg^{2+} (BRA 1960).

2.6.4 Stability and Phase Diagram

- Phase equilibrium in the system CaO–Iron oxide depends on oxygen pressure because of the stable existence of iron in the Fe, Fe^{2+} and Fe^{3+} oxidation states. When equilibria are studied in air or oxygen, it is possible to show the results without serious error.

- Similarly, when conditions are sufficiently reducing to cause the appearance of metallic Fe, the oxide compositions can be represented in the CaO–FeO system.
- At PO_2 value between these two extremes, it is more accurate to represent the equilibria in terms of the CaO FeO–Fe_2O_3 system.
- The main references for the CaO–Fe oxides system are shown in table 2.24.

2.6.4.1 Solidus in the System CaO–FeO–Fe_2O_3

- Data available in the literature are used to construct the subsolidus equilibria at a temperature around 1000 °C (figure 2.15).
- The phase assemblages at equilibrium are given by the following numbers on figure 2.15 (IML 1971–2, HAR 1984, TIM 1970, GUR 1950):

(1) Fe + FeO solid solution + CaO.
(2) CaO + FeO + C_2F.
(3) FeO + C_2F + Cf_3F.
(4) FeO + Cf_3F + Fe_3O_4.
(5) Fe_3O_4 + Cf_3F + CfF.
(6) C_2F + Cf_3F + CfF.
(7) C_2F + CfF + Fe_3O_4.
(8) C_2F + Fe_3O_4 + C_4fF_4.
(9) CF + C_4fF_4 + Fe_3O_4.
(10) Fe_2O_3 + Fe_3O_4 + C_4fF_8.
(11) CF + C_4fF_8 + Fe_2O_3.
(12) C_2F + CF + C_4fF_4.
(13) CF + Fe_2O_3 + C_4fF_4CF + Fe_2O_3 + C_4fF_4.

2.6.4.2 CaO–Fe_2O_3 System

- The Redlich Kister equation for the integral excess Gibbs energy has been applied to model the diagram CaO–Fe_2O_3.
- At liquidus temperature, in air, the CaO–Fe_2O_3 phase diagram is shown in figure 2.16.
- C_2F melts congruently.
- CF decomposes above 1216 °C by a peritectic reaction to C_2F and liquid (PHI 1960).
- CF_2 exists, but only within a fairly narrow temperature range. Above 1226 °C, it decomposes peritectically to liquid and Fe_2O_3 and below 1155 °C, it decomposes to CF and Fe_2O_3.
- Iron oxide melts at 1594 °C (in air) and 1583 °C (at $PO_2 = 1$ atm), magnetite being the crystalline phase stable at liquidus temperature. Liquidus temperature decreases when CaO is added and hematite becomes the stable phase.

TAB. 2.24 – Main references describing the stability behaviour in the CaO–Fe oxides system.

Experimental works	Experimental works	Modelling works
Sosman et al. (SOS 1916)	Martin et al. (MAR 1935)	Dufour et al. (DUF 1969)
Vogel et al. (VOG 1933)	Crook (CRO 1939)	Björkman (BJO 1984)
Tavashi (TAV 1936)	R. Hay et al. (HAY 1940)	Hillert et al. (HIL 1985, HIL 1990–1)
Gurry et al. (GUR 1950)	Burdese (BUR 1952–1 BUR 1952–2)	
Allen et al. (ALL 1955)	Phillips et al. (PHI 1958, PHI 1960)	Selesby et al. (SEL 1996)
Turkdogan (TUR 1961)	Reeve (REE 1966, REE 1967–1, REE 1967–2)	
Johnson et al. (JOH 1965)	Willshee et al. (WIL 1967)	
Trömel et al. (TRO 1966)	Timucin (TIM 1970)	
Obst et al. (OBS 1968 OBS 1969)	Aubry et al. (AUB 1970)	
Imlach et al. (IML 1971–2)	Scheel (SCH 1974)	
Schurman et al. (SCH 1976–1 SCH 1976–2)	Alota (ALO 1978)	
Lykasof et al. (LYK 1980)	Takeda et al. (TAK 1980)	

- Along any oxygen isobar, starting with pure iron oxide and increasing the CaO content, the ratio Fe_2O_3/FeO to the melt is seen to increase. CaO acts as a flux for Fe_2O_3 and vice versa (GUR 1950).

2.6.4.3 CaO–FeO System

- The binary CaO–FeO system is represented in figure 2.17 (HIL 1990–1).
- CaO and FeO form a range of solid-solutions. CaO solid-solution contains up to 10% FeO accompanied by a contraction from 4.307 to 4.76 Å with increasing substitution.
- Solid solution of CaO in FeO forms a phase known as Calico-Wüstite (26 wt % CaO) resulting in a shift of lattice parameter from 4.307 to 4.490 Å.
- At the T° of liquid formation (1190 °C), Fe_3O_4 dissolves about 14 mol % of CF (no solubility below 950 °C) (BUR 1952–1).
- CaO and Wüstite at its lowest oxidation limit can form a nearly binary system.
- At a temperature below 1070 °C, C_2F and iron are stable. Above this temperature, in the field Wüstite, calcium oxide and iron, these three phases exist in the solid state and the liquid appears at 1120 °C. In the field Wüstite, calcium oxide and C_2F, the liquid appears above 1070 °C. Two saturated solid-solutions

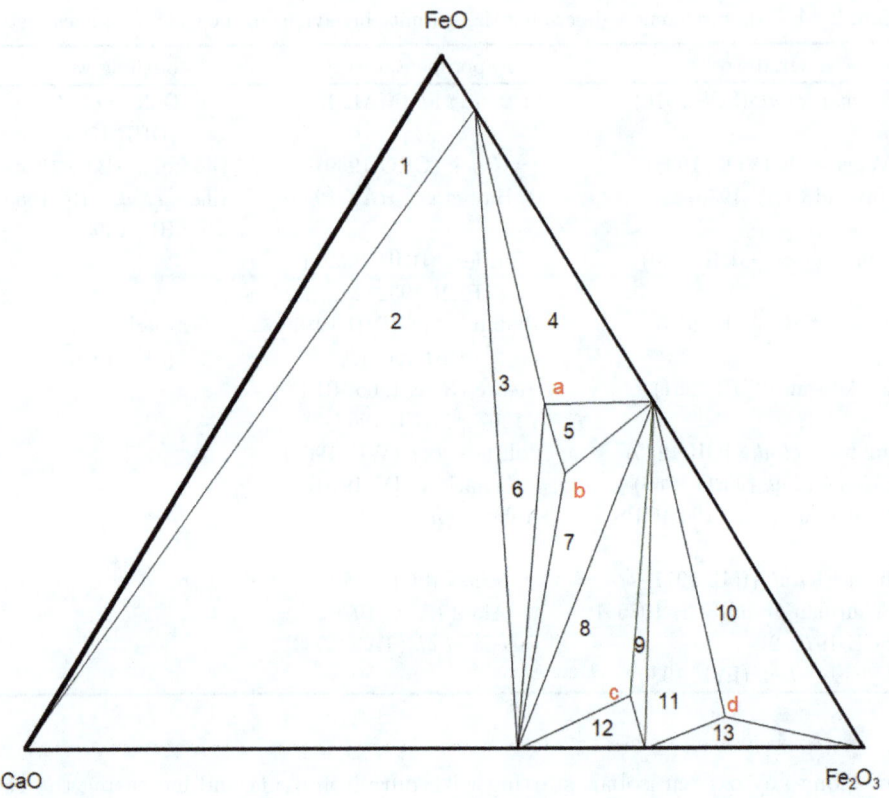

FIG. 2.15 – Isotherm at 1000 °C of the CaO–FeO–Fe_2O_3 system. Key: a = Cf_3F, b = CfF, c = C_4fF_4 and d = C_4fF_8.

between calcium oxide and Wüstite at its lowest oxidation limit and the compound C_2F can coexist in equilibrium with the liquid phase.
- Wüstite decomposes during cooling through a eutectoid reaction and results in a mixture of Wüstite, C_2F and metallic iron (AUB 1970, BUR 1952–2, BER 1959, CIR 1952–1).
- At 1130 °C, a peritectic reaction involves a liquid containing 34% CaO in equilibrium with solid solution $(Fe,Ca)O_x$ and $(Ca,Fe)O_y$ containing 40 and 60% CaO (ABB 1975).

2.6.4.4 Liquidus of the Ternary CaO–FeO–Fe_2O_3 System

- At liquidus temperature, in the ternary CaO–iron oxides system, C_4fF_8, C_4fF_4, CF, CF_2, C_2F, Hematite and Fe_3O_4 occur as primary phases.

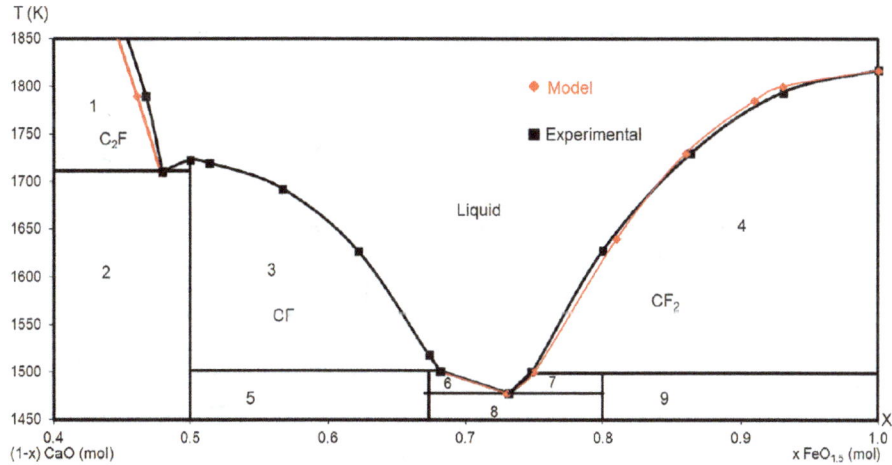

FIG. 2.16 – CaO–Fe$_2$O$_3$ phase diagram at liquidus. T° measured experimentally or by thermodynamic modelling. Key: (1) liquid + CaO, (2) CaO + C$_2$F, (3) liquid + C$_2$F, (4) CF$_2$ + liquid, (5) CF + C$_2$F, (6) CF + liquid (7) CF$_2$ + liquid, (8) CF + CF$_2$ and (9) CF$_2$ + FeO.

- Figure 2.18 represents the isotherm at 1450 °C of the CaO–FeO–Fe$_2$O$_3$ system.
- CfF melts incongruently at 1200 °C into Fe$_3$O$_4$ and liquid (BUR 1952–1).
- It is unclear if Cf$_3$F decomposes according to the reaction Cf$_3$F → f + CfF + C$_2$F or to f + CfF + liquid at 1380 °C (BER 1989, CHU 1988, SCH 1976–3).
- The ratio Fe^{3+}/Fe^{2+} depends on the PO$_2$ and lime content by the equation:

  log (Fe^{3+}/Fe^{2+}) + 0.170 log PO$_2$ + 0.018 CaO + 5500/T − 2.52 (TAK 1978, TAK 1980).

- The conditions of existence of multiphase equilibria in this system have been established for the reactions; Fe$_3$O$_4$ + f + CfF, C$_2$F + CfF$_3$ + CfF, C$_2$F + f + CfF, C$_2$F + Fe + CaO, C$_2$F + Fe + f with the equation, log PO$_2$ = A/T + B with A, respectively, 33 410, 32 700, 34 730, 29 810 and 26 700 and B, respectively, 13.62, 12.13, 13.14, 8.29 and 5.92 (BER 1988, BER 1989, BUR 1952–2, HAY1940, PHI 1958, SCH 1973, WIL 1967).
- The solubility of CaO in the melt is increased with increasing degree of oxidation. The ratio FeO/Fe$_2$O$_3$ increases in the lime saturation range as the temperature increases.
- The data Fe^{3+}/(Fe total) function of oxygen pressure have been used to compute the activities of Fe, FeO, Fe$_2$O$_3$ and CaO in slags of the ternary system. Activities of the first three have been obtained also for two quasi-ternaries involving fixed ratio (LAR 1953, LAR 1954).
- The compound energy model applied to the solid solutions with ionic species and the sub-lattice model to describe the liquid has been applied to the CaO–FeO–Fe$_2$O$_3$ system, but is limited by a lack of data concerning the ternary oxide (CfF,

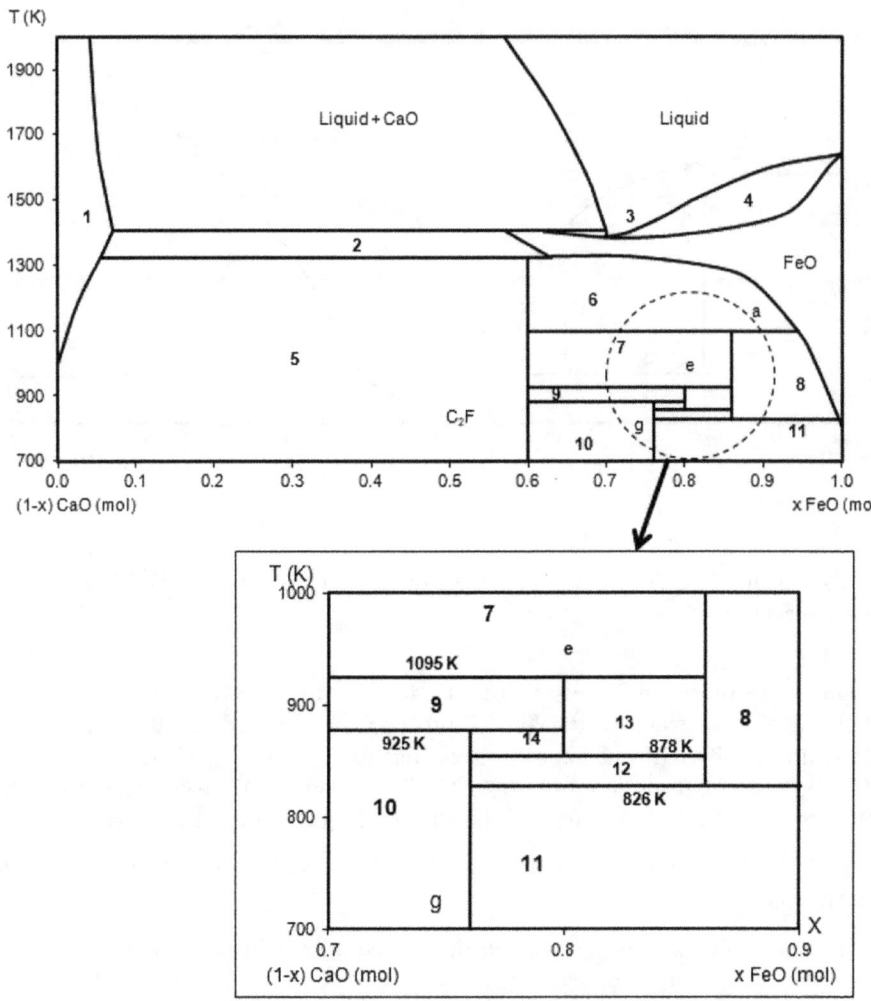

FIG. 2.17 – CaO–FeO system and detail corresponding to the circle. Key: a, e, g, represent respectively the compositions Cf_3F, CfF and C_4fF_4. (1) CaO, (2) CaO + calcio Wüstite, (3) FeO + liquid, (4) calcio Wüstite + liquid, (5) CaO + C_2F, (6) C_2F + calcio Wüstite, (7) C_2F + Cf_3F, (8) calcio Wüstite + Cf_3F, (9) C_2F + Cf_3F, (10) C_2F + C_4fF_4, (11) FeO + C_4fF_4, (12) C_4fF_4 + Cf_3F, (13) CfF + Cf_3F and (14) C_4fF_4 + Cf_3F.

Cf_3F, C_4fF_4 and C_4fF_8 are considered) (BJO 1984, HIL 1985, HIL 1990–1, SEL 1996).
- Figure 2.19 shows the flowchart of the ternary CaO–FeO–Fe_2O_3 system. Squares are used to show invariant equilibria with liquid that are linked by univariant lines.

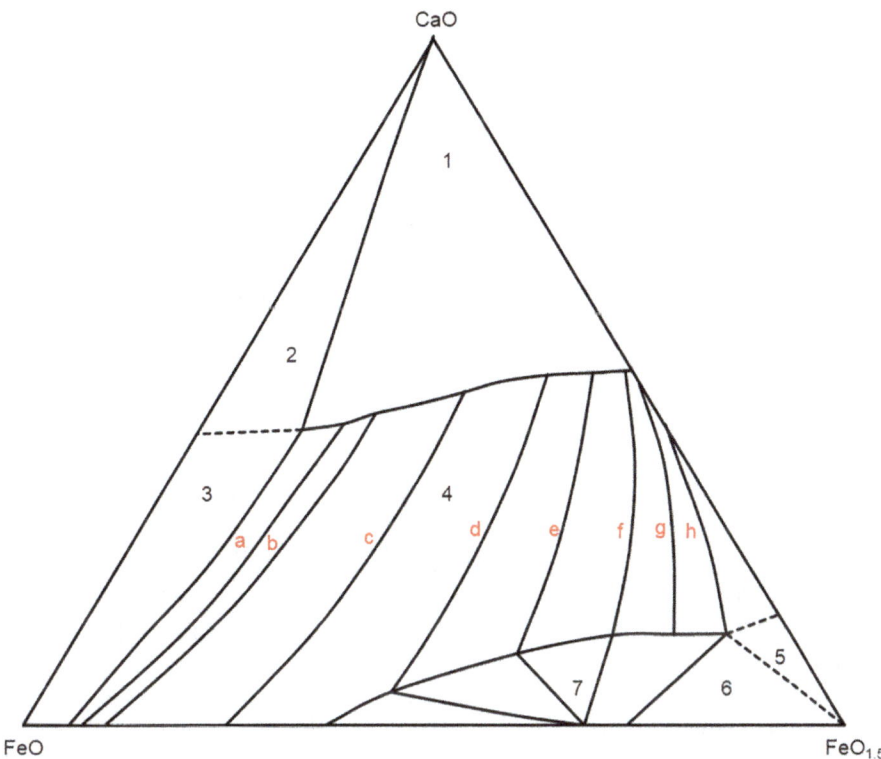

FIG. 2.18 – Isotherm at 1450 °C of the CaO–FeO–Fe$_2$O$_3$ system. The curves a, b, c, d, e, f, g and h represent the oxygen isobars (numbers −9.3/−9/−7/−5/−3.7/−2.9/−1.4 corresponding to the negative logarithm of the PO$_2$ (atm)). Key: (1) liquid + CaO solid solution, (2) liquid oxide + Fe liquid + CaO, (3) liquid oxide + Fe liquid, (4) liquid oxide, (5) Fe$_2$O$_3$ + liquid and (6) FeO + Fe$_2$O$_3$ + liquid.

2.6.5 Synthesis of Calcium Ferrite

- Complex metal oxides are synthesised by the thermal decomposition of solid solution precursors (formed by isomorphous compounds of component metals).
- First step is to prepare Ca$_{1-x}$Fe$_x$CO$_3$ by adding an aqueous solution of calcium and iron II nitrate to a large excess of ammonium carbonate in CO$_2$ atmosphere. The precursors are heated above their decomposition temperature (>400 °C). Decomposition of Ca$_2$Fe(CO$_3$)$_3$ at 750 °C for 6 h yielded a new phase CaFeO$_{3.5}$ containing iron in the 3+ state. The method has been extended to oxalate solid solution precursors (HOR 1978, VID1984).
- C$_2$F, CF and CF$_2$ are obtained by mixing calcium carbonate or oxalate and iron oxide and heated in air at 1195/1215 °C, 1200 °C and 1195/1250 °C, respectively, for CaO·2Fe$_2$O$_3$, CaO·Fe$_2$O$_3$ and 2CaO·Fe$_2$O$_3$.

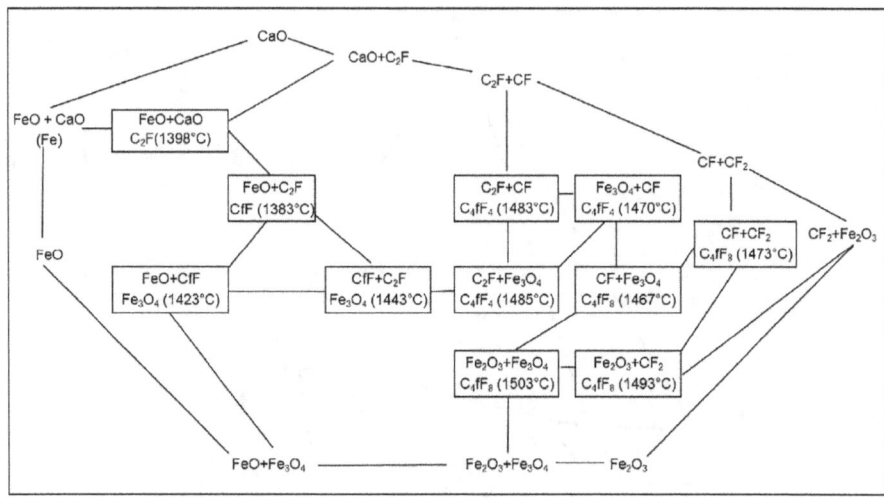

FIG. 2.19 – Flowchart of the ternary CaO–FeO–Fe$_2$O$_3$ system.

- C$_2$F is prepared by intimately mixing finely ground powders of calcium oxalate and iron III pressing the mixture to a cylindrical pellet (13 mm diameter and 2 mm thickness) and heated to 1100 °C for 3 h in a tube furnace through which oxygen was passed (BER 1971, HUG 1967).
- CF is also prepared by (DEC 1957, BRA1960, MIL 1986):

  (a) Solidification liquid (CaCO$_3$ + Fe$_2$O$_3$ at 1250 °C) and quenched.
  (b) Heating in air at 1200 °C a mixture of 2Fe$_2$O$_3$ and CaCO$_3$ during 24 h.

- A mono-crystal is obtained by heating at 1200 °C for 3 days in a sealed tube containing hydrated iron chloride.
- Synthesis of compounds containing Fe^{2+} are basically prepared in one of two ways; starting with FeO and fired in neutral atmosphere or by starting with Fe^{3+} and heating in controlled atmosphere CO/CO$_2$ or H$_2$O/H$_2$:

  ○ FeO is first synthesized mixed with CaO·2Fe$_2$O$_3$ at 1215 °C.
  ○ CaO·Fe$_2$O$_3$ is fired in a Pt crucible in an evacuated quartz tube.
  ○ CaO·3FeO·Fe$_2$O$_3$ is heated and kept at 1130 °C for 30 h.
  ○ CaO·FeO·Fe$_2$O$_3$ is heated and kept at 1000 °C during 30 h.
  ○ 4CaO·FeO·4Fe$_2$O$_3$ is heated and kept at 1200 °C during 60 h.
  ○ 3CaO·FeO·7Fe$_2$O$_3$ is heated and kept at 1200 °C during 60 h (ASA 1968).
  ○ CaO·FeO·Fe$_2$O$_3$ (CfF) is prepared by mixing Fe$_3$O$_4$ and CaO in Pt crucible and then heating and kept at 1373 °C under vacuum during 100 h (JAS 1975).
  ○ 2CaO·5FeO is heated and kept 8 days in H$_2$/H$_2$O atmosphere with 31.1% H$_2$ at 1030 °C (EVR 1977).
  ○ 2CaO·5FeO·2Fe$_2$O$_3$ is heated and kept under vacuum of CaO–2FeO–Fe$_2$O$_3$ and CaO–3FeO–Fe$_2$O$_3$ at 1120 °C during 48 h.
  ○ 2CaO·3FeO·2Fe$_2$O$_3$ is heated and kept under vacuum of CaO–FeO–Fe$_2$O$_3$ and CaO–2FeO–Fe$_2$O$_3$ at 1120 °C during 25 days (MAL 1981).

Chapter 3

Ternary Chemical Systems

3.1 General Introduction

- Ternary silicate systems are usually represented using an equilateral triangle with each apex representing 100% of one component. All possible compositions between the three components are thus represented into a triangle.
- Temperature may be plotted vertically on the triangular base to yield a solid model which is the triangular prism truncated by a series of liquidus surface.
- The liquidus temperatures for the several phases have been projected and are represented by contours of equal temperature (isotherm).
- Details of changes in composition of all solid and liquid phases with changing temperature may be followed by constructing a series of isotherm planes at selected temperature.

3.2 $CaO-Al_2O_3-SiO_2$ System

3.2.1 Ternary Constituents of the $CaO-Al_2O_3-SiO_2$ System

- Six ternary compounds occur in the $CaO-Al_2O_3-SiO_2$ system; Anorthite ($CaAl_2Si_2O_8$), Gehlenite ($Ca_2Al_2SiO_6$), Grossularite ($Ca_3Al_6Si_3O_{12}$), and Calcium Tschermac's molecules ($CaAl_2SiO_6$) are the main compounds found in the $CaO-Al_2O_3-SiO_2$ system.
- Besides these compounds, some others have been found in more unusual conditions, by devitrification of glass ($Ca_3Al_6Si_2O_{16}$), under high pressure and temperature ($Ca_2AlSiO_{5.5}$), as mineral from the moon ($Ca_{7.5}Al_{15}SiO_{32}$), in refractory inclusions or by calcium exchange and dehydration of zeolite such as Wairakite, Chabazite, Mordenite and Faujasite.

DOI: 10.1051/978-2-7598-2480-9.c003
© Science Press, EDP Sciences, 2020

3.2.2 Mineralogy

3.2.2.1 Anorthite ($CaAl_2Si_2O_8$)

- It is found in magmatic and metamorphic rocks at Monte Somma and Valle di Fassa (Italy), Val Pasmeda (Austria) named from the Greek *an* plus *orthos* (not upright).
- Two compounds of the same chemical composition exist:

  ○ Dmisteinbergite with a hexagonal structure is prepared under a high temperature microscope.
  ○ Svyatoslavite with an orthorhombic structure is found as a sublimate (700/900 °C) in a burning coal dump at Chelyabinsk coal basin (Russia) named from Svyatoslav Nestorovich Ivanov (Russian geologist).

3.2.2.2 Gehlenite ($Ca_2Al_2SiO_6$)

- It is an end member of the Melilite group of minerals consisting of the solid solution between Gehlenite and Åkermanite ($Ca_2MgSi_2O_7$).
- It occurs in nature, in the zone of contact of metamorphosed limestone and in basic igneous rocks with a high CaO content.
- It is found at Trentino (Italy) and named after the German chemist A.F. Gehlen.
- Natural Gehlenite is tetragonal but a cubic form can be prepared in special conditions.

3.2.2.3 Grossularite ($Ca_3Al_6Si_3O_{12}$)

- It is found in the Isle of Mull (Scotland) and named after the Latin botanical name of gooseberry, which is *Ribes grossularia* in reference to the green garnet of this composition that is found in Siberia.
- It belongs to the group of Garnet and characteristics of metamorphic rocks but is also found in some igneous types and detrital grains in sediments.

3.2.2.4 Calcium Tschermack's Molecule ($CaAl_2SiO_6$)

- Calcium Tschermack's molecule ($CaAl_2SiO_6$) is a calcium Pyroxene (also known as Calcium Tschermak's pyroxene (CaTs)). It is found in igneous metamorphic rocks and in meteorites (CHI 1989).

3.2.2.5 Kushiroite ($CaAlSiO_6$)

- It belongs to the Pyroxene group and occurs in refractories' inclusions (Ca, Al-rich inclusions) in carbonaceous chondrites.

3.2.2.6 Yoshiokaite ($Ca_{7.3}Al_{15}SiO_{32}$)

- It is found on the moon, near the Apollo 14 landing site, and named after the mineralogist T. Yoshioka.

3.2.2.7 Zeolitic Compounds

- Compounds of formula $Ca_xAl_ySi_zO_{48}$ with x from 5.6 to 6.13, y from 11.5 to 12.3 and z from 12 to 12.5, are obtained by calcium exchanged, dehydration of artificial or natural zeolite such as:
  - Wairakite ($Ca_8[Al_{16}Si_{32}O_{96}] \cdot 16H_2O$).
  - Scolecite ($Ca_8[Al_{16}Si_{24}O_{80}] \cdot 24H_2O$).
  - Heulandite ($Ca_4[Al_8Si_{28}O_{72}] \cdot 24H_2O$).
  - Stellerite ($Ca_8[Al_{16}Si_{56}O_{144}] \cdot 56H_2O$).
  - Gismondine ($Ca_4[Al_8Si_8O_{32}] \cdot 16H_2O$).
  - Epistilbite ($Ca_3[Al_6Si_{18}O_{48}] \cdot 16H_2O$).
  - Faujasite ($Ca_{3.5}Al_7Si_{17}O_{48} \cdot 32H_2O$).
  - Mordenite ($CaAl_2Si_{10}O_{24} \cdot 7H_2O$).
  - Chabazite ($Ca_2Al_4Si_8O_{24} \cdot 12H_2O$).
  - Levynite ($Ca_3[Al_6Si_{12}O_{36}] \cdot 18H_2O$).
  - Laumonite ($Ca_4[Al_8Si_{16}O_{48}] \cdot 16H_2O$).
  - Yugawaralite ($Ca_4[Al_8Si_{20}O_{56}] \cdot 16H_2O$).

3.2.3 Structure and Solid Solutions

- The crystallographic characteristics of the main compounds of the CaO–Al_2O_3–SiO_2 system are shown by table 3.1.
- The other polymorphic forms are shown in table 3.2.

3.2.3.1 Anorthite ($CaAl_2Si_2O_8$)

- Anorthite is part of the Plagioclase Feldspar family of minerals of general formula: $XAl_{(1-2)}Si_{(3-2)}O_8$. When X is K or Na, the formula contains one Al and three Si. If the formula contains Ca, then the formula contains two Al and two Si leading to the formula of Anorthite ($CaAl_2Si_2O_8$).
- It crystallises with triclinic symmetry with space group P-1. Three polymorphic varieties are known; monoclinic (Svyatoslavite), hexagonal (Dmisteinbergite also named Dmistweinbergite), and orthorhombic. These phases are metastable and can grow steadily prior to the nucleation of the stable phase when the supercooling or the cooling rate is large. From refining the structure on single crystals, the space groups, monoclinic C2 (or pseudo hexagonal), and monoclinic P21 (or pseudo orthorhombic) were determined for these polymorphs in the feldspar

TAB. 3.1 – Crystallographic characteristics of the main compounds of the CaO–Al$_2$O$_3$–SiO$_2$ system.

Chemical formula	CaAl$_2$Si$_2$O$_8$	Ca$_2$Al$_2$SiO$_7$	Ca$_3$Al$_2$Si$_3$O$_{12}$	CaAl$_2$SiO$_6$
Oxide formula	CaO·Al$_2$O$_3$·2SiO$_2$	2CaO·Al$_2$O$_3$·SiO$_2$	3CaO·Al$_2$O$_3$·3SiO$_2$	CaO·Al$_2$O$_3$·SiO$_2$
Cement formula	CAS$_2$	C$_2$AS	C$_3$AS$_3$	CAS
Mineral	Anorthite	Gehlenite	Grossularite	CaTs
Lattice	Triclinic	Tetragonal	Cubic	Monoclinic
S PG	P-1	P-42$_1$m	Ia-3d	C2/c
Z	8	2	8	4
a (Å)	8.184	7.69	11.851	9.619
b (Å)	12.865			8.659
c (Å)	14.15	5.05		5.278
V (Å^3)	1338		1663	422
d (g/cm^3)	2.75	3.07	3.57	3.43
JCPDS N°	12-301	9-216	39-368	31-249
	41-1486	35-755		70-2129
Ref	ANG 1990	RAA 1930	DEE 1982	HAY 1966

structure of Anorthite, Si and Al tetrahedral alternate so that each O atom has one Si and one Al neighbour: there is no Si/Al disorder. Si–O and Al–O bound lengths show variation within the same tetrahedron, the average value of each increasing as the number of Ca neighbours of the O atoms increases from zero to two. There are 4 independent Ca atoms 6 or 7 coordinated. There is no disorder of Ca position.

- Structural variation arises from changes in composition and cation order. The structural effects depend upon the model chosen to represent Anorthite structure.
- The Raman spectrum of crystalline CaAl$_2$Si$_2$O$_8$ shows characteristic bands of Feldspar.
- The presence of weak band corresponding to ν_{as}(Si–O–Al) modes in addition to the sharp and well defined ν_{as}(Si–O–Al) bands in the spectrum of well crystallised Anorthite implies a small degree of Si–Al disorder.
- The similarity in the Raman spectra of crystalline and glassy Anorthite composition confirms that the random network of the glass contains predominantly four membered rings of TO$_4$ tetrahedra similar to those present in crystalline Anorthite. A high degree of order (far from perfect) exists in the glass.
- Anorthite can take about 10% Al$_2$O$_3$ into solid solution created by crystallisation of glass below 1150 °C. There are several mechanisms to explain these substitutions:

TAB. 3.2 – Other polymorphic forms of the main compounds of the CaO–Al$_2$O$_3$–SiO$_2$ system.

Chemical formula	CaAl$_2$Si$_2$O$_8$		
Oxide formula	CaO–Al$_2$O$_3$ 2SiO$_2$		
Cement formula	CAS$_2$		
Mineral	Dmisteinbergite	Svyatoslavite	Svyatoslavite
Lattice	Hexagonal	Monoclinic	Orthorhombic
S PG	$P6_3/mcm$	$P12_11$	$P2_12_12$
Z	2	2	
a (Å)	5.113	8.228	8.224
b (Å)		8.621	8.606
c (Å)	14.743	4.827	4.836
V (Å^3)			
d (g/cm^3)	2.768	2.7	
JCPDS N°	31-248	46-1266	5-528
	31-247	71-788	
Ref	TAK 1959	CHE 1989	DAV 1952

Chemical formula	Ca$_2$Al$_2$SiO$_7$	Ca$_3$Al$_2$Si$_3$O$_{12}$	CaAl$_2$SiO$_6$
Oxide formula	2CaO·Al$_2$O$_3$·SiO$_2$	3CaO·Al$_2$O$_3$·3SiO$_2$	CaO·Al$_2$O$_3$·SiO$_2$
Cement formula	C$_2$AS	C$_3$AS$_3$	CAS
Mineral	Gehlenite	Grossularite	CaTs
Lattice	Cubic	Tetragonal	Hexagonal
S PG			
Z			
a (Å)	14.58	9.943	6.264
b (Å)			
c (Å)		7.332	8.228
V (Å^3)		287	
d (g/cm^3)		2.6	
JCPDS N°	34-1236	34-1417	25-1456
Ref	LIU 1979	LIU 1979	KIR 1973

(a) Al^{3+} may replace Si^{4+} with concomitant stuffing of excess Al^{3+}.
(b) Al^{3+} may replace Si^{4+} with concomitant stuffing of excess Ca^{2+}.
(c) Al^{3+} may replace Ca^{2+} with the excess positive charge balanced by omission of Ca^{2+}.

- On increasing temperature anorthite undergoes a displacive phase transition at 514 °K to a structure with I1bar symmetry.
- On increasing pressure (2.6 GPa) at room temperature, phase transition P1bar to I1bar also occurs. Upon increasing pressure up to 2.3 GPa, Raman bands shift towards higher frequency without any major change in the spectrum. Between 2.9 and 9.6 GPa, the major change involves the shifts in frequency of most of the

bands towards higher frequencies. Upon compression some of the T–O–T bond angles decrease considerably while other are relatively insensitive to pressure leading to a strong distortion of the aluminosilicate framework. Increasing up to 10.3 GPa induces a significant change in all regions of the spectrum involving the T–O–T modes (shortening of the T–O lengths).
- Along the join $CaAl_2Si_2O_8$–SiO_2, Anorthite and silica are not pure substance; 8 wt % of excess silica is present in Anorthite at the eutectic and a maximum of about 5% of a complementary $CaAl_2Si_2O_8$ is present in Cristobalite.

3.2.3.2 Gehlenite ($Ca_2Al_2SiO_6$) (LIS 1981)

- The general formula of Melilite is $X_2ZT_2A_7$ in which X represents large cations (Ca, Na, Sr…), Z represents cations in tetrahedral coordination (Mg, Fe^{2+}, Fe^{3+}, Al), T represents smaller cations in tetrahedral coordination and A represents the anions (O, F, S).
- The Melilite structure is based on tetragonal lattice. Its structure consists of alternate layers of large calcium polyhedra with eight vertices joined via edges and faces and layers of tetrahedral of two sorts, T1 and T2.
- Gehlenite structure is deduced from the Åkermanite structure by substituting 2 Al for Mg and Si.
- In Åkermanite, the Mg atoms occupy the larger tetrahedral sites (T1) which lie at the corners and face centres of the unit cell, while the two Si atoms occupy two equivalent tetrahedral sites (T2). There are twice as many T2 sites as T1 sites.
- Two limiting cases are possible:

   (1) One Al occupies a T1 sites and one Al and Si occupy the equivalent T2 site.
   (2) One Si occupies a T1 site and two Al are placed in two T2 sites.

   It seems that the most probable distribution would be Al in T1 and Al/Si in T2 but the silicon–aluminium distribution is temperature dependent.

- Strontium can substitute Ca and there is the same ordering with Al in T1 tetrahedra and Al/Si in T2 ones as in natural Gehlenite.
- Silica deficient Gehlenite is obtained by devitrification of the appropriated composition. The deficiency increases with cooling rate.
- Gehlenite exhibits a cubic polymorph (table 3.1).
- The Raman spectrum of crystalline $Ca_2Al_2SiO_7$ shows characteristic bands of Melilite.
- In Gehlenite, the presence of half the Al^{3+} ions as tetrahedrally coordinated network modifiers at the 4 bars in the structure decreases the intensities of the symmetrical stretching bands of non-bridging oxygen compared with those of the corresponding bands in the spectrum of non Al bearing Melilite (SHA 1983).
- There is no similarity between the Raman spectra of crystallised and glass of Gehlenite.
- The presence of v_s(T–O–T) band in the spectrum of Gehlenite glass at much lower frequency than the corresponding band in the spectrum of crystallised

Gehlenite has been attributed to predominantly tetrahedrally coordinated Al^{3+} charge balanced by Ca^{2+} in the aluminate network of the glass. The 0.33 non bridging oxygen per tetrahedron required by the stoichiometry of Gehlenite glass exists largely in the form of SiO_4 monomers in the polymerized network. Because of these monomeric silicate species Ca^{2+} has a minimum dispersion in the glass.
- The transition temperature from unmodulated high-temperature Melilite to the incommensurately modulated structure depends on the chemistry of these phases and is less than 24 °C as measured by TEM.

3.2.3.3 Grossularite ($Ca_3Al_6Si_3O_{12}$)

- The general formula of garnet is $X_3Y_2Z_3O_{12}$. The formula of grossularite is $Ca_3Al_2Si_3O_{12}$ ($3CaO \cdot Al_2O_3 \cdot 3SiO_2$).
- It crystallises as b–c cubic or tetragonal (table 3.1).
- The structure consists in alternating ZO_4 tetrahedra and YO_6 octahedra which share corners to form a three-dimensional framework. Within these, there are cavities that can be described as distorted cube or alternatively as triangular dodecahedra of 8 oxygen atoms which contain the X ions. Each oxygen atom is coordinated by one Z, one Y and two X cations. The X, Y, Z, cations occupy each special fixed positions in the unit cell so that the only variables to describe the structure are the cell edge and a single set (x, y, z) of oxygen atom coordinates.
- Two edges of each tetrahedron and six of each octahedron are shared with dodecahedra and four dodecahedral edges are shared with other dodecahedra. The high percentages of shared edges lead to a tightly packed structure accounting for the high density and refractive index of garnet.

3.2.3.4 Calcium Tschermack's Molecule ($CaAl_2SiO_6$)

- The general formula of Pyroxene is $(M2,M1)(Si,Al)_2O_6$. Calcium Tschermak's molecule is a calcic Pyroxene of formula $CaAl_2SiO_6$.
- It crystallises as monoclinic.
- In a calcic Pyroxene, Ca occupies more than 2/3 of the M2 sites. The structure of Pyroxene is based on SiO_4 tetrahedra, linked by sharing two of four corners to form continuous chain of composition $(SiO_3)_n$. The chain is linked laterally by cations which occupy the site M1 and M2. M1 atoms lie principally between the apices of SiO_3 chains, while M2 atoms lie principally between their bases.
- The Raman spectra of crystalline $CaAl_2SiO_6$ show characteristic bands of Pyroxene. The presence of half the Al^{3+} ions in six-fold coordination enhances the intensities of the band resulting from the non-bridging oxygen in the 900–1200 cm^{-1} region of the spectrum.
- There is no similarity between the Raman spectra of crystallised and glass of $CaAl_2SiO_6$.
- In the spectrum of glass, the v_s(T–O–T) bands appear at a lower frequency than the corresponding bands in the spectrum of crystalline $CaAl_2SiO_6$ indicating that the structure of the glass is more polymerized than that of the crystal.

- The increase in the relative intensity of the v_s(T–O–T) bands is due to non-bridging oxygen atoms. On the basis of the relative intensity of the Raman bands associated with the anti-symmetry and symmetric stretching motion of the bridging oxygen, it is proposed that in $CaAl_2SiO_6$ glass, Al^{3+} remains predominantly in tetrahedral coordination and only a fraction <1% mol % of that Al^{3+} ions have five- or six-fold coordination and acts as a network modifier.
- A hexagonal polymorph is synthesised from a glass of the same composition in the temperature range of 950–1050 °C.

3.2.4 Stability and Phases Diagrams (At One Atmosphere)

3.2.4.1 Solidus

RUM 1981

- Table 3.3 shows the main contributions to the $CaO–Al_2O_3–SiO_2$ system.
- Figure 3.1 represents the ternary $CaO–Al_2O_3–SiO_2$ system at the sub solidus temperature (around 1000 °C and absence of liquid phase) and the phase assemblages are:

(1) $C–C_3S–C_3A$	(5) $C_2S–CA–C_2AS$	(9) $C_2AS–CAS_2–CA_6$	(13) $CS–C_2AS–S$
(2) $C_3S–C_2S–C_3A$	(6) $C_2S–C_3S2–C_2AS$	(10) $CA_6–C_2AS–A$	(14) $CAS_2–C_2AS–CS$
(3) $C_2S–C_3A–C_{12}A_7$	(7) $CA–CA_2–C_2AS$	(11) $C_2AS–A_3S_2–A$	(15) $C_3S_2–C_2AS–CS$
(4) $C_2S–C_{12}A_7–CA$	(8) $CA_2–CA_6–C_2AS$	(12) $CAS_2–A_3S_2–S$	

3.2.4.2 Liquidus

- The phase diagram of the $CaO–Al_2O_3–SiO_2$ system, in presence of a liquid phase, has been studied on subsystems or specific joins; $CAS_2–SiO_2$, $CA–C_2S$, $CA–CS$, $C_2S–C_2AS$, $CAS_2–A$, $C_2AS–CAS_2–A$, CS plus eutectic composition C_2AS/CAS_2, $CAS_2–S$, $CS–CAS_2$, $CA_6–CAS_2–C_2AS$, $CA_2–A–S$, $C–C_2S–C_{12}A_7$ and $C_2S–C_{12}A_7$.
- From the results obtained in these partial systems, it is possible to construct the liquidus of the overall diagram (figure 3.2).
- C, C_3S, C_2S (Pseudowollastonite), C_3S_2, CS, CAS_2 (Anorthite), S (Tridymite, Cristobalite), A_3S_2 (Mullite), A (Corundum), C_2AS (Gehlenite), CA_6, CA_2, $C_{12}A_7$, and C_3A occur as primary phases.
- In the flowchart (figure 3.3), the invariant points are represented in square boxes, linked by univariant lines which can join binary systems (in circles).
- The invariant points are reported in table 3.4.

TAB. 3.3 – Main contributions to the CaO–Al$_2$O$_3$–SiO$_2$ system.

Experimental works	Experimental works	Modelling works
Osborn et al. (OSB 1941)	Schairer et al. (SCH 1947)	Baisanov (BAI 1986)
Filonenko et al. (FIL 1950)	Langenberg et al. (LAN 1956)	Barry et al. (BAR 1993)
Goggi (GOG 1961)	Gentile et al. (GEN 1963)	Wang (WAN 1989)
Criado et al. (CRI 1991)	Dragoi et al. (DRA 1970)	Mao et al. (MAO 2006)
	Teoreanu et al. (TEO 1985)	

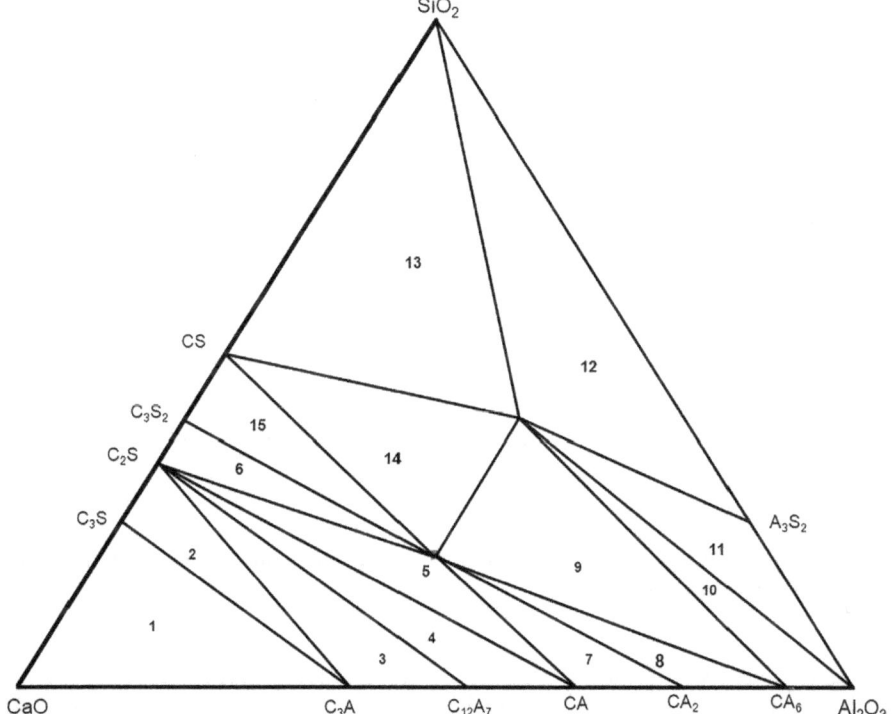

FIG. 3.1 – Ternary CaO–Al$_2$O$_3$–SiO$_2$ system at the sub solidus temperature (around 1000 °C and absence of liquid phase).

- Along the join CaAl$_2$Si$_2$O$_8$–SiO$_2$, Tridymite is not a stable phase at liquidus temperature because it admits less Anorthite in solid solution than does Cristobalite. The Cristobalite–Tridymite inversion is placed at 1244 and 1200 °C, respectively, in the presence of Anorthite and near the eutectic. Concentrations of the excess silica in natural feldspar are much lower than in synthetic Anorthite.
- Along the join Ca$_2$SiO$_4$–Ca$_{12}$Al$_{14}$O$_{33}$, when α-C$_2$S, with a large amount of impurity is cooled from its stable region, it decomposes into a liquid and a solid solution α'H-C$_2$S which has a lower impurity concentration (re-melting reaction). The reaction started at 1395 °C and finished at 1348 °C. At 1395 °C, the liquid

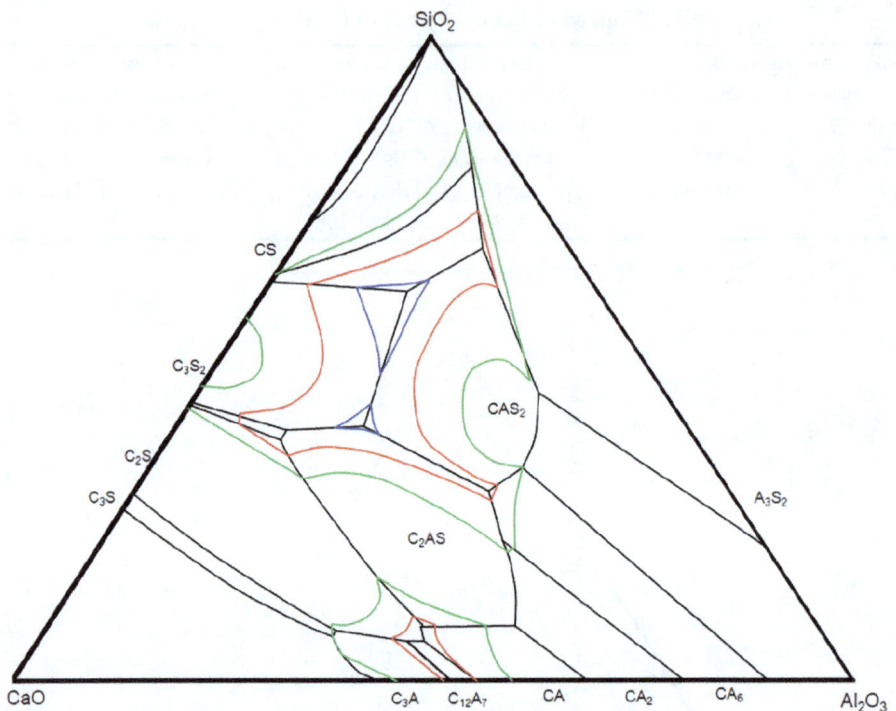

FIG. 3.2 – Liquidus of the overall diagram of the ternary $CaO–Al_2O_3–SiO_2$ system. Isotherms; 1300 °C in blue (c), 1400 °C in red and 1500 °C in green.

exsolved was rich in $C_{12}A_7$ and with decreasing temperature the $C_{12}A_7$ content in the exsolved liquid increases.

3.2.4.3 Structure of Liquid

- Density measurements performed up to 1800 °C on liquids in the $CaO–Al_2O_3–SiO_2$ system especially along the join $CA–SiO_2$ and $C_3A–SiO_2$ show that the molar volume of calcium aluminosilicate melts at low SiO_2 content does not behave linearly as function of the composition.
- The deviation from ideal of the molar volume of calcium aluminosilicate melts might result from coordination change in Si and/or Al.
- This deviation from ideal of the physical properties has been explained by the dual structural role of Al. As a network former in four-fold coordination, it can substitute for Si in tetrahedral positions (AlO_4). It can also be a network modifier element in six-fold coordination acting like alkaline or alkaline earth element to break up the tetrahedral network in octahedral position (AlO_6) when the excess metal cation is over that required for charge balance.

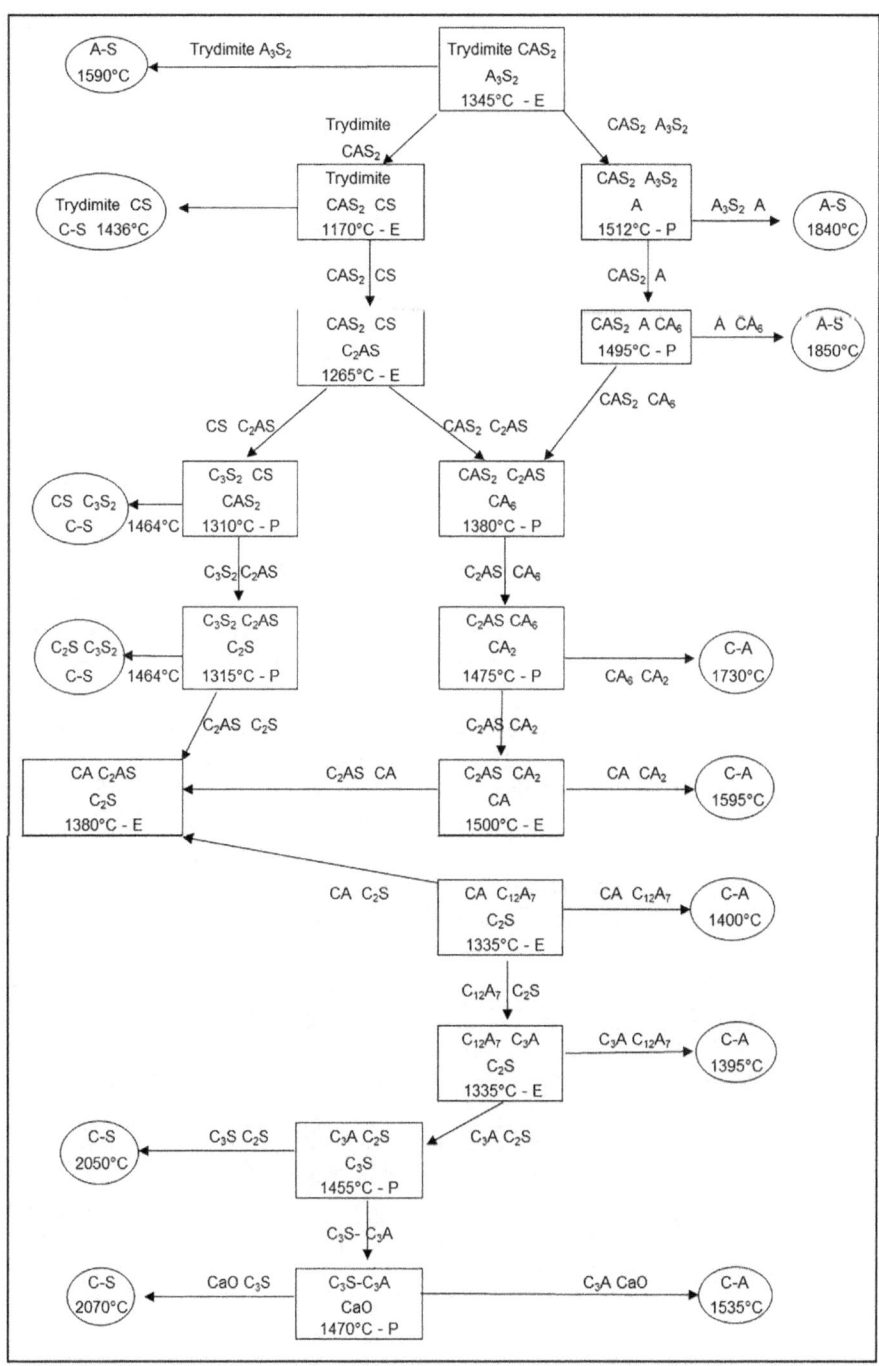

FIG. 3.3 – Flowchart of the CaO–Al$_2$O$_3$–SiO$_2$ system.

TAB. 3.4 – Invariant points of the ternary CaO–Al$_2$O$_3$–SiO$_2$ system (E for eutectic, P for peritectic).

	T°C			T°C	
Trydimite–A$_3$S$_2$–CAS$_2$	1345	E	CAS$_2$–C$_2$AS–CA$_6$	1380	P
Tridymite–CAS$_2$–CS	1170	E	CAS$_2$–A–CA$_6$	1495	P
CAS$_2$–A$_3$S$_2$–A	1512	P	CAS$_2$–CA$_6$–CA$_2$	1475	P
CAS$_2$–CS–C$_2$AS	1265	E	C$_2$AS –CA$_2$–CA	1500	E
C$_3$S$_2$–CS–CAS$_2$	1310	P	CA–C$_2$AS–C$_2$S	1380	E
C$_3$S$_2$–C$_2$AS–C$_2$S	1315	P	CA–C$_{12}$A$_7$–C$_2$S	1335	E
C$_3$A–C$_{12}$A$_7$–C$_2$S	1335	E	C$_3$A–C$_3$S–C$_2$S	1455	P
C$_3$S–C$_3$A–C	1470	P			

- Existence of Al sites in the melt structurally distinct from those well defined in the glass suggests the coexistence of different coordination numbers from Al atoms (IV, V and VI) (COT 1992).
- Significant structural changes occur with increasing temperature in the super-cooled and stable liquid above the glass transformation range.
- By Raman spectroscopic studies, it is shown that the principal low frequency band remains nearly constant with increasing temperature, indicating little change in the T–O–T angle and that the angle bending vibration is quite harmonic. Above Tg, intensity changes in the 560–590 cm^{-1} region indicate configurational changes in the supercooled liquids associated with formation of additional Al–O–Al linkages or three membered (Al, Si) containing ring. Additional intensity at 800 cm^{-1} reflects also rearrangement of the Si–O–Si network.
- The high temperature behaviour of Anorthite is distinguished by the increasing degree of non-Arrhenius viscosity behaviour as the silica content is decreased.
- In liquid silicates, silica exists as large silicate ions associated with oxygen ions. Thus, the self-diffusion coefficient of silicon tends to be smaller than that of other cations.
- Oxygen atoms exist as silicate but also as free O^{2-} ions (existence of free O^{2-} ions have been verified by ESCA). The possibility of free O^{2-} ions is also suggested from the self-diffusion coefficient of oxygen which is the largest of all constituents of the system.
- The magnitude of self-diffusion of all four constituent elements is in the order oxygen, calcium, aluminium, and silicon.
- A model based on diffusing species formed by multi-atomic ions with associated oxygen, calcium aluminium, and silicon plus free oxygen ions shows that the thermodynamic non-ideality cannot be explained by these diffusing units. Non-ideality of concentrated ionic solution cannot be fully described as an ideal solution of the actual diffusion units.

3.2.4.4 Viscosity

- Viscous flow is a momentum transport process in the fluid phase. A liquid flows when external forces such as gravity, are applied. On a molecular scale, viscous flow involves relative movement of the structural element of the liquid. Two conditions must be realised for such displacement; an energetic condition (jumping probability) and a geometric condition (free volume probability). With these two conditions, it is possible to obtain an expression between viscosity (μ) and temperature (T). The following equation is the more often used; $\mu = A \cdot \exp^{(B/T)}$. A and B depend on the chemical composition of the liquid. They are related by the following relation established for silicate and aluminosilicate; $\text{Log } A = 11.571 + 0.293B$.

3.2.4.5 Structure of Glass

- The addition of silica to calcium aluminate glass broadens the range of CaO/Al_2O_3 ratio which will form a glass. As the ratio CaO/Al_2O_3 increases, the number of non-bridging oxygen increases causing a decrease in the glass transition temperature and an increase in the refractive index. The role of silica in the aluminate structure depends on the silica content and the C/A ratio.
- The replacement of calcium oxide by alumina in glasses of constant SiO_2 content results in a decrease in thermal expansion coefficient.
- The replacement of silica by alumina results in the following different effects. The thermal expansion for low % (>CaO 20%) is almost independent of the composition while that of glasses containing more CaO becomes more composition sensitive as CaO is added to glass. The thermal expansion coefficient for glasses lying on lines of constant Al_2O_3 to CaO ratio increases with decreasing silica content with a monotonic decrease in the thermal expansion for glasses of equal silica content with increasing Al_2O_3 to CaO ratio. Refractive index varies monotonically with composition with a trend toward increasing with increasing CaO content.
- The interaction between CaO and Al_2O_3 in sections with 50 and 60 mol % of SiO_2 is accompanied by positive deviations from additive in the molar volume of glass. The maximum of dependence occurs for a ratio equal to 1 and characterizes the interaction between CaO and Al_2O_3 of being irreversible.
- The dissociation of CAS_2 under high pressure is associated with the destruction of the order of alternation of the Al and Si positions. These fluctuations in concentration of Al and Si appear in the isomorphism of Tschermack's calcium molecule CAS.
- The $xCaO \cdot 5Al_2O_4(1-x)SiO_2$ glasses show a definite minimum in molar volume at composition near $x = 0.5$ where maximum energetic stabilization is observed by calorimetry.
- The coordination number of Si in glasses and melts is generally assumed to be four. However, Si in coordination V is not excluded. Si coordination increases with increasing Al_2O_3 content in the SiO_2–Al_2O_3 join and is about 4.7 close to Al_2O_3 pole.
- Al is tetrahedrally coordinated in the silicate glass whereas the six-fold coordination polyhedra AlO_6 is to be discussed (^{27}Al NMR).

3.2.4.6 Crystallisation from Melts and Glass

- By devitrification of glass of composition in the area of composition SiO_2 between 14.3 and 76.9 mol %, Al_2O_3 from 9.0 to 43 mol %, CaO from 10 to 50 mol %, it is shown that the glass transition temperature decreases with the ratio Ca/(Al + Si). This is related to a progressive modification of the glass structure which moves from a Quartz like network with Al in fourfold coordination to a fragmented structure due to the modifier effect of Ca.
- Crystallisation of glass forming melts in the C_2AS/C_2S and C_2AS/CAS_2 systems can be explained by the characterization of nucleation and growth of crystals. In the range 10–30% of C_2S in the C_2S/C_2AS system, heterogeneous nucleation occurs on the surface of the heater and gives rise to the crystals with the leading phase β-C_2S, then crystals of C_2AS appear. Homogeneous nucleation occurs at lower temperatures giving Gehlenite crystals in the bulk.
- The growth rate and crystal formation could be concentrated in a single figure: a linear growth rate equal to the growth rate in the direction of the greatest growth rate of the crystal. The measurements are carried out by annealing method without or with observation made in a high-temperature microscope (depending on the intensity of the growth rate $>>$ or <50 μm/s). In the region 100–60% Anorthite of the system CAS_2/C_2AS, two modifications of Anorthite crystal growth occur: α-Anorthite (triclinic) and, in spontaneous nucleation, hexagonal Anorthite also appears. In the region 47%, only Gehlenite grows at the initial stage.
- Triclinic Anorthite grows in the shape of spherulites on supercooling temperatures greater than 50–100 °C and as acicular form, but non-facetted crystals, on supercooling temperatures smaller than 10 °C.
- In the region of diagram, where C_2AS precipitates, the growth of the induced triclinic Anorthite phase separates over a period with growth of Gehlenite.
- The results obtained by XRD, IR spectrometry and thermal analysis during the crystallisation process of a glass of molar composition 7.04 CaO, 30.29 Al_2O_3, and 62.67 SiO_2 indicates the presence of $C_{12}A_7$ and OH groups.
- The TG curve shows an increase in weight during the crystallisation process assuming the occurrence of a reaction between amorphous $C_{12}A_7$ and water. The kinetic parameters of the crystallisation process can be explained by the assumption of a change in mechanism from 3 to 2 dimensional growths with increasing heating rate. By comparing the observed and theoretical weight changes, the degree of crystallinity of the final product has been calculated. It was found to be about 15% with maximum of 18.5% for the heating rate of 5 K min^{-1}. The activation energy is found to be 92 and 86 kcal/mole.
- Anorthite never nucleates as the first phase in the supercooled melt. The metastable polymorphs nucleate prior to Anorthite at T° lower than 1250 °C where the supercooling for Anorthite exceed about 300 °C and for the metastable polymorph about 130 °C. Before the appearance of Anorthite, the metastable polymorphs behave as if they were stable phases: they can grow steadily below the respective metastable liquidus but dissolve above this temperature.

The liquidus T° of the pseudo-hexagonal polymorph was determined to be about 1400 °C, about for pseudo-orthorhombic polymorph, about 1420 °C for Corundum 1480 °C and about 1420 °C for Mullite. As soon as Anorthite appears in the liquid, the metastable polymorphs become unstable and transform to Anorthite in a short time.
- At initial stage of the dissolution process of alumina in a CaO–Al$_2$O$_3$–SiO$_2$ melt, a C$_2$AS rich boundary layer occurs at Al$_2$O$_3$ surface. As the reaction continued, a reaction zone consisting of a coherent layer of CA$_2$ is formed at the melt/alumina interface. With increasing temperature and time, CA$_2$ reacts with Al$_2$O$_3$ to form CA$_6$ at the interface. Because of the large difference in the molar volume of CA$_6$ and A, the interface disintegrates and allows the liquid to penetrate to the bulk of Al$_2$O$_3$. The liquid leads to the crystallisation of CAS$_2$ and the dissolution process stops.

3.2.5 Compounds Obtained in Special Conditions

- It includes the products obtained by devitrification, dehydration, high pressure and temperature. The crystallographic properties are shown in table 3.5.

3.2.5.1 Yoshiokaite ($Ca_{7.3}Al_{15}SiO_{32}$)

- The general formula of Yoshiokaite is $Ca_{(8-x/2)}V_{(x/2)}Al_{(16-x)}Si_xO_{32}$ (V for vacancies). Its composition can be considered in the join CaAl$_2$O$_4$ (CA) – CaAl$_2$SiO$_6$ (caTs).
- It is a metastable phase formed by devitrification of glass at 950/1200 °C in a part of CaO–SiO$_2$–Al$_2$O$_3$ system where CA$_6$ and Al$_2$O$_3$ are the stable liquidus phase. The synthetic phase and the natural collected on the moon are equivalent.
- It crystallises as trigonal with a Nepheline like structure with a space group P3c1 or P-3c1. By single crystal analysis, it is shown that it has a very distorted stuffed Tridymite structure.

3.2.5.2 $Ca_3Al_6Si_2O_{14}$ ($C_3A_3S_2 = C_2AS + CA_2$)

- It is obtained by devitrification at 1000 °C of glass having the same composition (SUG 1978).

3.2.5.3 $Ca_2AlSiO_{5.5}$

- It is a metastable phase, synthesised at high pressure (16 GPa) and temperature (1700 °C).
- It has a structure analogous to CaSiO$_3$ Perovskite but half the Si atoms replaced by Al and charge balanced provided by vacancies in the oxygen sub lattice. The unit cell has a lattice parameter $a = 3.706$ Å based on a simple Perovskite structure. By electron diffraction, a superstructure has developed so that the cell can be described formally as rhombohedral with $a = 11.12$ Å and $\alpha = 27.27°$.

TAB. 3.5 – Crystallographic properties of products obtained by devitrification, dehydration, high pressure and temperature.

	Yoshiokaite	Devitrified	Devitrified	16 GPa 1700 °C
Chemical formula	$Ca_{7.5}Al_{15}SiO_{32}$	$Ca_3Al_6Si_2O_{16}$	$Ca_5Al_{10}Si_2O_{22}$	$Ca_2AlSiO_{5.5}$
Lattice	Trigonal	Hexagonal	Hexagonal	Rhombohedral
Space group	$P3c1$ or $P\text{-}3c1$			R
Z	1	2		
a (Å)	9.927	9.96	5.022	5.24
b (Å)				
c (Å)	8.22	8.22	8.17	32
V (Å^3)	701	711	178	763
d (g/cm^3)	2.755	2.77	7.92	0.486
JCPDS N°	80-1547	23-105	24-179	47-699
Ref	STE 1989	YOS 1970	YOS 1970	FIT 1991

3.2.6 Stability Relative to Temperature or Pressure and Both

3.2.6.1 $CaAl_2Si_2O_8$ (Anorthite)

- Anorthite melts congruently at 1553 °C and 1 atm pressure and incongruently at pressure above 9 kbars to corundum plus liquid.
- The upper stability of the stability field of Anorthite, determined experimentally over the range 890/1250 °C and 19–28 kbars by the reaction:

$$CaAl_2Si_2O_8 \rightarrow Ca_3Al_2Si_3O_{12} + Al_2O_3 \cdot SiO_2 \text{ (Kyanite)} + \text{Quartz}$$

is given by the relation 22.80 kbars and temperature 1093 °C (or at higher temperature) by the reaction:

$$CaAl_2Si_2O_8 \rightarrow CaAl_2SiO_6 + \text{Quartz} \rightarrow Ca_3Al_2Si_3O_{12} + Al_2O_3 \cdot SiO_2$$
(Kyanite) + Quartz.

- The open framework structure of Anorthite in which aluminium exits in fourfold coordination with oxygen collapses under elevated pressure and temperature 950–1640 °C to a denser assemblage of minerals with 6-coordinated aluminium.

3.2.6.2 $Ca_2Al_2SiO_6$ (Gehlenite)

- Gehlenite melts congruently at 1760, 1695, 1645 and 1590 °C, respectively, at 25, 16, 8 kbars and 1 atm.

3.2.6.3 $Ca_3Al_6Si_3O_{12}$ (Grossularite)

- Grossularite transforms into a new tetragonal form at pressures greater than about 250 kbars but the stability field for the garnet solid solution extends to pressure up to about 300 kbars (LIU 1979).
- Grossularite is stable as a phase up to at least 855 °C and one atmosphere. Above 855 °C, Grossularite breaks down to Gehlenite, Anorthite and Wollastonite.
- At pressures below about 250 kbar, the assemblage of grossularite plus corundum is stable for compositions containing more than 25 mol % Al_2O_3. Above 250 kbars, phase assemblages for the latter composition are truncated by those in the join $CaAl_2O_4$–SiO_2.
- Garnet ss are stable between about 10 and 25 mol % Al_2O_3. By increasing the pressure, the larger eight-coordinated polyhedra compress more than the aluminium octahedra which compress more than the silicon tetrahedra, and distorted polyhedra tend to become more regular at high pressure.

3.2.6.4 $CaAl_2SiO_6$ (Ca Tschermack's Molecule)

- It is not stable below 1160 °C and has a restricted pressure stability field (figure 3.4).
- The stability field of $CaAl_2SiO_6$ is defined by the reactions:

$CaAl_2Si_2O_8 + Ca_2Al_2SiO_7 + Al_2O_3 \rightarrow 3CaAl_2SiO_6$.
P (bars) = 12 500 + 9.9(T°C−1250).
$3CaAl_2SiO_6 \rightarrow Ca_3Al_2Si_3O_{12} + 2Al_2O_3$.
P (bars) = 17 500 + 63.8(T°C−1250).

- Activities of $CaAl_2SiO_6$ by the reaction $CaAl_2Si_2O_8 + CaAl_2SiO_6 + SiO_2$ are a simple function of the Clinopyroxene composition.

3.2.7 Thermodynamic Models

- Thermodynamic model permits to confirm experimental data and to predict phase equilibria in regions where experiments are difficult to carry out. They are based on an assumption of a model supported by thermodynamic data (*i.e.* activity, excess free energy measurements...).
- Thermodynamic analysis of CaO–Al_2O_3–SiO_2 system in the subsolidus regions of reactions shows significant influence of the initial mixture composition but a slight influence of the sintering temperature.
- The liquidus of the join Wollastonite/Anorthite calculated by the application of the Schröder-Le Chatelier equation is in good agreement with the experimental data; the calculated values for the invariant point CS/C_2AS are $X_{CAS2} = 0.277$ and $T_{eut} = 1307.7$ °C.

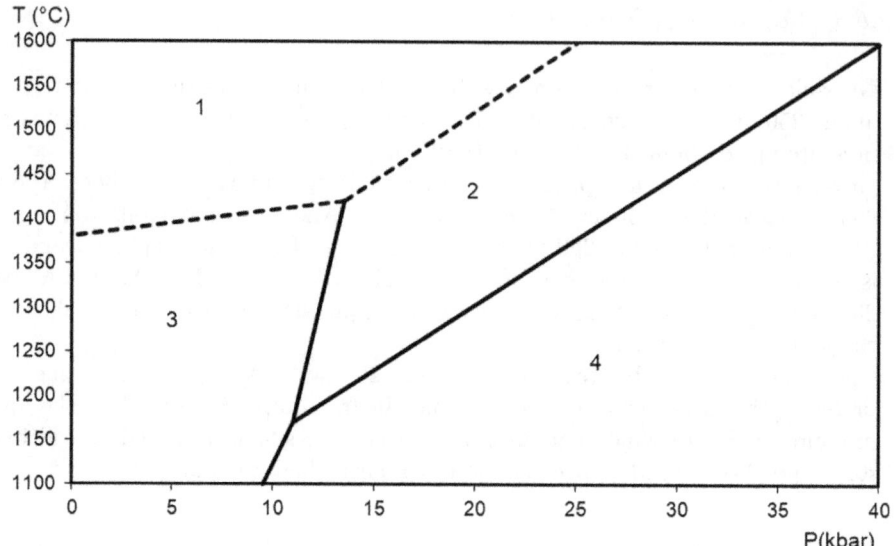

FIG. 3.4 – Stability field of Ca Tschermak's molecule. Key: (1) Corundum + liquid, (2) Ca Tschermak's molecule, (3) Anorthite + Gehlenite + Corundum, (4) Grossularite + Corundum.

- From the data obtained with equilibrium of CaO–SiO$_2$–Al$_2$O$_3$ melts and liquid or gasses containing sulphur, it is possible to assess the activity of lime in this melt:

  Activity of CaO = (N CaO)/(N SiO$_2$ + N Al$_2$O$_3$/3).

- Activity of silica can be derived from the data by the equilibrium SiO$_2$ + 3C → SiC + CO and by application of the Gibbs–Duheim relationship; the model based on the molecular interaction volume reduced to the Flory–Huggins equation predicts values of activity of CaO. Al$_2$O$_3$ and SiO$_2$ in agreement with experimental values for a range of concentration which is <0.25 for CaO, between 0.05 and 0.55 for Al$_2$O$_3$ and between 0.03 and 0.85 for SiO$_2$. The Flory Huggins equation links the molar excess Gibbs energy, Gm, to the molar fraction, xi, and the molar volume fraction of components in the system (TAO 2008):

  $$\Phi = xi * Vmi/Vm.$$

  Vmi and Vm being the molar volume of component and the system, respectively,

  $$Gm = RT \sum xi \ln (\Phi/xi).$$

- Enthalpy of mixing of glasses in the SiO$_2$/CaAl$_2$O$_4$ series measured by calorimetry study of heats of solution in molten 2PbO·B$_2$O$_3$ at 985 K is given by the equation:

  $$\Delta Hmix \text{ (calories per 2 oxygen mole)} = x(1-x)(-254 - 28551x + 22673x^2).$$

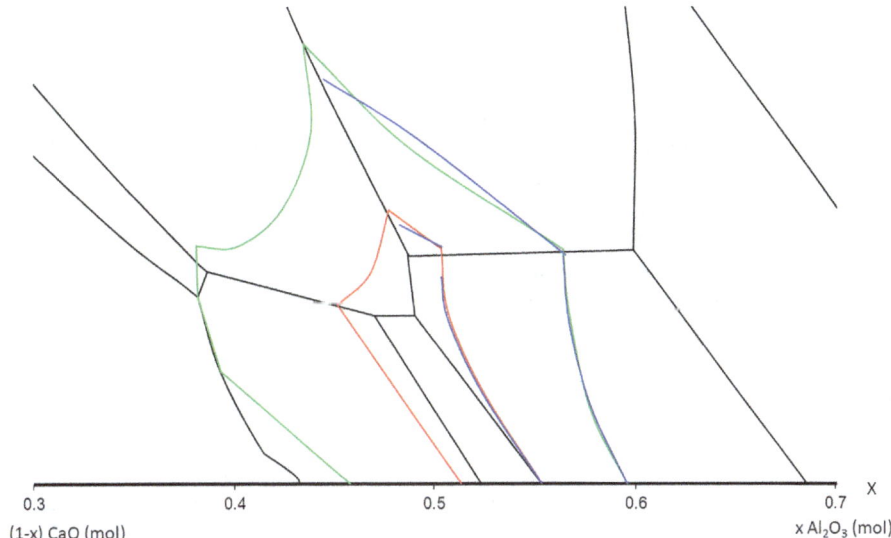

FIG. 3.5 – Part of the diagram relevant to $C_3A/C_{12}A_7/CA/C_2AS/CS$. Key: (Green) isotherm 1500 °C, (red) isotherm 1400 °C, (blue) model of isotherm at 1300 °C in the primary field of CA and C_2AS.

The shape of the heat of mixing curve between $x = 0$ and 0.4 shows a pronounced tendency towards immiscibility, and is roughly symmetrical about $x = 0.5$. That can be explained in terms of a glass structure by considering essentially random substitution of Si an Al on a continuous three-dimensional tetrahedral framework with stabilizing arising from electrostatic interactions between aluminium and the non-framework cations balancing the destabilizing effects arising from perturbation of the aluminosilicate framework by the non-framework cation (NAV 1982). See figure 3.5.

- A model of liquidus of the system CaO–Al_2O_3–SiO_2 that has been calculated on the basis of the stoichiometric 'Margules solution model' is sufficient to fit the CaO–Al_2O_3–SiO_2 liquidus relationships. To account for the non ideality in the liquid a Margules solution is derived in a generalised form which can be extended to systems of any number of components and for polynomial of any degree. Equations are presented for calculations of both the excess Gibb free energy of solution and the component activity coefficients. G excess can be approximated by a variable degree polynomial expansion.

3.2.8 Synthesis of Calcium Silicoaluminate

3.2.8.1 $CaAl_2Si_2O_8$ (Anorthite)

- It is obtained by sintering. During firing mixture of kaolinite and $CaCO_3$, C_2AS is first formed (870 °C), later hexagonal CAS_2 (890 °C) and finally triclinic CAS_2

(950/1000 °C). At 1200 °C, triclinic CAS_2 and traces of C_2AS, Cristobalite and Mullite are observed.
- The use of mineralisers (sodium chloride or carbonate, borax ($Na_2B_4O_7$)) allows us to decrease the firing temperature (1000/1050 °C).
- A glass of CAS_2 composition was heated at 2000 °C and quickly cooled to 1200 °C, where it was allowed to crystallize.
- Hexagonal CAS_2 can be prepared by hydrothermal processing of monocalcium aluminate and quartz at 2000 °C.
- CAS_2 can be prepared using chemical synthesis route (the PVA steric entrapment route) and crystallise at 1000 °C. The PVA content and its degree of polymerisation had significant influence on the synthesis behaviour and final powder.

3.2.8.2 $Ca_2Al_2SiO_6$ (Gehlenite)

- Tetragonal Gehlenite is obtained by sintering from a mixture of $CaCO_3$ with alumina and silica between 1200 and 1400 °C or by devitrification of a glass of the same composition.
- Cubic Gehlenite is obtained by quenching from 150/200 kbars and 1000 °C.

3.2.8.3 $Ca_3Al_6Si_3O_{12}$ (Grossularite)

- Grossularite is prepared by subjecting C_3AS_3 to 280 kbars and 1000 °C and quenched in room conditions, in piston cylinder apparatus at about 25 kbar and 1000 °C using pure component oxides as the starting material.
- Grossularite was synthesised from stoichiometric glass at 1200 °C and 26 kbars.

3.2.8.4 Calcium Tschermack's Molecule

- It is prepared by crystallisation of glass of $CaAl_2SiO_6$ composition at 20 kbar and 1400 °C.

3.2.8.5 C_3A_3S

- It is prepared by devitrification of a glass having a composition C_2AS/CA_2 (equimolar) of the solid solution in the system C_2AS/CA_2. Its area of crystallisation lies in the triangle $C_2AS/CA/CA_2$.
- At about 1100 °C, it decomposes into stable minerals (SUG 1978).

3.2.8.6 $Ca_{1.82}Al_{3.64}Si_{0.36}O_8$

- It is obtained by short heating of glass at 1000 °C.

3.2.8.7 $Ca_2AlSiO_{5.5}$

- It is synthesised by coprecipitated gel as starting materials fused by heating it in iridium strip at 1600 °C and then quenching. The product was subjected to 16 GPa pressure and 1700 °C for 15 min in a multiple anvil.

3.3 Al_2O_3–SiO_2–Fe Oxides

3.3.1 Introduction

- Two ternary compounds occur in the Al_2O_3–SiO_2–Fe oxides system: Almandine ($Fe^{2+}_3Al_2Si_3O_{18}$ or $Fe^{2+}_3Al_2(SiO_4)_3$ (f_3AS_3 in shorthand notation)) and iron Cordierite also named Sekaninaite ($Fe^{2+}_2Al_4Si_5O_{18}$ or $Fe^{2+}_2Al_4(SiO_2)_5$).
- Binary compounds occurring in the Al_2O_3–SiO_2–Fe oxides system are described in their respective binary systems. They are: Mullite, Spinel (Hercynite), Fayalite, $AlFeO_3$, Silica as Tridymite and Cristobalite, Hematite, Corundum and Wüstite.

3.3.2 Mineralogy

3.3.2.1 Almandine ($Fe^{2+}_3Al_2Si_3O_{12}$)

- Almandine takes its name after the locality Alabanda (Asia Minor) is found in metamorphic and pegmatitic rocks.

3.3.2.2 Sekaninaite ($Fe^{2+}_2Al_4(SiO_2)_5$)

- Sekaninaite belongs to a solid solution of iron Cordierite. It is named after J. Sekanna (Czech geologist). It is found at Koln Bory in Moravia (Czech Republic) at a site close to pegmatitic rock outcrops.

3.3.3 Structure and Solid Solution

3.3.3.1 Almandine ($Fe^{2+}_3Al_2Si_3O_{12}$)

- It belongs to the garnet family and crystallises in the cubic system (table 3.6) (LIU 1975).

3.3.3.2 Sekaninaite

- It is the end member of the solid solution of iron into cordierite. It crystallises in the orthorhombic system (table 3.6).

TAB. 3.6 – Crystallographic characteristics of almandine and sekaninaite.

Chemical formula	$Fe_3^{2+}Al_2Si_3O_{12}$	$Fe_2^{2+}Al_4Si_5O_{18}$
Oxide formula	$3FeO \cdot Al_2O_3 \cdot 3SiO_2$	$2FeO \cdot 2Al_2O_3 \cdot 5SiO_2$
Cement formula	$3feAS_3$	$f_2A_2S_5$
Mineral	Almandine	Sekaninaite
Lattice	Cubic	Orthorhombic
Space group	$Ia3d$	$Pccm$
Z	8	4
a (Å)	11.47	17.1
b (Å)		9.8
c (Å)		9.2
V (Å^3)	1509	1570
d (g/cm^3)	4.32	2.6
JCPDS N°	41-1423	9 473
Ref	LIU 1975	RIC 1949

- Cordierite is a silicate with a tetrahedral framework structure. Cordierite can be written as follows: $(M)_2(T_23)_2(T_21)_2(T_26)_2(T_11)_2O_{18}(Ch0, CH¼)$ where M represents an octahedrally coordinated metal ion (Fe^{2+}, Mg^{2+},...). T represents a tetrahedral position; T_2 tetrahedra build six membered rings and T_1 tetrahedra cross link these units to form a framework. T_11 and T_26 are occupied by Al whereas all the other tetrahedra are occupied by Si. The six membered rings produce endless channels parallel to c. Ch0 designs the channel position at (0, 0, 0) in the centre of the six membered ring. CH¼ designs the channel position at (0.0.¼) in the middle of a larger cavity in the six membered ring.
- Electronic absorption and ^{57}Fe Mössbauer spectra confirm Fe^{2+} in two different structural positions. The major fraction of Fe^{2+} occupies the octahedral sites while some occupy the larger of the two channel sites and a little of iron Fe^{3+} replaces Al^{3+}.
- At high temperatures, a polymorph of magnesium rich cordierite exists – Indialite which is isostructural with Beryl. It has a random distribution of Al in the $(Si,Al)_6O_{12}$ rings.
- No evidence of polymorphism has been found for iron cordierite.
- Mullite and ferric oxide form a limited solid solution when heated between 1000 and 1300 °C (18 mol % of ferric oxide). Ferrous oxide does not form a solid solution but form ferrous aluminate and ferrous silicates (figure 3.6).

3.3.4 Stability and Phase Diagram

- Table 3.7 reports the list of the principal contributors of the findings in the Al_2O_3 SiO_2–Fe oxides system.
- Figure 3.7 represents the ternary Al_2O_3–SiO_2–Fe oxides system.

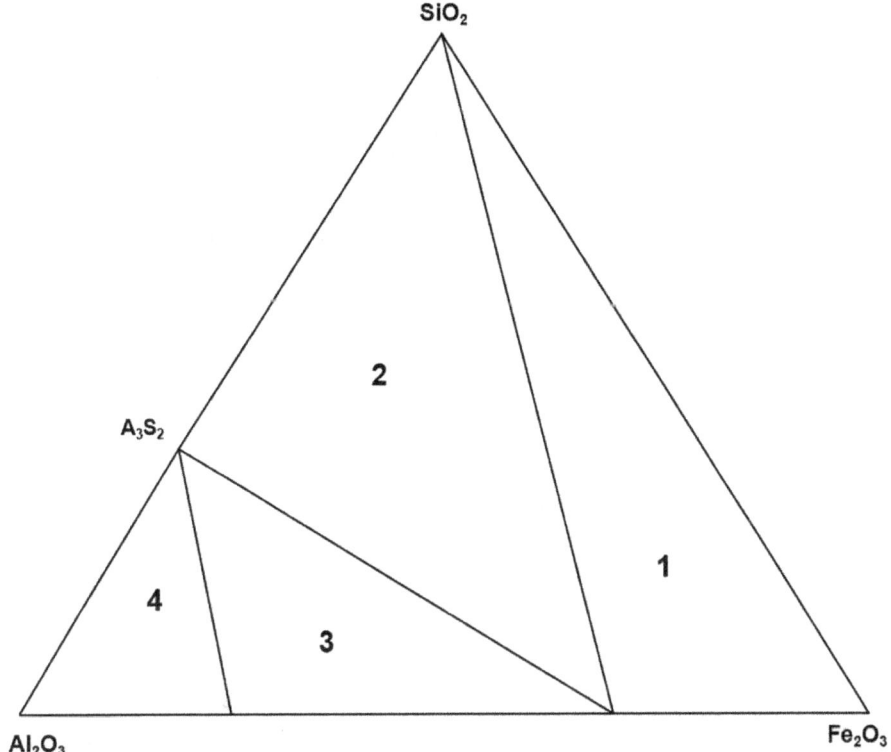

FIG. 3.6 – Equilibrium phase diagram at 1000 °C in the Al_2O_3–SiO_2–Fe_2O_3 solid solution system. Key: (1) SiO_2 + Fe_2O_3 ss, (2) Mullite + SiO_2 + Fe_2O_3 ss, (3) Mullite + Al_2O_3 + Fe_2O_3, (4) Mullite + Al_2O_3.

- When the pressure of oxygen is high enough to keep *i.e.* as Fe^{3+}, Silica, Hematite, Corundum (ss), $Al_2Fe_2O_6$, Mullite and Tridymite crystallise as primary phases.
- In air, ($PO_2 = 0.21$ atm), solid solution of Corundum, Spinel, Mullite, Tridymite and Cristobalite crystallise as primary phases.
- Two assemblages (Corundum + Spinel + Mullite and Spinel + Mullite and Tridymite) form invariant points at equilibrium with liquid in the Al_2O_3–SiO_2–FeO oxides system (figure 3.7).
- The phase compatibilities *versus* pressure and temperature are described by mean the flowchart presented by figure 3.8 wherein:

  ○ At $PO_2 = 0.5$ atm, $Al_2Fe_2O_6$ occurs as primary phases and the invariant points are:

   (a) Spinel + Corundum + $Al_2Fe_2O_6$.
   (b) Corundum + Mullite + $Al_2Fe_2O_6$.
   (c) Hematite + Spinel + $Al_2Fe_2O_6$.
   (d) Hematite + Spinel + Mullite.

TAB. 3.7 – Principal contributors of the Al_2O_3 SiO_2–Fe oxides system.

Experimental	Experimental
Greig (GRE 1927–2)	Yoder (YOD 1955)
Hay et al. (HAY 1937)	Muan (MUA 1957-1 MUA 1957-2)
Snow et al. (SNO 1942)	Brownell (BRO 1958)
Nowotny et al. (NOW 1951)	Shulter et al. (SHU 1989)
Schairer (SCH 1952)	Woerman (WOE 1991)

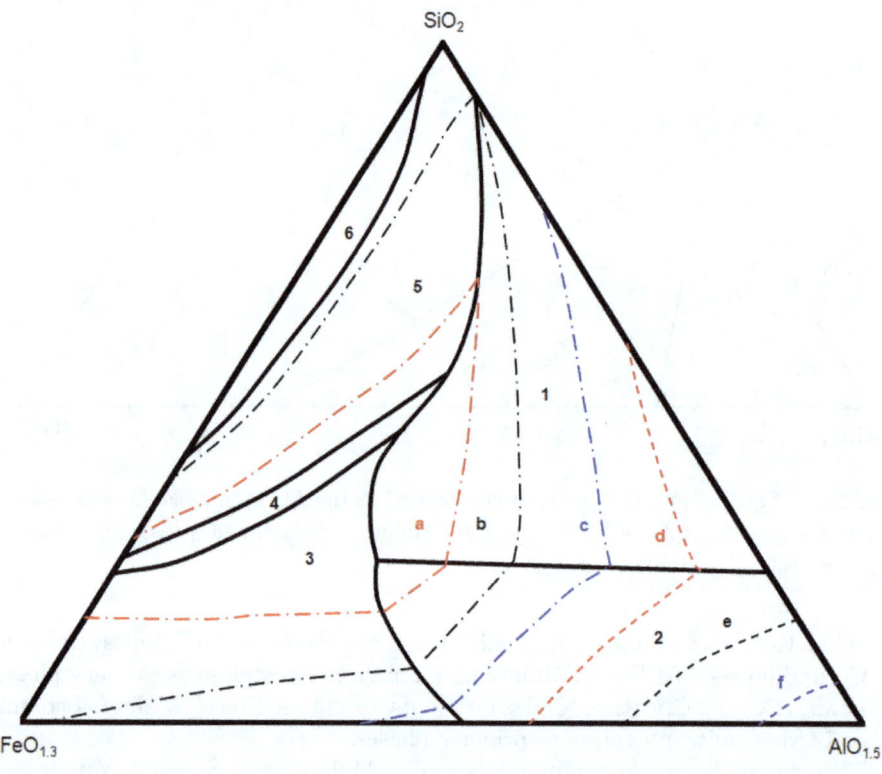

FIG. 3.7 – Phase diagram of the Al_2O_3–SiO_2–Fe oxides system. $FeO_{1.3}$ corresponds to $FeO + Fe_2O_3 = Fe_3O_4$. Key: (1) Mullite, (2) Corundum, (3) Spinel, (4) Tridymite, (5) Cristobalite, (6) Tridymite a, b, c, d, e, f isotherms at respectively 1500, 1600, 1700, 1800, 1900 and 2000 °C.

(e) Hematite + Tridymite + Mullite.
(f) Hematite + Spinel + Tridymite.

○ When PO_2 decreases to 0.4 atm, Hematite crystallises as primary phases, with Spinel, Mullite, Tridymite and Cristobalite while invariant points are:

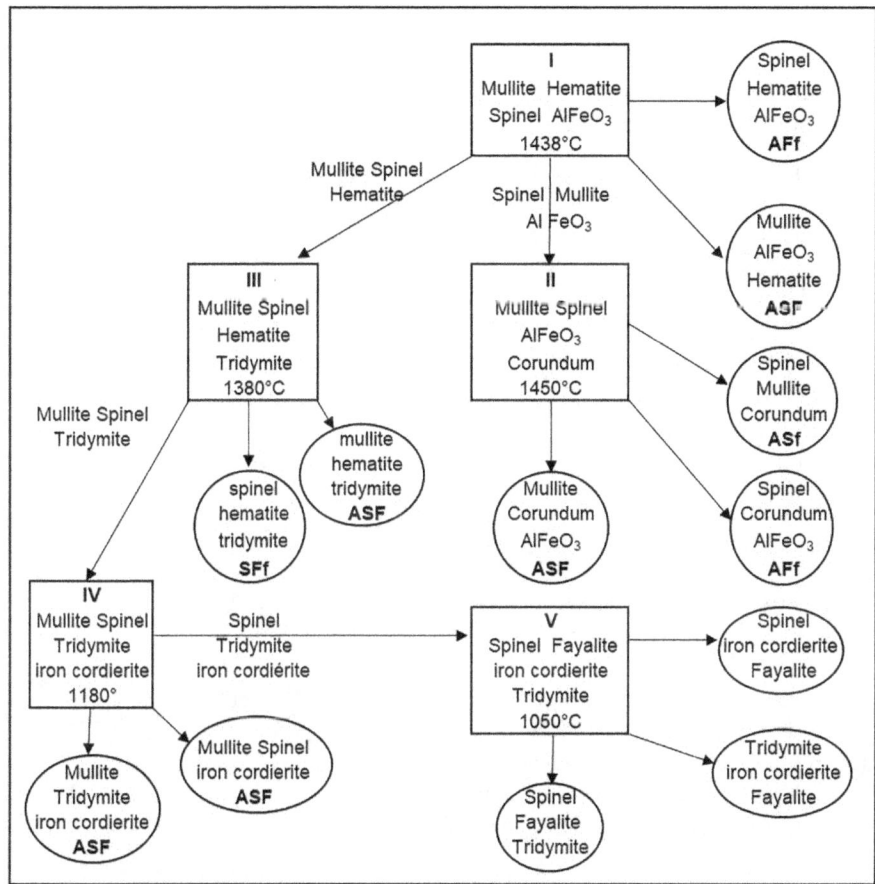

FIG. 3.8 – Flowchart of the phase compatibilities *versus* pressure and temperature for the Al_2O_3–SiO_2–Fe oxides system.

(a) Corundum + Spinel + Mullite.
(b) Hematite + Spinel + Mullite.
(c) Hematite + Tridymite + Mullite.
(d) Hematite + Spinel + Tridymite.

○ Under strongly reducing conditions in the Al_2O_3–SiO_2–Fe oxides system, iron Cordierite crystallises as primary phase and the following invariant points occur in the plane A–S–F:

(a) Fayalite + Spinel (Hercynite) + Wüstite (1148 °C).
(b) Fayalite + Spinel + Iron Cordierite (1088 °C).
(c) Iron Cordierite + Tridymite + Fayalite (1083 °C).

(d) Spinel + iron Cordierite + Mullite (1210 °C).
(e) Iron Cordierite + Mullite + Tridymite (1205 °C).

- Five quaternary liquidus invariant points are found. Three of these are located in the volume between the $PO_2 = 1$ and 0.2 atm isobaric surface.

  (1) 1438 °C (I); Mullite ss, Hematite ss, Spinel, $AlFeO_3$, and a liquid of the composition 20% SiO_2, 24% Al_2O_3, 15% FeO, 41% Fe_2O_3 by weight and a gas of $PO_2 = 0.9$ atm coexist in equilibrium.
  (2) 1450 °C (II); present together in equilibrium; Mullite ss, Spinel ss, $AlFeO_3$, Corundum and a liquid composed of composition 20% SiO_2, 25% Al_2O_3, 17% FeO, 41% Fe_2O_3 by weight and a gas of $PO_2 = 0.4$ atm.
  (3) 1380 °C (III); Tridymite, Mullite, Hematite and Spinel coexist with a liquid of 40% SiO_2, 19% Al_2O_3, 16% FeO, 25% Fe_2O_3 by weight and a gas phase of $PO_2 = 0.2$ atm.
  (4) 1180 °C (IV); Tridymite, Mullite, Spinel, iron Cordierite, liquid and gas coexist.
  (5) 1050 °C (V); Tridymite, Spinel, iron Cordierite, Fayalite, liquid and gas coexist.

- Hematite and Fe_2O_3–Al_2O_3 phases can only accommodate iron Fe^{3+} in their structure, and are therefore stable at liquidus temperature only under relative oxidizing conditions.
- Any Almandine ($Fe^{2+}_3Al_2Si_3O_{18}$) crystallises from the melt at any temperature: natural Almandine decomposes at about 900 °C giving a mixture of Hercynite (Spinel), iron Cordierite and Fayalite.
- $Fe^{2+}_3Al_2Si_3O_{18}$ is involved in equilibrium limiting the stability of Hercynite (Spinel) plus Quartz and Hercynite plus Sillimanite:

$$FeAl_2O_4 + Quartz \rightarrow Fe_3Al_2Si_3O_{12} + Sillimanite\ (Al_2SiO_5).$$
$$FeAl_2O_4 + Sillimanite \rightarrow Fe_3Al_2Si_3O_{12} + Corundum.$$

- By oxidation:

$$Fe_3Al_2Si_3O_{12} + oxygen \rightarrow FeAl_2O_4 + Fe_3O_4 + silica.$$
$$Fe^{2+}_2Al_4Si_5O_{18} + oxygen \rightarrow Fe_3O_4 + Al_2SiO_5\ (Mullite).$$

- The $Fe_4Si_4O_{12}$ – $Fe_3Al_2Si_3O_{12}$ system shows a high-pressure transformation; the maximum solubility of Ferrosilite ($FeO \cdot SiO_2$) in Almandine forming a homogeneous Garnet solid solution is 40 mol % at 93 kbars and 1000 °C.

3.3.5 Synthesis

- Almandine is produced under 12.1 KPa at 1400–1800 °C.
- Iron Cordierite is obtained only by crystallisation in sealed tube after several days or weeks of heating in the temperature range between 800 and 1050 °C.

3.4 CaO–SiO$_2$–Fe Oxides

3.4.1 Introduction

- The following compounds and solid solutions are found in the CaO–SiO$_2$–Fe oxides system:
    - CaFe^{2+}Si$_2$O$_6$ (Hedenbergite).
    - Ca$_3$Fe$^{3+}_2$Si$_3$O$_{12}$ garnet solid solution (Andradite).
    - CaFe^{2+}SiO$_4$ (Kirschsteinite).
    - (CaO$_{(1-x)}$FeO$_x$)$_2$ SiO$_2$ Olivine solid solution (Calcic-fayalite).
    - (Ca Fe^{2+})SiO$_3$ (Iron-wollastonite).
    - (Ca Fe^{2+})Si$_2$O$_6$ (Ferrobustamite).
    - Ca$_2$Fe^{2+}Si$_2$O$_7$ (Iron-åkermanite).
    - Ca$_{2.82}$Fe$_{10.4}$Si$_{0.79}$O$_{20}$ (SiO$_2$ substituted Ca ferrite ss Ca$_2$Fe$_{12}$O$_{20}$ denoted as SFC ss.
    - Iron-gehlenite (Ca$_2$(Al$_x$Fe$^{3+}_{(1-x)}$)$_2$SiO$_7$) exists in the system C–S–A–F but not in the system C/S/F because the hypothetical end member ($x = 0$) Ca$_2$Fe$^{3+}_2$SiO$_7$ decomposes into α′-C$_2$S and Hematite below 1315 °C.

3.4.2 Mineralogy

3.4.2.1 Hedenbergite

- Hedenbergite (CaFe^{2+}Si$_2$O$_6$ or CaO·FeO·2SiO$_2$) is named after the Swedish mineralogist, M. A. L. Hedenberg and it was first found at Nordmark (Sweden). It is present in both igneous and metamorphic rocksand.

3.4.2.2 Andradite

- Andradite (Ca$_3$Fe$^{3+}_2$Si$_3$O$_{12}$) is named for the Brazilian mineralogist José Bonifácio de Andrada e Silva and was originally found at Magnet Cove (Arkansas) at the environment of igneous and metamorphic rocks formed in oxidising conditions.

3.4.2.3 Kirschsteinite

- Kirschsteinite [CaFe^{2+}SiO$_4$ or CaO·FeO·SiO$_2$ (iron Monticellite)] is named for Egon Kirstein, German mineralogist, a pioneer in the geological exploration of Kivu Province Congo (Zaire). It is found in the environment of Melilite-Nepheline lava.

3.4.2.4 Calcic-Fayalite

- Calcic-fayalite (2FeO$_x$·CaO$_{(1-x)}$·SiO$_2$). The iron end member of the Olivine group is Fayalite, named for the locality Fayal (Island) and is found in ultramafic silica poor igneous rock.

3.4.2.5 Iron-Wollastonite

- Iron-wollastonite [$(Fe^{2+}Ca^{2+})SiO_3$] is found in endogenous zone of dolerite contact, in alkali gabbro, in xenoliths in andesite lava, in skarn at dolomite contact.

3.4.2.6 SFC ss

- The compound SFC ss is found industrially as a binder in iron ore sinter.

3.4.3 Structure and Solid Solution

- Table 3.8 summarizes the crystallographic characteristics of $CaFe^{2+}Si_2O_6$ (Hedenbergite), $Ca_3Fe^{3+}{}_2Si_3O_{12}$ (Andradite) and $CaFe^{2+}SiO_4$ (Kirchsteinite).

3.4.3.1 Hedenbergite

- Hedenbergite forms a series of solid solutions which extends to 80% by weight $FeSiO_3$ and 20% by weight $CaSiO_3$.
- It crystallizes in the monoclinic system (table 3.8) with the Pyroxene structure. The Pyroxenes of structural formula (M2)(M1)(Si,Al)$_2$O$_6$ can be considered to be members of the four component system $CaMgSi_2O_6$–$CaFeSi_2O_6$–$Mg_2Si_2O_6$–$Fe_2Si_2O_6$.
- The M1 site octahedrally coordinated is occupied by Fe^{2+}. The larger M2 site is either 6 or 8 coordinated and is occupied by Ca^{2+}. There is only one type of tetrahedral site completely occupied by silicon. There are 3 crystallographic non-equivalent oxygen atoms, O(1), O(2), and O(3). O(3) is referred to as a bridging oxygen that is bonded to two silicon atoms whereas O1 and O2 are non-bridging. The angle O(3)–O(3)–O(3) is used to measure the extension of the silicate chains. The increase in the Si–O interatomic distance with the temperature is not significant. The mean M–O distance and polyhedral volume of both M1 and M2 site increase regularly with increasing temperature.
- The average distance Fe–O (2.00 Å) and average coordinence (4.5) are determined by EXAFS and molecular dynamic.

3.4.3.2 Andradite

- Andradite belongs to the family of garnet and crystallises in the cubic system. It forms a solid solution with Skyagite ($Fe_3{}^{2+}Fe_2{}^{3+}Si_3O_{12}$).
- Most of the Garnets crystallise in the body centered cubic space group Ia-3d. There are $Z = 8$ formula units, $X_3Y_2Si_3O_{12}$, per unit cell of a silicate Garnet (X = Ca^{2+}, Fe^{2+} and Y = Al^{3+}, Fe^{3+}). The structure can be described as consisting of chains with corner sharing SiO_4 tetrahedra and YO_6 octahedra

TAB. 3.8 – Crystallographic characteristics of Hedenbergite, Andradite and Kirschsteinite.

Chemical formula	$CaFe^{2+}2SiO_6$	$Ca_3Fe^{3+}2Si_3O_{12}$	$CaFe^{2+}SiO_4$
Oxide formula	$CaO·FeO·2SiO_2$	$3CaO·Fe_2O_3·3SiO_2$	$CaO·FeO·SiO_2$
Cement formula	CfS_2	C_3FS_3	CfS
Mineral	Hedenbergite	Andradite	Kirsteinite
Lattice	Monoclinic	Cubic	Orthorhombic
Space group	$C2/c$	$Ia3d$	$Pbmn$
Z	4.00	8.00	4.00
a (Å)	9.84	12.05	4.88
b (Å)	9.03		11.17
c (Å)	5.25		6.45
V (Å^3)	453.00	1 753.00	350.00
d (g/cm^3)	3.5–3.68	3.86	3.43/3.56
JCPDS N°	41-1372	10-288	34-98
Ref	CAM 1972	GUS 1974	FOL 1997
	CAM 1973	NOV 1971	

alternate while each octahedron is corner linked to six different tetrahedra. A third type of coordination polyhedron is the triangular dodecahedron occupied by X cations

3.4.3.3 Calcic-Fayalite

- Calcic-fayalite and Kirschsteinite belong to the family of Olivine and crystallize in the orthorhombic system (table 3.8).
- Fayalite and Kirschsteinite form a complete series of solid solution (Ca–Fe Olivine) (BOW 1933).

3.4.3.4 Iron-Åkermanite

- Iron-åkermanite ($Ca_2Fe^{2+}Si_2O_7$) belongs to the Melilite family. In iron-åkermanite, the replacement of Mg^{2+} by Fe^{2+} occurs at a limited extend (80% by weight in Åkermanite at 1170 °C and 1 atm).
- The extent of solid solution decreases with increasing pressure and oxygen fugacity.

3.4.3.5 Iron-Gehlenite

- Iron-gehlenite ($Ca_2(Al_xFe_{(1-x)})_2SiO_7$) belongs to the Melilite family.

3.4.3.6 SFC ss

- SFC ss forms a solid solution along a trend line between the theoretical end-members $CaFe_6O_{10}$ and $Ca_4Si_3O_{10}$. The maximum solid solution range occurs with compositions containing approximately 7 through to 11.7% $Ca_4Si_3O_{10}$. The solid solution range is valid between 1060 and 1240 °C.

3.4.4 Stability and Phase Diagrams in Air

- In silicate melts, iron is usually present as ferric and ferrous ions depending on the temperature and the partial pressure of the gas phase in equilibrium with the mixture. It is therefore necessary to control the oxygen partial pressure. Experiments have been carried out in different ways: in contact with iron metal or in equilibrium with oxygen atmosphere (air or different PO_2).
- Table 3.9 gives the main references of the $CaO-SiO_2-Fe$ oxides system.

3.4.4.1 Solidus in the $CaO-SiO_2-Fe_2O_3$ System in Air

- In the $CaO-SiO_2-Fe_2O_3$ system, the evolution of the phase assemblages *versus* temperature is shown in figure 3.9 (T° below 1067 °C), figure 3.10 (between 1067 and 1155 °C) and figure 3.11 (between 1155 and 1192 °C).
- The phase relations at subsolidus temperatures below 1155 °C which is the lower limit of the stability of $CaFe_4O_7$ are shown in figure 3.9.
- Between 1067 and 1155 °C, the occurrence of the solid solution SFC ss yields a modification of the phase relation $Ca_2SiO_4-CaFe_2O_4-SFC$ ss (5) and three new assemblages are observed; (8) $CaFe_2O_4$-SFC ss, (9) hematite–SFC ss and (10) $CaFe_2O_4$–hematite–SFC ss (figure 3.10).
- At a temperature between 1155 and 1192 °C, SFC ss remains stable and $CaFe_4O_7$ becomes stable. Phase relation involving $CaFe_4O_7$ are generated; (10) $CaFe_2O_4-CaFe_4O_7$–SFC ss and (11) $CaFe_4O_7$–hematite–SFC ss (figure 3.11).

3.4.4.2 Liquidus in the $CaO-SiO_2-Fe_2O_3$ System

- In the presence of a liquid phase, the impact of the PO_2 takes a fundamental importance; modification of the ratio Fe^{3+}/Fe^{2+}, of the liquidus surfaces and the phase assemblages.
- On the extreme side of low and high oxygen potentials, the system can be considered as the pseudo ternaries $CaO-SiO_2-Fe_2O_3$ and $CaO-SiO_2-FeO$. Between these two cases, PO_2 is controlled by a mixture of CO_2/CO according to the reaction:

$$CO + 1.2O_2 \rightarrow CO_2 \text{ with } \Delta G(J) = -279\,710 + 84.08 \times \text{Temperature}.$$

- Besides the diagram described as $CaO-SiO_2-Fe_2O_3-FeO$ and $CaO-SiO_2$, the system has been studied through different joins; $CaSiO_3-Ca_2Fe_2O_5$, $CaSiO_3-Fe_2O_3$, $Ca_2SiO_4-Ca_2Fe_2O_5$ and $Ca_4Si_3O_{10}-CaFe_6O_{10}$ in oxidizing conditions and $CaSiO_3-FeSiO_3$, $Ca_2SiO_4-Fe_2SiO_4$, $CaSiO_3-FeO$ and Ca_2SiO_4-FeO.

Tab. 3.9 – Main references of the CaO–SiO$_2$–Fe oxides system.

System	Experimental works	Modelling works
CaO–SiO$_2$–Fe$_2$O$_3$	Iwase et al. (IWA 1937)	
	Phillips et al. (PHI 1959)	
CaO–SiO$_2$–FeO	Days et al. (DAS 1906)	Rasim (RAS 1982)
	Johnson et al. (JOH 1967)	
Ca–Si–Fe–O	Timucin et al. (TIM 1970)	Danek (DAN 1984)
	Toguzov (TOG 1987)	Davidson et al. (DAV 1985)
	Canha et al. (CAN 1986)	Nikolic et al. (NIK 2008)
	Pownceby et al. (POW 1998)	
	Henaeo (HEN 2005)	
	Matsura et al. (MAT 2009)	

3.4.4.3 'Binary' Systems CaO.SiO$_2$ Fe$_2$O$_3$ in Air

- Figure 3.12 summarizes the joins that have been studied in the ternary system CaO–SiO$_2$–FeO$_{1.5}$.

3.4.4.4 CaSiO$_3$–Ca$_2$Fe$_2$O$_6$ System

- The diagram representing the join CaSiO$_3$–Ca$_2$Fe$_2$O$_6$ is given in figure 3.13. It has been constructed from the corresponding ternary diagram CaO–FeO–SiO$_2$ and by data following the path of crystallisation for selected mixtures located along the line connecting the compositions CaSiO$_3$ and Ca$_2$Fe$_2$O$_6$ (PHI 1959).

3.4.4.5 CaSiO$_2$–FeO$_{1.5}$ System

- The join CaSiO$_2$/Fe$_2$O$_3$ is given in figure 3.14.
- Pseudo-wollastonite (α-CaSiO$_3$), Hematite and Magnetite occur as primary phases.
- The assemblage α-CaSiO$_3$ and Hematite begins to melt at 1280 °C. α-CaSiO$_3$, Hematite and Magnetite solid solution with Fe$_2$O$_3$ appear as primary phases along the join α-CaSiO$_3$–Fe$_2$O$_3$. Hematite, CaSiO$_3$ and liquid coexist over a range of temperatures. The non-invariant coexistence of these 3 phases implies that this system has more than 3 components and that CaSiO$_3$ is not an independent component of this system.
- Hematite transforms to Magnetite in air at 1385 °C.
- Andradite is stable up to 1137 °C in air. Above 1137 °C, Andradite breaks down to Pseudo-wollastonite, Hematite and a Garnet containing a small amount of Skyagite (Fe$_3^{2+}$Fe$_2^{3+}$Si$_3$O$_{12}$).
- Ca$_3$Fe$_2^{3+}$Si$_3$O$_{12}$ bulk composition is completely melted at a temperature of 1343 °C Two reactions define the stability limit of Andradite:

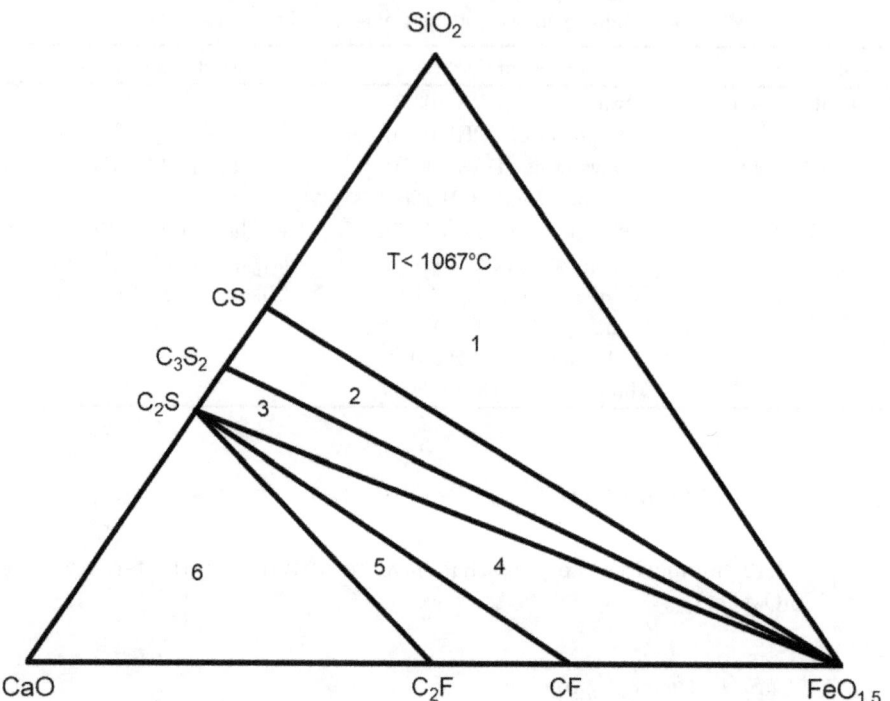

FIG. 3.9 – Subsolidus in the CaO–SiO$_2$–Fe$_2$O$_3$ system at T° below 1067 °C (PHI 1959). Key: (1) SiO$_2$–CaSiO–hematite, (2) CaSiO$_3$–Ca$_3$Si$_2$O$_7$–hematite, (3) Ca$_3$Si$_2$O$_7$–Ca$_2$SiO$_4$–hematite, (4), Ca$_2$SiO$_4$–CaFe$_2$O$_4$–hematite, (5) Ca$_2$SiO$_4$–Ca$_2$Fe$_2$O$_5$–CaFe$_2$O$_4$, (6) CaO–Ca$_2$SiO$_4$–Ca$_2$Fe$_2$O$_5$.

$$\text{Andradite} \rightarrow \text{Magnetite} + \text{Pseudo-wollastonite} + O_2.$$
$$\text{Andradite} \rightarrow \text{Kirschsteinite} + \text{Pseudo-wollastonite} + O_2.$$

- Kirschsteinite is not stable in the presence of excess SiO$_2$.
- Under reducing condition, Kirschsteinite decomposes into Wollastonite and Fe (BRI 1965).
- Iron åkermanite is stable below 775 °C and decomposes into CaSiO$_3$ and Kirschsteinite.
- Kirschsteinite [analogous as iron Monticellite (CaO–FeO–SiO$_2$)] melts congruently at 1208 °C (BOW 1933).

3.4.4.6 Ca_2SiO_4–$Ca_2Fe_2O_5$ System

- The join Ca$_2$SiO$_4$–Ca$_2$Fe$_2$O$_5$ is given in figure 3.15. It has been constructed from data taken from the literature.
- Ca$_2$SiO$_4$ and Ca$_2$Fe$_2$O$_5$ occur as primary phases and the system presents a eutectic point.

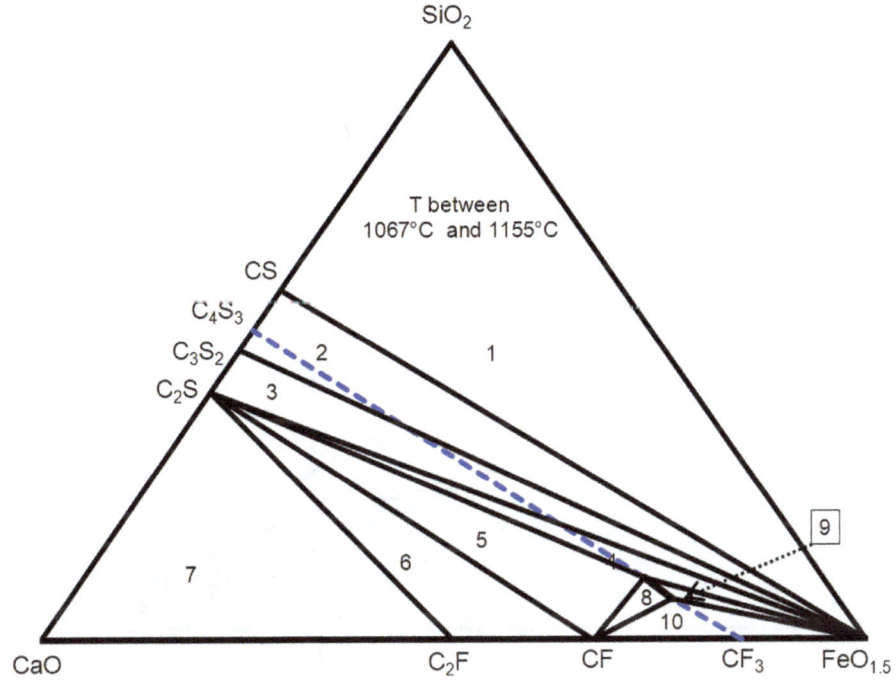

FIG. 3.10 – Subsolidus in the $CaO-SiO_2-Fe_2O_3$ system at T° below between 1067 and 1155 °C. Key: (1) SiO_2–$CaSiO$–hematite, (2) $CaSiO_3$–$Ca_3Si_2O_7$–hematite, (3) $Ca_3Si_2O_7$–Ca_2SiO_4–hematite, (4) Ca_2SiO_4–hematite, SFC ss, (5) Ca_2SiO_4–$CaFe_2O_4$–SFC ss, (6) Ca_2SiO_4–$Ca_2Fe_2O_5$ $CaFe_2O_4$ (7) Ca_2SiO_4–$Ca_2Fe_2O_5$ CaO, (8) $CaFe_2O_4$–SFC ss, (9) hematite–SFC ss and (10) $CaFe_2O_4$–hematite–SFC ss.

3.4.4.7 $Ca_4Si_3O_{10}$–$CaFe_6O_{10}$ System

- The phase diagram for the $CaFe_6O_{10}$ rich part of this system in air is shown in figure 3.16. The end member composition $Ca_4Si_3O_{10}$–$CaFe_6O_{10}$ does not exist as discrete phase and the composition of some of the crystalline and liquid phases cannot be expressed in terms of the chosen components and therefore deviate from the binary phase.

3.4.4.8 Ternary $CaO-SiO_2-Fe_2O_3$ System in Air

- The diagram shown in figure 3.17 that is the projection of the true composition on the plane $CaO-SiO_2-Fe_2O_3$. It has the appearance of a ternary diagram although the actual compositions are located on the 0.21 atm O_2 isobaric surface in the tetrahedron representing the $CaO-SiO_2-FeO-Fe_2O_3$ system.
- There is no ternary phase in equilibrium with liquid. CaO, Ca_3SiO_5, Ca_2SiO_4, $CaSiO_3$, Hematite (Fe_2O_3), Magnetite (Fe_3O_4), SiO_2 and calcium ferrites

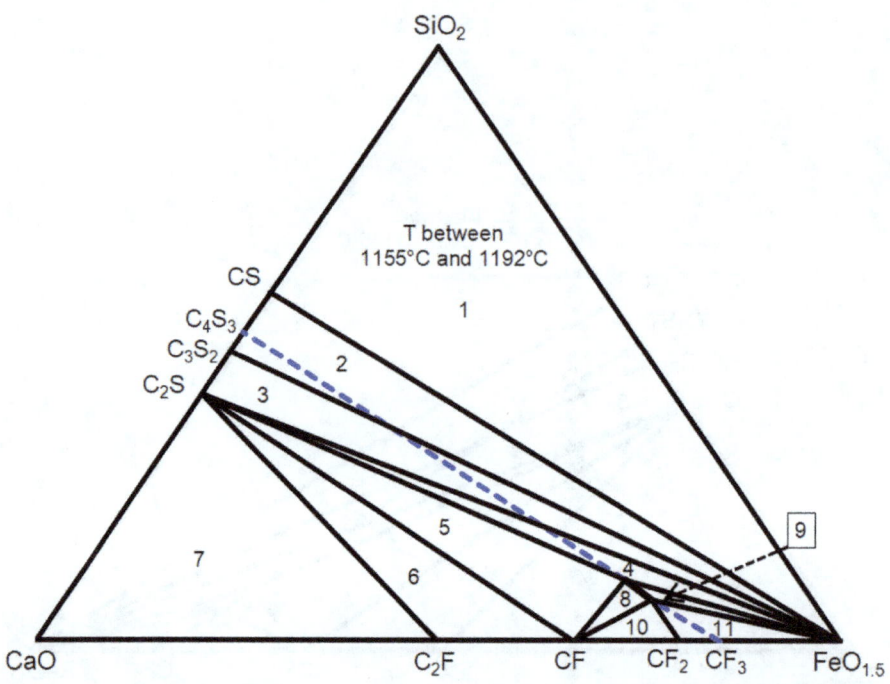

FIG. 3.11 – Subsolidus in the CaO–SiO$_2$–Fe$_2$O$_3$ system at T° below between 1155 and 1192 °C. Key: (1) SiO$_2$–CaSiO–hematite, (2) CaSiO$_3$–Ca$_3$Si$_2$O$_7$–hematite, (3) Ca$_3$Si$_2$O$_7$–Ca$_2$SiO$_4$–hematite, (4), Ca$_2$SiO$_4$–hematite, SFC ss (5) Ca$_2$SiO$_4$–CaFe$_2$O$_4$–SFC ss (6) Ca$_2$SiO$_4$–Ca$_2$Fe$_2$O$_5$ CaFe$_2$O$_4$ (7) Ca$_2$SiO$_4$–Ca$_2$Fe$_2$O$_5$ CaO, (8) CaFe$_2$O$_4$–SFC ss, (9) SFC ss–hematite, (10) CaFe$_2$O$_4$–CaFe$_4$O$_7$–SFC ss and (11) CaFe$_4$O$_7$–hematite–SFC ss.

(Ca$_2$Fe$_2$O$_5$, CaFe$_2$O$_4$, CaFe$_4$O$_7$) crystallise as primary phases in the CaO–SiO$_2$–Fe$_2$O$_3$ system. Olivine [(Ca,Fe)$_2$SiO$_4$] and Wüstite (FeO) replace Hematite and Magnetite as primary phases while calcium ferrites disappear.

- In the CaO–SiO$_2$–Fe$_2$O$_3$ system, invariant points involved the following assemblages in contact with liquid and in equilibrium with air (E for eutectic, P for peritectic) – see figure 3.18:

    ◦ SiO$_2$ + Hematite + Magnetite (P, 1390 °C).
    ◦ SiO$_2$ + CaSiO$_3$ + Hematite (E, 1204 °C).
    ◦ CaSiO$_3$ + Ca$_3$Si$_2$O$_7$ + Hematite (E, 1214 °C).
    ◦ Ca$_2$SiO$_4$ + Ca$_3$Si$_2$O$_7$ + Hematite (P, 1230 °C).
    ◦ Ca$_2$SiO$_4$ + CaFe$_4$O$_7$ + Hematite (P, 1216 °C).
    ◦ Ca$_2$SiO$_4$ + CaFe$_2$O$_4$ + CaFe$_4$O$_7$ (P, 1192 °C).
    ◦ Ca$_2$SiO$_4$ + CaFe$_2$O$_4$ + Ca$_2$Fe$_2$O$_5$ (P, 1192 °C).
    ◦ Ca$_3$SiO$_5$ + Ca$_2$SiO$_4$ + Ca$_2$Fe$_2$O$_5$ (E, 1412 °C).
    ◦ CaO + Ca$_2$SiO$_4$ + Ca$_2$Fe$_2$O$_5$ (P, 1405 °C).

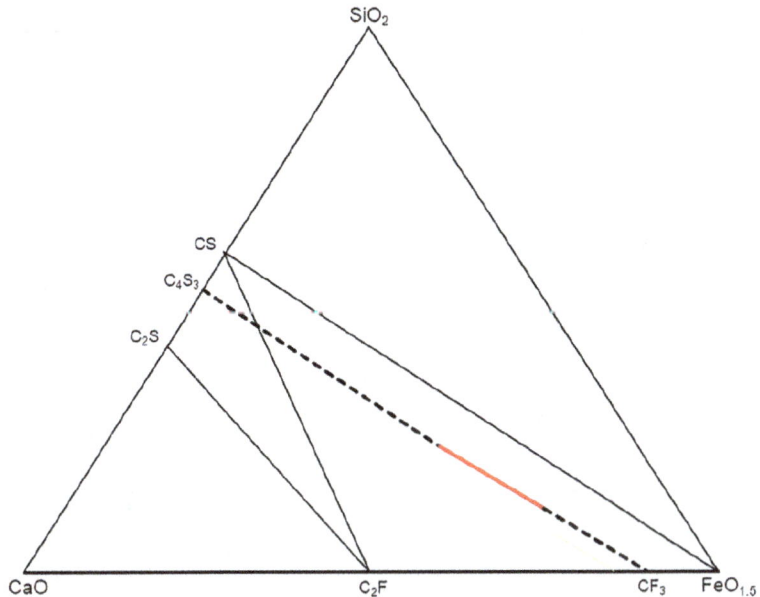

FIG. 3.12 – Pseudo binary systems in the CaO–SiO_2–$FeO_{1.5}$ system. $CaSiO_3$–$FeO_{1.5}$(CS–$FeO_{1.5}$); $CaSiO_3$–$Ca_2Fe_2O_6$(CS–C_2F); Ca_2SiO_4–$Ca_2Fe_2O_6$(C_2S–C_2F); $Ca_4Si_3O_{10}$–$CaFe_6O_{10}$ (C_4S_3–CF_3).

- The rich portion Fe_2O_3 of the diagram CaO–Fe_2O_3–SiO_2 has been explored in more details at 1 atm in air between 1250 and 1300 °C. A phase called SFC phase crystallises as a primary phase up to 1250/1255 °C after which it decomposes giving Hematite + liquid or Hematite + Ca_2SiO_4 + liquid. The isotherm at 1240 °C shows the presence of SFC phase (figure 3.19).
- Chemically, the SFC ss phase is closely related to the SiO_2 substituted CF_3 ferrite phase. Average calculated formulae are $Ca_{2.98/2.66}Fe_{10.02/10.79}Si_{1.0/0.56}O_{20}$ close to the ideal formula $Ca_2Fe_{12}O_{20}$ ($M_{14}O_{20}$). Thus, the solid solution is located between the two end members ($Ca_4Si_3O_{10}$) and ($CaFe_6O_{10}$).

3.4.5 Stablity and Phase Diagram in the CaO–SiO_2–FeO System

3.4.5.1 Solidus in the CaO–SiO_2–FeO System

- The solidus of the CaO–SiO_2–FeO system at 1080 °C is given in figure 3.20.
- In the area 1/3/7/8/10, two crystalline phases coexist in equilibrium of which at least one phase is a solid solution of variable composition. Three phase triangles border these phase areas.

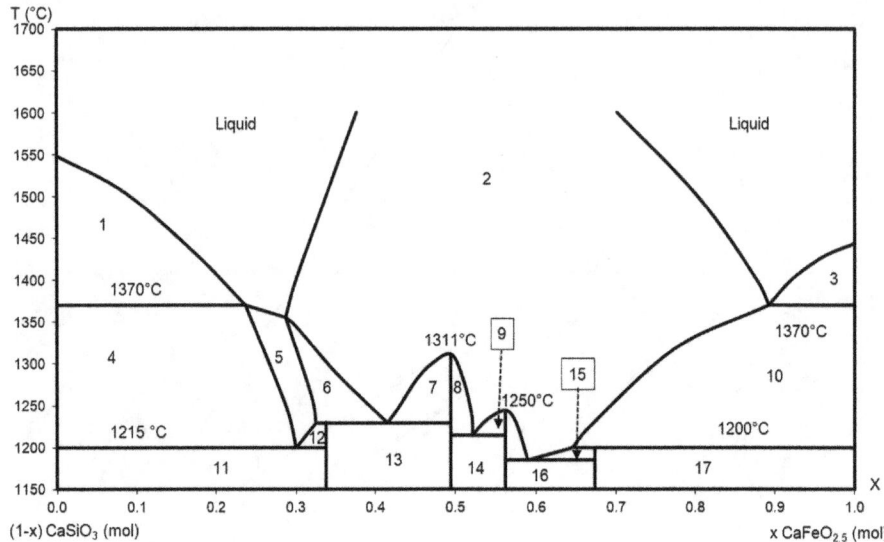

FIG. 3.13 – Pseudo binary diagram of the $CaSiO_3$–$Ca_2Fe_2O_5$ system. Key: (1) $CaSiO_3$ + Liquid, (2) Ca_2SiO_4 + liquid, (3) $Ca_2Fe_2O_5$ + liquid, (4) $CaSiO_3$ + $Ca_3Si_2O_7$ + liquid, (5) $Ca_3Si_2O_7$ + liquid, (6) $CaSiO_3$ + $Ca_3Si_2O_7$ + liquid, (7), (8) Ca_2SiO_4 + hematite + liquid, (9) Ca_2SiO_4 + $CaFe_4O_7$ + liquid, (10) Ca_2SiO_4 + $Ca_2Fe_2O_5$ + liquid, (11) $CaSiO_3$ + $Ca_3Si_2O_7$ + hematite, (12) $Ca_3Si_2O_7$ + hematite + liquid, (13) Ca_2SiO_4 + $Ca_3Si_2O_7$ + hematite, (14) Ca_2SiO_4 + $CaFe_4O_7$ + hematite, (15) $CaFe_4O_7$ + Ca_2SiO_4 + liquid, (16) Ca_2SiO_4 + $CaFe_2O_4$ + $CaFe_4O_7$, (17) Ca_2SiO_4 + $CaFe_2O_4$ + $Ca_2Fe_2O_5$.

- In the CaO–SiO_2–FeO system, there are two series of metasilicate solid solution (from pure Wollastonite to 76% $FeSiO_3$ and from Hedenbergite to 80% of $FeSiO_3$) and a orthosilicate series (Ca, Fe Olivine) extending from Fayalite $FeSiO_4$ through $CaFeSiO_4$ to 59% of Ca_2SiO_3.
- Hedenbergite is stable at temperatures below 965 °C. If it is heated above this temperature, it inverts to a homogeneous solid phase of the same composition. It is one of the β-Ca_2SiO_3 solid solution.
- Hedenbergite ($CaO \cdot FeO \cdot 2SiO_2$) decomposes at subsolidus temperature (800–965 °C) into a Pyroxenoids similar to a solid solution of iron into Wollastonite. This compound has been recognized to be structurally analogue to Bustamite ($CaMn_2SiO_6$).
- The compound $2CaO \cdot FeO \cdot 2SiO_2$ is stable only below 775 °C.
- Along the join Hedenbergite-$FeSiO_3$ at one atmosphere in contact with metallic iron, the Pyroxenes are stable only for compositions that are close to $CaFeSi_2O_6$.
- Pure $FeSiO_3$ is not stable below 12/13 kbars in the temperature range 800/1000 °C Hedenbergitic clinopyroxenes invert to ferriferous Wollastonite solid solution upon heating to temperatures below the solidus. The assemblage Fayalite + Tridymite is stable relative to Ferrosilite at 1 atm.

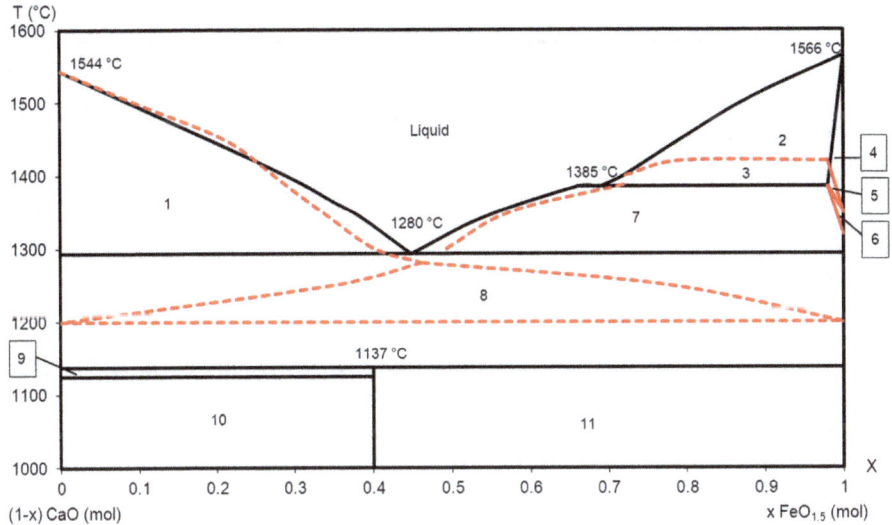

FIG. 3.14 – Pseudo binary diagram $CaSiO_3$–Fe_2O_3. Key: (1) α-$CaSiO_3$ + liquid, (2) Magnetite ss + liquid, (3) Magnetite ss + Hematite + liquid, (4) Magnetite + α-$CaSiO_3$, (5) Magnetite + Hematite, (6) Hematite, (7) Hematite + Liquid, (8) α-$CaSiO_3$ + Hematite + liquid, (9) α-$CaSiO_3$ + Hematite, (10) Andradite + α-$CaSiO_3$, (11) Andradite + Hematite.

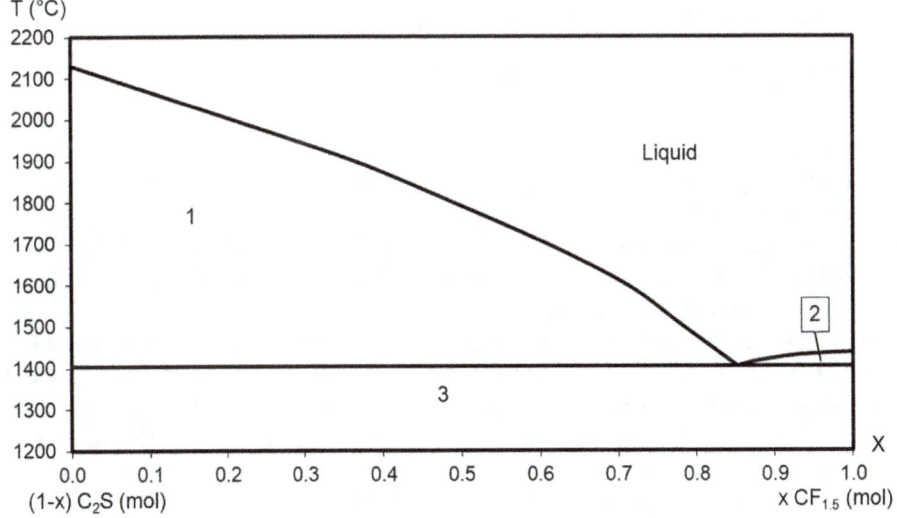

FIG. 3.15 – Pseudo binary diagram of the Ca_2SiO_4–$Ca_2Fe_2O_5$ system. Key: (1) Ca_2SiO_4 + liquid, (2) $Ca_2Fe_2O_5$ + liquid, (3) Ca_2SiO_4 + $Ca_2Fe_2O_5$.

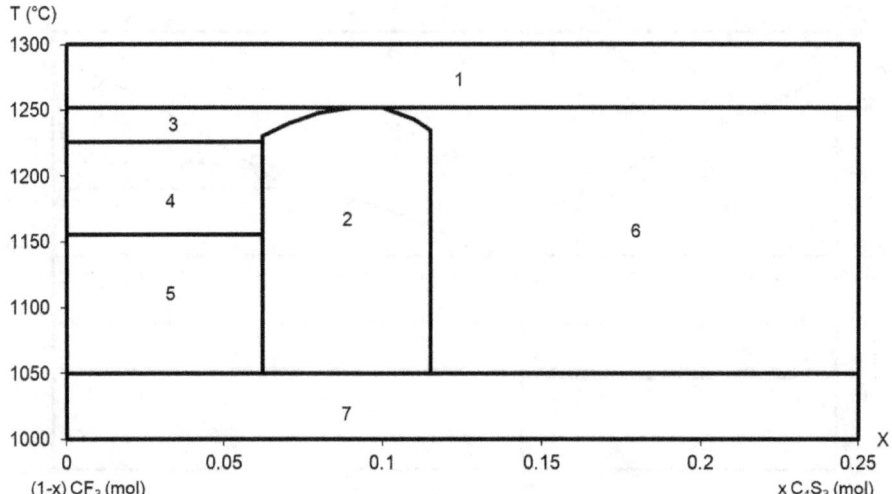

FIG. 3.16 – Pseudo binary diagram of the $Ca_4Si_3O_{10}$–$CaFe_6O_{10}$ system. Key: (1) Hematite + liquid, (2) SFC ss, (3) Hematite + SFC ss + liquid, (4) Hematite + SFC ss + $CaFe_4O_7$, (5) Hematite + SFC ss + $CaFe_2O_4$, (6) Hematite + SFC ss + Ca_2SiO_3, (7) Hematite + $CaFe_2O_4$ + Ca_2SiO_3.

- Along the join Hedenbergite–Ferrosilite ($FeSiO_3$), Hedenbergitic clinopyroxenes are stable at low pressure and over the compositional range from $Ca_{0.5}Fe_{0.5}SiO_3$ to a maximum Fe content $Ca_{0.4}Fe_{0.6}SiO_3$. More iron rich compositions are represented by the following three phase assemblages:

  ○ Hedenbergite ss + Fayalite(f2S) + SiO_2 below 940 °C.
  ○ Wollastonite ss + Fayalite + SiO_2 above 940 °C.

- Pyroxenoids that are richer in iron than approximately $Ca_{0.37}Fe_{0.63}SiO_3$ are unstable at low pressure (<2 kbar). By increasing the pressure, the stability of Pyroxenoids that are more iron-rich increases.
- Pyroxenoids of composition $Ca_{0.15}Fe_{0.85}SiO_3$ have been obtained at 12.5 kbars and 1175 °C. Wollastonite is one such pyroxenoid and is able to accept a limited amount of iron as a solid solution.
- When the percentage of iron in Wollastonite increases, several parameters indicate that the structure is modified for a composition at around $Ca_{0.9}Fe_{0.1}SiO_3$; deviation of optical properties, variation of unit cell parameters, an X ray powder line cannot fit with the triclinic Wollastonite structure and the IR spectrum is similar to the spectrum of Bustamite.
- Possible miscibility gap exists between Iron Wollastonite ($Ca_{0.83}Fe_{0.17}SiO_3$) and Wollastonite. $Ca_{0.83}Fe_{0.17}SiO_3$ could be a stable phase in the $CaSiO_3$–$CaFeSi_2O_6$ join. The temperature of inversion of Wollastonite (β-$CaSiO_3$) to Pseudo-wollastonite (α-$CaSiO_3$) is raised by solid solution of $FeSiO_3$ in β-Wollastonite up to 9% of $FeSiO_3$.

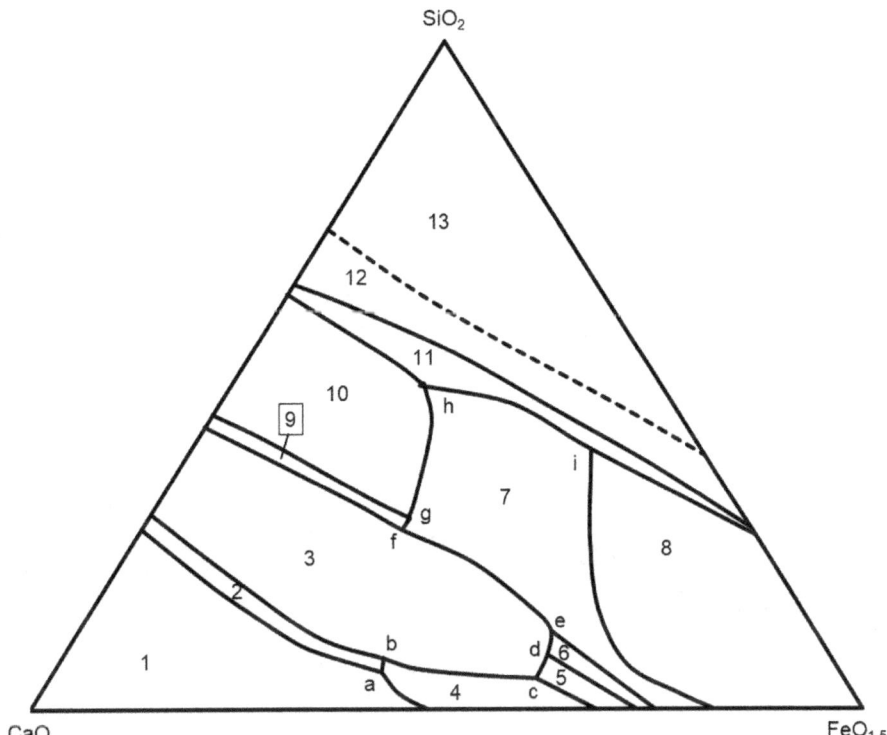

FIG. 3.17 – Ternary diagram of the CaO–Fe$_2$O$_3$–SiO$_2$ system in equilibrium with air. Key: (1) CaO, (2) Ca$_3$SiO$_5$, (3) Ca$_2$SiO$_4$, (4) Ca$_2$Fe$_2$O$_5$, (5) CaFe$_2$O$_4$, (6) CaFe$_4$O$_7$, (7) Hematite, (8) Magnetite, (9) Ca$_3$Si$_2$O$_7$, (10) α-CaSiO$_3$, (11) Tridymite (12/13) Cristobalite with a boundary curve in dashed lines.

3.4.5.2 Liquidus in the System CaO–SiO$_2$–FeO (in Equilibrium with Metallic Iron)

- Figure 3.21 summarizes the joins that have been studied in the ternary CaO–SiO$_2$–FeO system.

3.4.5.2.1 CaSiO$_3$–FeSiO$_3$ Pseudo Binary System
- The CaSiO$_3$–FeSiO$_3$ join (metasilicate join) given in figure 3.22 does not represent a true binary system in the sense that the end member, FeSiO$_3$, does not exist but is decomposed in Olivine and Tridymite.
- A liquid of the composition C begins to crystallise at 1308 °C as pure α-CaSiO$_3$ (pseudo Wollastonite). At 1285 °C, β-CaSiO$_3$ (Wollastonite) begins to form at the expense of α-CaSiO$_3$.

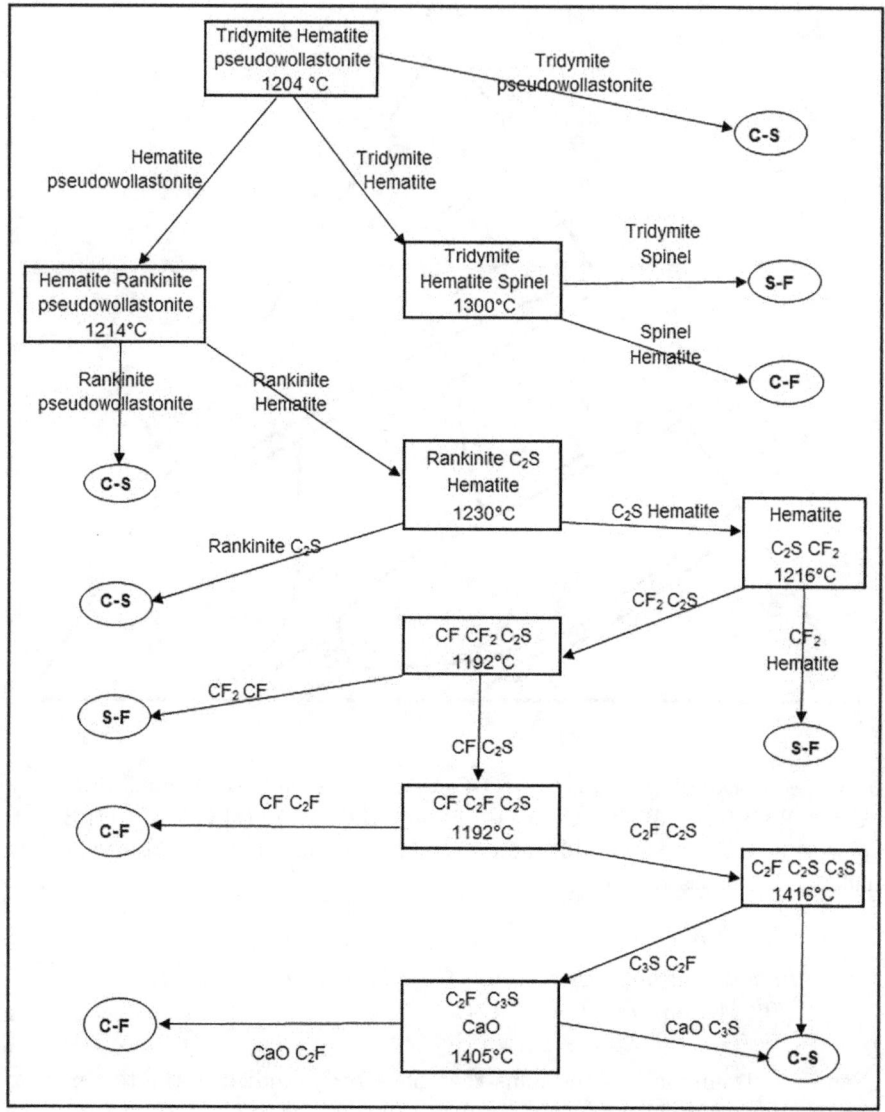

FIG. 3.18 – Flowchart of the ternary CaO–SiO_2–Fe_2O_3 system in air.

- With further cooling, the compositions of liquid and solid change, respectively, along MZ and BC. At 1160 °C, the mass consists entirely of mixed crystals of composition C.
- Liquid C has the highest content of $CaSiO_3$ that behaves in a binary manner.

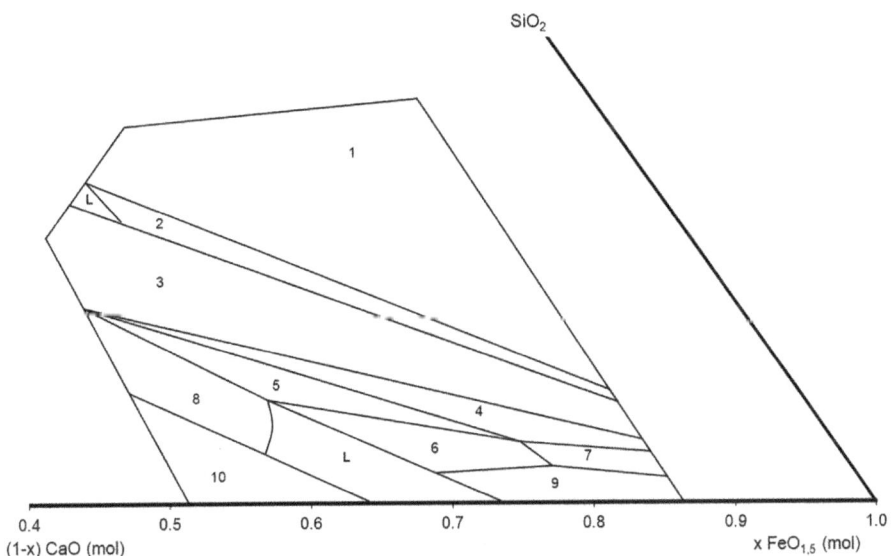

FIG. 3.19 – Isotherm at 1240 °C of the iron-rich part of the $CaO-SiO_2-Fe_2O_3$ system in air. Key: (1) Hematite + $CaSiO_3$ + liquid, (2) Hematite + liquid, (3) Hematite + Ca_2SiO_4 + liquid, (4) Hematite + Ca_2SiO_4 + SFC, (5) Ca_2SiO_4 + SFC + liquid, (6) SFC, (7) Hematite + SFC, (8) Ca_2SiO_4 + liquid, (9) Hematite + SFC + liquid, (10) Ca_2SiO_4 + $Ca_2Fe_2O_5$ + liquid, (L) liquid.

- Liquids of compositions between $CaSiO_3$ and B crystallise at 1285 °C as α-$CaSiO_3$ and β-$CaSiO_3$ ss of composition B.
- A liquid of the composition of Hedenbergite begins to crystallize at 1207 °C as β-$CaSiO_3$ solid solution containing 25% of $FeSiO_3$.
- With further cooling (until at 1160 °C) the compositions of liquid and solid change until point Z (liquid) and C (solid).
- At this temperature, Tridymite separates and the phases cannot be read from the diagram. At 1118 °C, liquid and Tridymite disappear and the solid is a $CaSiO_3$ solid solution having the composition of Hedenbergite.
- A liquid of the composition $CaSiO_3$ 30% and $FeSiO_3$ 70%, begins to crystallize at 1240 °C as Tridymite. At 1106 °C, Tridymite and liquid disappear and the solid is β-formed by $CaSiO_3$ ss containing 70% $FeSiO_3$.
- A liquid of the composition $CaSiO_3$ 20% and $FeSiO_3$ 80%, begins to crystallize as Tridymite at 1350 °C. At 1105 °C, the solid is a mixture of β-Ca_2SiO_2 ss, Olivine and Tridymite.
- The curve with its maximum at the composition of Hedenbergite (H) (figure 3.22) is analogous to the liquid curve of a melting point diagram except that the phase lying above is a solid solution instead of a liquid solution.

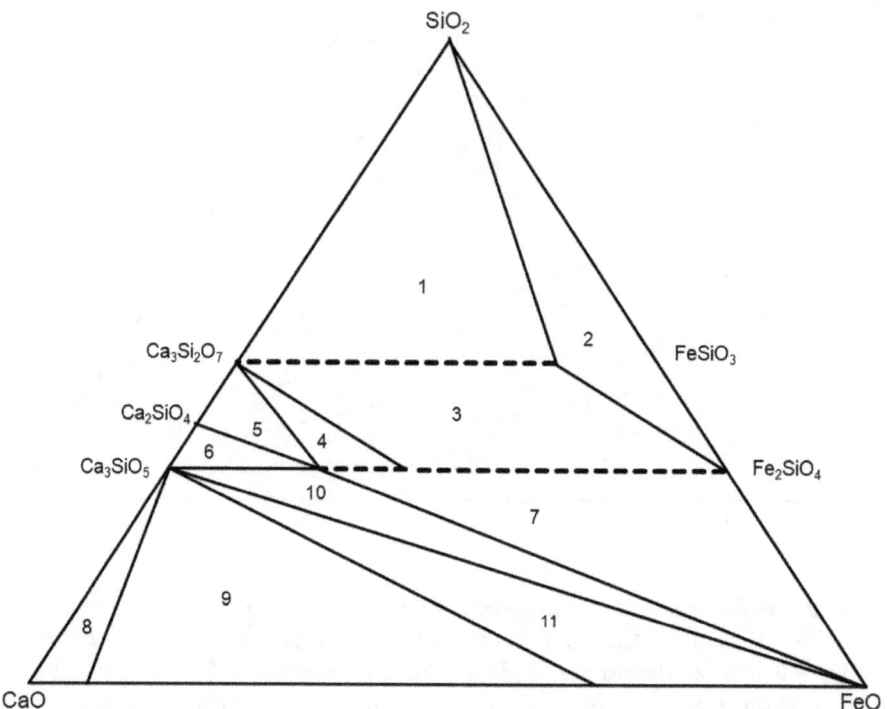

FIG. 3.20 – Solidus of the CaO–SiO$_2$–FeO system at 1080 °C. The dashed lines represent the solid solution FeSiO$_3$–CaSiO$_3$ and Fe$_2$SiO$_4$–Ca$_2$SiO$_4$. Key: (1) CaSiO$_3$ ss + SiO$_2$, (2) SiO$_2$ + CaSiO$_3$ ss + Fe$_2$SiO$_4$, (3) CaSiO$_3$ ss + Fe$_2$SiO$_4$, (4) CaSiO$_3$ + Ca$_3$Si$_2$O$_7$ Fe$_2$SiO$_4$, (5) Ca$_2$SiO$_4$ + Ca$_3$Si$_2$O$_7$ + Fe$_2$SiO$_4$, (6) Ca$_2$SiO$_4$ + Fe$_2$SiO$_4$ + FeO, (7) Fe$_2$SiO$_4$ + FeO, (8) CaO ss + Ca$_2$SiO$_4$, (9) Ca$_2$SiO$_4$ + CaO ss + (CaFe)O, (10) Ca$_2$SiO$_4$ + (CaFe)O.

3.4.5.2.2 Ca$_2$SiO$_4$–Fe$_2$SiO$_4$ Pseudo Binary System

- The Ca$_2$SiO$_4$–Fe$_2$SiO$_4$ pseudo binary system represents the orthosilicate join in the CaO–FeO–SiO$_2$ system (figure 3.23).
- Ca$_2$SiO$_4$ and Fe$_2$SiO$_4$ form a compound CaFeSiO$_4$ that melts congruently at 1208 °C.
- A complete solid solution exists between Fe$_2$SiO$_4$ and CaFeSiO$_4$ with a minimum at 1117 °C. The solid solution extends through the side Ca$_2$SiO$_4$ and CaFeSiO$_4$ without a eutectic. The limits are 10% and 41% Fe$_2$O$_4$.

3.4.5.2.3 CaSiO$_3$–FeO Pseudo Binary System

- The Wollastonite/Wüstite join (CaSiO$_3$–FeO) is given in figure 3.24. It shows the existence of a ternary compound, Ca$_2$FeSi$_2$O$_7$, with a limited temperature of stability (<775 °C) and above this temperature, it decomposes into β-Ca$_2$SiO$_3$ and CaFeSiO$_4$.

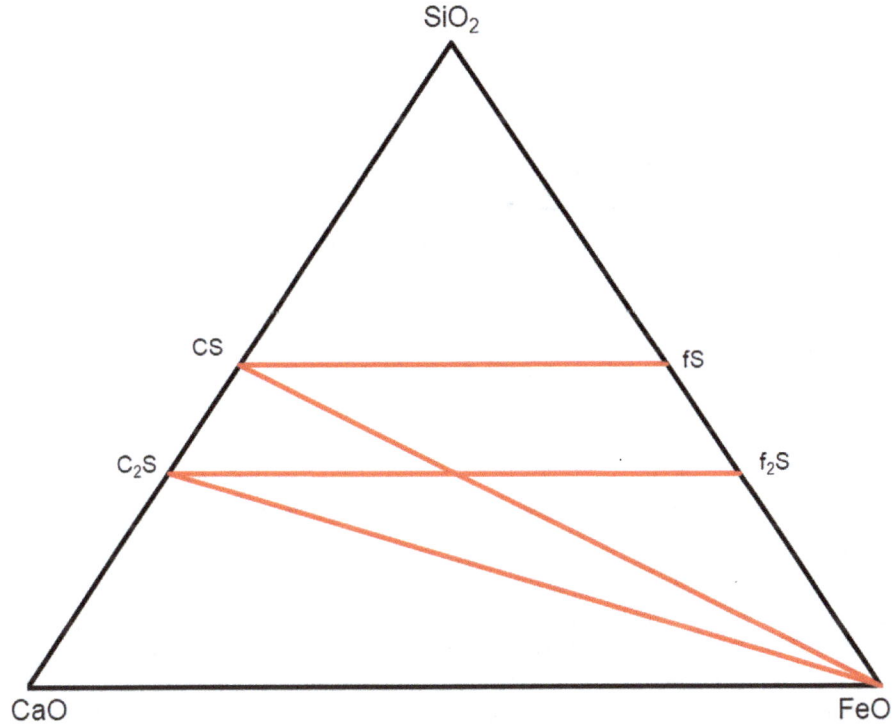

FIG. 3.21 – Joins that have been studied in the ternary CaO–SiO_2–FeO system. Pseudo binary systems; $CaSiO_3$–$FeSiO_3$, Ca_2SiO_4–Fe_2SiO_4, $CaSiO_3$–FeO, Ca_2SiO_4–FeO.

3.4.5.2.4 Ca_2SiO_4–FeO Pseudo Binary System

- The Ca_2SiO_4–FeO join has been constructed from the data of the ternary system CaO–SiO_2–FeO (figure 3.25).
- Ca_2SiO_4 and Wüstite occur as primary phases.
- The part that has been explored presents a eutectic point.

3.4.5.3 Ternary CaO–SiO_2–FeO System

- The liquidus of the ternary CaO–SiO_2–FeO system is shown in figure 3.26; CaO, Ca_3SiO_5, Ca_2SiO_4, $Ca_3Si_2O_7$, $CaSiO_3$, Olivine, and Tridymite occur as primary phases.
- In reducing atmosphere (in contact with iron), Fe_2SiO_4 (Fayalite) transforms into Wollastonite and the following invariant points are obtained (figure 3.27):
    (a) (P 1200 °C) CaO–Ca_3SiO_5–$Ca_2SiO_4^{(2)}$.
    (b) (P 1375 °C) $Ca_2SiO_4^{(1)}$–$Ca_2SiO_4^{(2)}$–Wüstite.
    (c) (E 1200 °C) CaO–$Ca_2SiO_4^{(2)}$–Wustite.
    (d) (P 1283 °C) $Ca_2SiO_4^{(1)}$–$Ca_2SiO_4^{(2)}$–Wustite.

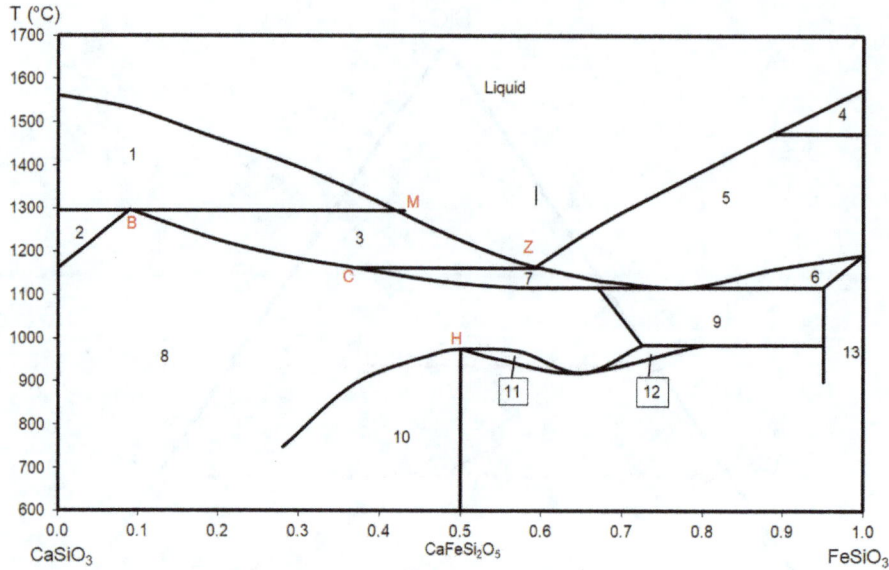

FIG. 3.22 – Equilibrium diagram of metasilicate join ($CaSiO_3$–$FeSiO_3$). Key: (1) α-$CaSiO_3$ + liquid, (2) α-$CaSiO_3$ + β-$CaSiO_3$ ss, (3) β-$CaSiO_3$ ss + liquid, (4) Cristobalite + liquid, (5) Tridymite + liquid, (6) Tridymite + Olivine + liquid (7), β-$CaSiO_3$ ss + Tridymite + liquid, (8) β-$CaSiO_3$ ss, (9) β-$CaSiO_3$ ss + Tridymite + liquid, (10) β-$CaSiO_3$ ss + Hedenbergite, (11) Hedenbergite ss, (12) Hedenbergite ss + Tridymite + Olivine, (13) Olivine + Tridymite. Letters in red are defined in the text.

(e) (P 1223 °C) $Ca_2SiO_4^{(1)}$–Wustite–Olivine.
(f) (P 1227 °C) $Ca_3Si_2O_7$–Olivine β-$CaSiO_3$.
(g) (P 1220 °C) $Ca_2SiO_4^{(1)}$–$Ca_3Si_2O_7$–Olivine.
(h) (P 1193 °C) Pseudo-wollastonite Wollastonite, Olivine.
(i) (P 1272 °C) Pseudo-wollastonite, Wollastonite, Tridymite.
(j) (P 1105 °C) Olivine, Wollastonite, Tridymite.

- The mineral Ca_2SiO_4 is recognizable in polished section by its rounded shape. $Ca_2SiO_4^{(1)}$ is characterized by a cross twinned structure corresponding to a polymorph of Ca_2SiO_4 (α or α') while $Ca_2SiO_4^{(2)}$ is characterized by a fine parallel twining.

3.4.6 CaO–SiO_2–FeO_x System at Various Oxygen Pressures

- The CaO–SiO_2–FeO_x system has been studied at PO_2 pressure from 10^{-1} to 10^{-8} Pa at 1250 and 1300 °C in its iron rich part.
- The ratio Fe^{3+}/Fe^{2+} depends on PO_2 and the chemical environment. The effect of PO_2 on the oxidation degree of the melt is quantified by the equation:

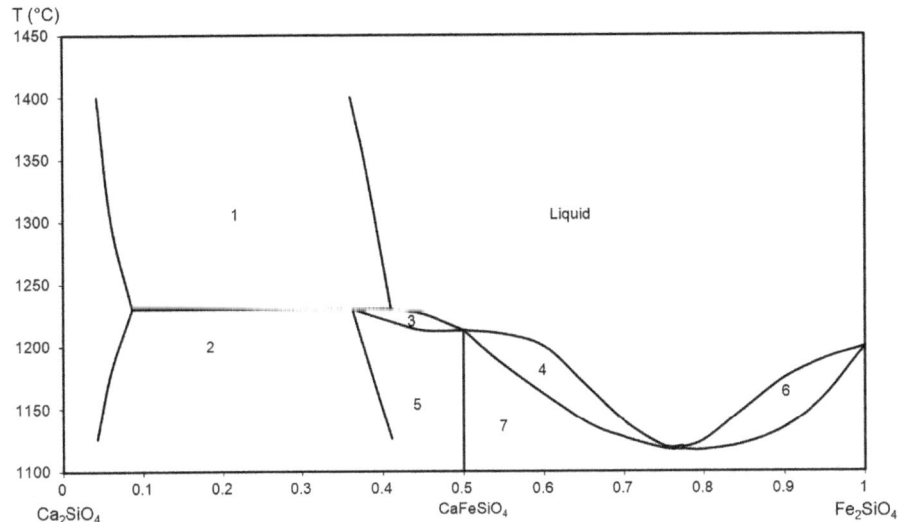

FIG. 3.23 – Pseudo binary diagram of Ca_2SiO_4–Fe_2SiO_4 system. Key: (1) β-Ca_2SiO_4 ss + liquid, (2) β-Ca_2SiO_4 ss + Fe/Ca Olivine, (3), (4) and (6) Fe/Ca Olivine + liquid, (5) and (7) Fe/Ca Olivine.

$$\log (Fe^{3+}/Fe^{2+}) = \log (a_{feO}/a_{FeO1.5}) + \tfrac{1}{4} \log PO_2 - (7441.14/T - 2.57)$$

where $a_{feO}/a_{FeO1.5}$ are the activity coefficients.

- High oxygen potentials involve an increase in the dissolution of silica. The solubility of silica along the isotherm 1200 and 1300 °C is estimated at 1 and 1.5%, respectively. Increasing the equilibrium temperature results in higher solubilities of silica.
- In the CaO–SiO_2–Fe_2O_3 system in air, the solubility of silica is constant around 5%.
- A decrease in temperature leads to a decrease in the ferric/ferrous ratio of the activity coefficients and PO_2 remains constant.
- In the case of iron silicate melts (acidic), increasing the silica content along the isopotential lines results in decreasing the ferric/ferrous ratio.
- In the case of iron lime melts (basic), increasing the silica content along the isopotential lines results in increasing the ferric/ferrous ratio.
- The degree of oxidation of the melt is related to the CaO content:

$$\log (Fe^{3+}/Fe^{2+}) = 0.170 \log PO_2 + 0.018(\% \, CaO) + 5500/T - 2.52.$$

- It is possible to estimate the effect of PO_2, T° and CaO content upon the Fe^{3+}/Fe^{2+} ratio; at the lowest PO_2, silica decreases the ferric to ferrous ratio. As the PO_2 increases, the Fe^{3+}/Fe^{2+} ratio increases.

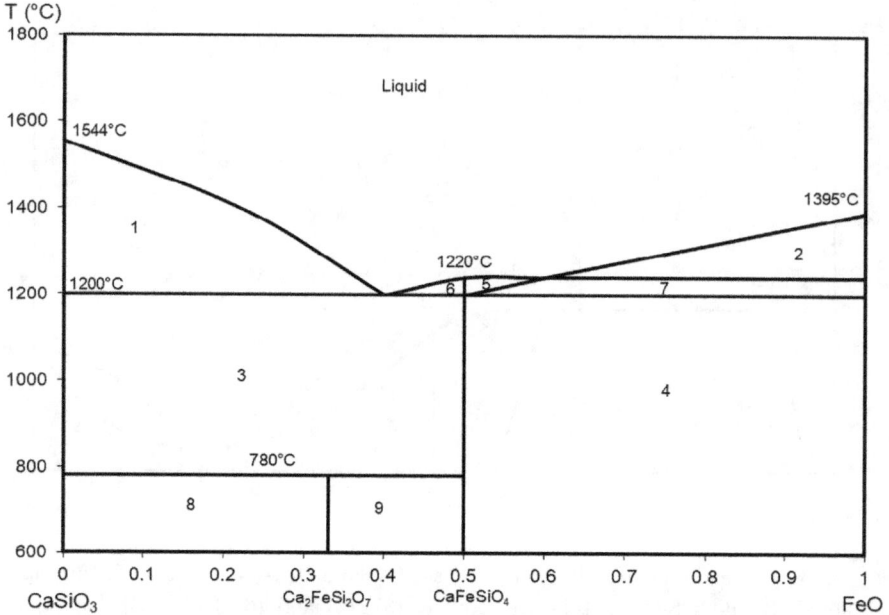

FIG. 3.24 – Pseudo binary diagram of CaSiO$_3$–FeO system. Key: (1) α-CaSiO$_3$ + liquid, (2) Wüstite + liquid, (3) β-CaSiO$_3$ + CaFeSiO$_4$, (4) CaFeSiO$_4$ + Wüstite, (5) Olivine + liquid, (6) CaFeSiO$_4$ + liquid, (7) Wüstite + olivine + liquid, (8) β-CaSiO$_3$ + Ca$_2$FeSi$_2$O$_7$, (9) Ca$_2$FeSi$_2$O$_7$ + CaFeSiO$_4$.

- The stability of the liquidus line and the primary phases has been determined in the CaO–SiO$_2$–FeO–Fe$_2$O$_3$ system at 1450 and 1550 °C for 5, 10 and 15% silica contents and 5, 10, 20 and 30% silica and 30% of CaO at 1300 °C.
- It is shown that PO$_2$ changes from 10^{-7} to 10^{-10} Pa from 1623 to 1373 °C, respectively, for a mixture of composition 46.3% FeO, 6.7% Fe$_2$O$_3$, 37% SiO$_2$ and 10% CaO while CO$_2$/CO is almost constant. Therefore, phase diagrams at constant CO$_2$/CO are an interesting alternative in the quantification of the liquidus surface when the PO$_2$ changes during cooling. This type of diagram gives directly the effect of silica, iron, lime and PO$_2$ on the liquidus surface.

3.4.6.1 *Liquidus Lines*

- Liquidus lines have been studied in the FeO–CaO–SiO$_2$ system at 1300, 1350 and 1550 °C and PO$_2$ between 3.63×10^{-3} and 10^{-9} Pa.
- There are considerable discrepancies in the position of the liquidus surface at 1300 °C, under various PO$_2$ and between the experimental data. Nevertheless, the following results can be assessed:

  ○ The homogeneous liquid area decreases with increasing oxygen partial pressure.

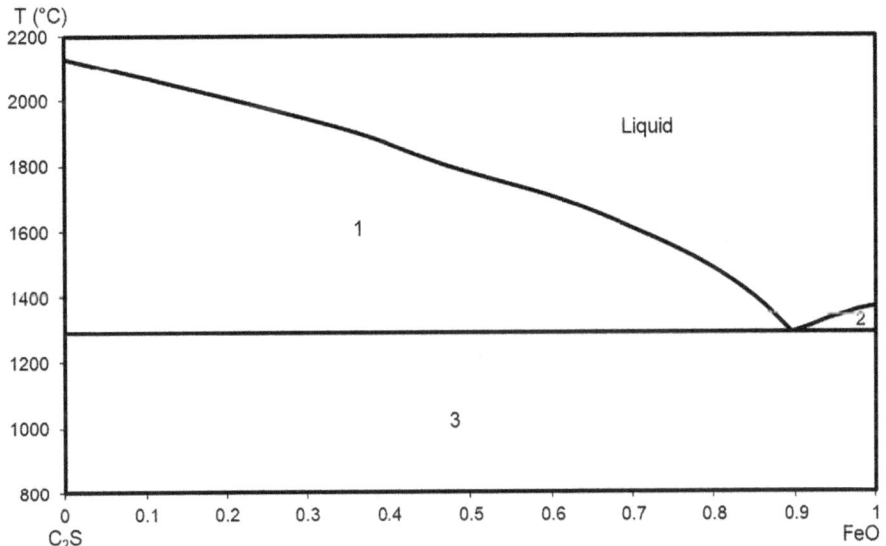

FIG. 3.25 – Pseudo binary diagram of Ca_2SiO_4–FeO system. Key: (1) Ca_2SiO_4 + liquid, (2) Wüstite + liquid, (3) Ca_2SiO_4 + Wüstite.

○ At 1300 °C, the liquid region is located in one continuum region or in two separated regions depending on the PO_2. Liquid area shrinks with increasing PO_2.
○ For the liquidus saturated with Spinel at 1300 °C, decreasing the PO_2 from 3.63×10^{-4} to 10^{-6} Pa resulted in a displacement of the liquidus line to the FeO_x–SiO_2 side without any appreciable extension of the liquid area, except for liquid region near $2CaO \cdot Fe_2O_3$ compound, liquid area shrinks with decreasing temperature.
○ Increasing CaO content leads to a decrease in melt liquidus in the SiO_2 primary phase field.
○ The liquidus increases with the CaO content in the Magnetite primary phase field.
○ The liquid boundary saturated with SiO_2 and $CaSiO_3$ slide phases is not affected by PO_2.

3.4.6.2 Magnetite

- The liquidus surface at $PO_2 = 10^{-7}$ and 10^{-6} Pa shows that magnetite, Ca_2SiO_4 and $CaSiO_3$ precipitate as primary phase.
- The stability of Magnetite decreases with increasing temperature and decreasing PO_2 from 0.21 to 10^{-4} Pa and with increasing silica.

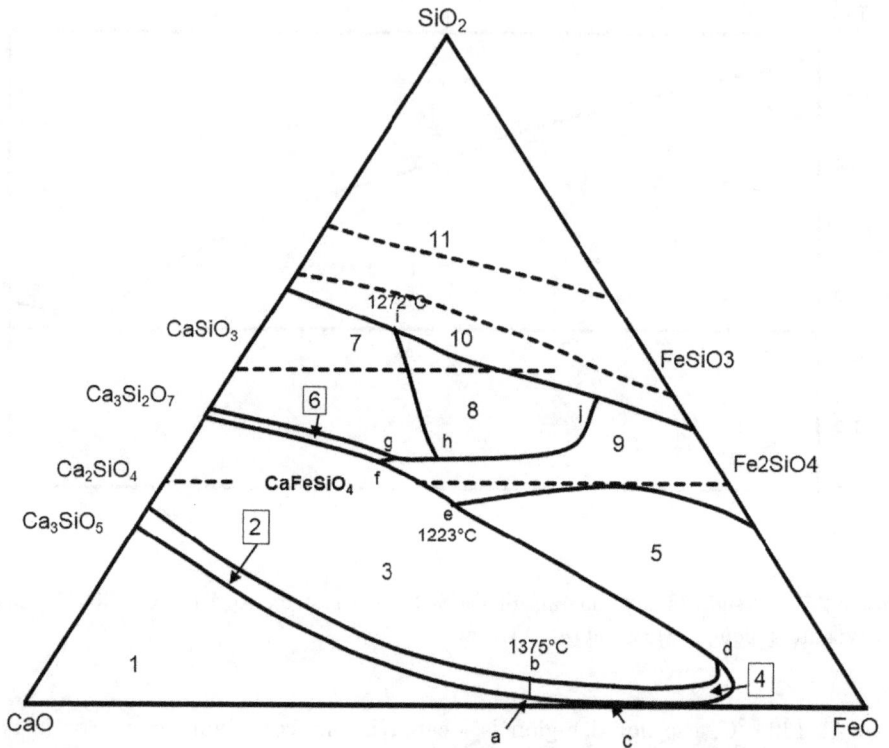

FIG. 3.26 – CaO–SiO$_2$–FeO system in equilibrium with metallic iron. Key: (1) CaO, (2) Ca$_3$SiO$_5$, (3) Ca$_2$SiO$_4^{(1)}$, (4) Ca$_2$SiO$_4^{(2)}$, (5) Wüstite, (6) Ca$_3$Si$_2$O$_7$, (7) Pseudo Wollastonite, (8) Wollastonite, (9) Olivine ((Fe,Ca)$_2$SiO$_4$), (10) Tridymite, (11–12) Cristobalite with boundary curve inferred.

3.4.6.3 Wüstite

- At 1300 °C, in the range of PO$_2$ where Wüstite appears, an increase in PO$_2$ has a large effect on the stable region of Wüstite.
- It was found that the two phases (solid Wüstite and liquid) region in the CaO–SiO$_2$–FeO system (1300 °C) increases remarkably by changing PO$_2$ between 10^{-4} and 10^{-3} Pa.
- It was also found that the two phases (solid Spinel and liquid) region in the CaO–SiO$_2$–Fe$_3$O$_4$ system (1300 °C) at PO$_2$ of 10^{-2} Pa was a little larger than that in the CaO–SiO$_2$–FeO system at PO$_2$ of 10^{-3} Pa, but it was extended considerably when PO$_2$ was increased to 10^{-1} Pa.
- Furthermore, it was found that the two phases region in the CaO–SiO$_2$–Fe$_3$O$_4$ system at PO$_2$ of 10^{-1} Pa decreases remarkably with increasing temperature.

Ternary Chemical Systems

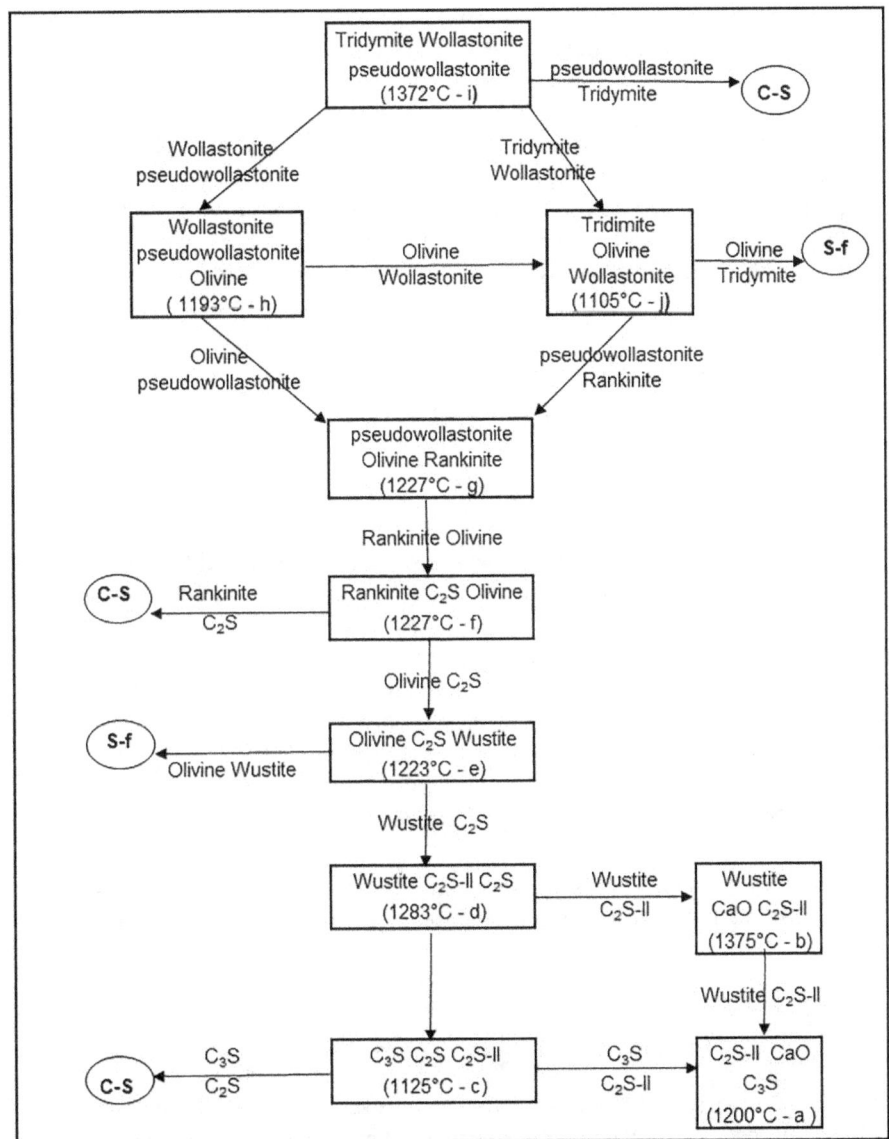

FIG. 3.27 – Flowchart of the ternary CaO–SiO$_2$–FeO system (in equilibrium with Fe).

3.4.7 Model

- Activities of CaO and FeO have been determined at 1450 and 1550 °C and pressure PO$_2$ from 1 to 10^{-11} atm. Several models based on thermodynamic data have tried to explain the experimental results.

- Multiphase equilibrium (MPE) based on the cell model was used to quantify the liquidus surface of the $CaO-FeO_x-SiO_2$ system.
- The effect of PO_2 on the phase constitution of a typical sinter mixture at 1300 °C (5% SiO_2, 12% CaO and 83% FeO_x) is quantified by model calculation.
- The quantity of melt can be calculated at 1300 °C and various PO_2 but also calculated at equilibrium (figure 3.28).
- Liquidus surface for $CaSiO_3$ and Ca_2SiO_4, liquidus boundary for Ca_2SiO_4 and Fe_2SiO_4 and isotherms have been calculated under the assumption that Fe^{2+} is at tetrahedral coordination in the melts. The model estimates the activity of $CaSiO_3$ and uses the following equation with the assumption of structural analogy of $CaSiO_3$ in the crystalline and molten state. The liquidus temperature is given by the equation:

$$T_{liquidus} = \Delta H_f \cdot T_f / (\Delta H_f - R \cdot T_f \ln(a_{CS})).$$

where ΔH_f, T_f, a_{CS} are, respectively, the enthalpy (constant), the temperature of fusion and the activity of $CaSiO_3$.

- Thermodynamic properties have been described using a liquid model that is the ionic two sublattice models and the compound energy model to describe all the solid solution phases (Wollastonite, Olivine, and Hedenbergite).
- The determination of the diagram is based on the system of semi empirical equations which take into account the concentration and temperature dependence of the solution properties of ordered system.

3.4.8 Preparation and Synthesis

- Synthesis of the compounds in the system CaO, SiO_2, Fe oxides are carried out by high temperature and low-pressure sintering, hydrothermal reaction or devitrification of glass.

3.4.8.1 Sintering

- Studies of mixtures of Hematite, Quartz, and Lime show that the assimilation of Quartz proceeds as (1) formation of calcium ferrite at 1000 °C by reaction of lime and hematite, (2) calcium ferrite melts at about 1160 °C and Quartz dissolves into the melt as follows; formation of molten slag with low basicity around Quartz grains, molten silicate slag reacts with calcium ferrite and Hematite precipitates at the boundary of the two melts, Hematite is divided into particles by molten silicate slag and the reaction between the two melts continue until assimilation completes at low temperatures; for basicity lower than 1.6, glassy silicate secondary hematite from calcium ferrite, for basicity higher than 2, calcium ferrite and calcium silicate.

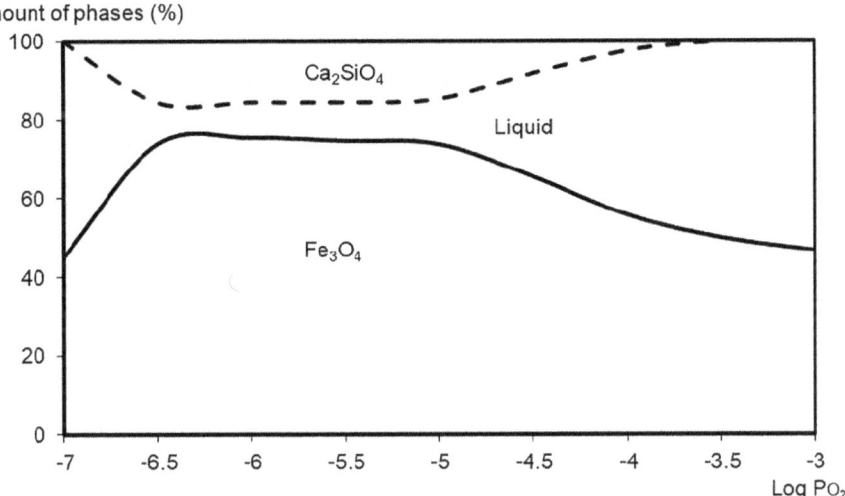

FIG. 3.28 – Calculation of the liquid quantity at 1573 K and of PO_2 for a composition 5% SiO_2, 12% CaO and 83% FeO_x.

- Andradite is synthesized dry from oxides at 1 atm by means of a lithium molybdate flux.

3.4.8.2 Sol Gel

- Andradite is synthesised from a gel of Andradite bulk composition at 1050 °C and 1 atm pressure.

3.4.8.3 Hydrothermal Reaction

- Hedenbergite is prepared by hydrothermal synthesis. The gel is prepared from $CaCO_3$ + Fe dissolved in nitric acid after evaporation and addition of TEOS and precipitation by ammonium hydroxide. The gel was treated at 400–600 °C under 1 kbar of methane in Ag–Pd crucible. Hedenbergite was also synthesized at 980 °C and 20 kbars for 8 h.
- Mössbauer spectroscopy of Hedenbergite prepared by hydrothermal synthesis reaction shows that the synthesis is completed in a time much shorter than evidenced by X ray.
- Andradite is synthesized hydrothermally in different conditions:

    ○ At 500 °C and 11 days from a glass of appropriate composition.
    ○ Crystallized hydrothermally at >150 atm, >480 °C from appropriate mixture of $CaCO_3$, $Fe_2O_3 \cdot xH_2O$ and $SiO_2 \cdot xH_2O$ of an Andradite bulk composition.
    ○ Synthesized hydrothermally at 850 °C and 20 000 psi in a cold sealed pressure vessel from a gel of nitrates of ferric iron and calcium and ethyl silicates.

3.4.8.4 Devitrification of Glass

- Homogeneous glass of the composition of Hedenbergite is crystallised at 1135 °C followed by 56 days of successive grindings.
- Andradite was crystallized from one batch of finely powdered glass of the composition $Ca_3Fe_2Si_3O_{12}$ at 1135 °C and 1 atm pressure. 56 days associated to 10 grindings were necessary for a complete dissolution of $CaSiO_3$ in Andradite.

3.5 CaO–Al$_2$O$_3$–Fe Oxides

3.5.1 Introduction

- A ternary compound $Ca[Al_{2x}Fe_{2(1-x)}]_3O_{10}$, called 'T phase' or AFC or $C[A_x F_{(1-x)}]_3$ and the following solid solutions are found in the CaO–Al_2O_3–Fe_2O_3 system:
  - $Ca_3Al_{2x}Fe_{2(1-x)}O_6$, $(C_3(A_xF_{(1-x)}))$.
  - $Ca_{12}Al_{14x}Fe_{14(1-x)}O_{33}$, $(C_{12}(A_xF_{(1-x)})_7)$.
  - $CaAl_{2x}Fe_{2(1-x)}O_4$, $(CA_xF_{(1-x)})$.
  - $CaAl_{4x}Fe_{4(1-x)}O_7$, $C(A_xF_{(1-x)})_2$.
  - $CaAl_{12x}Fe_{12(1-x)}O_{19}$, $C(A_xF_{(1-x)})_6$.
  - $Ca_4Al_{2x}Fe_{2(1-x)}O_{10}$, $C_4A_xF_{(1-x)}$.

- Several ternary compounds have been reported but not confirmed:
  - $Ca_{15/16}Al_2Fe_{20/22}O_{48/57} \cdot (C_{15/16}AF_{10/11})$ (TAV1 1936).
  - $Ca_4(AlFe)_{18}O_{31} \cdot Ca_4(AlFe)_{20}O_{33}$ $(C_4(AF)_{9/10})$ (MAC 1937).
  - $Ca_{13}Al_2Fe_{18}O_{43}$ $(CA_{13}F_9)$.
  - $Ca_{12/13}Al_2Fe_{18}O_{42}$ $(C_{12}AF_9)$ (APP 1972).
  - $Ca_{10.1}Al_{18.9}Fe^{3+}_{49.1}O_{112}$ and $Ca_{9.6}Al_{17.7}Fe^{3+}_{39.8}O_{96}$ (HAM 1969).
  - $Ca_{7.12}Al_{8.18}Fe_{0.88}Fe_{23.82}O_{56}$ $(C_7fF_{13}A_4)$ (ARA 1990).
  - $CaAl_4Fe_8O_9$ (CA_2F_4) similar to $CaAl_{12x}Fe_{12(1-x)}O_{19}$ (KUL 1994).
  - $Ca_2Al_{0.74}Fe_{9.26}O_{17}$ $(SFCA_1C_2F_{4.6}A_{0.37})$ (MUM 2003).

3.5.2 Mineralogy

3.5.2.1 Brownmillerite – $Ca_4Al_2Fe_2O_{10}$ (C_4AF)

- Brownmillerite is named after the chemist L.T. Brownmiller (of Easton, Pennsylvania USA).
- Brownmillerite is found as mineral in thermally metamorphosed limestone blocks included in volcanic rocks, but the composition C_4AF is the most often found.
- It is part of the solid solution between $Ca_2Fe_2O_5$ and a hypothetical $Ca_2Al_2O_5$ found at high pressure.

3.5.3 Structure and Solid Solutions

3.5.3.1 $Ca_3Al_{2x}Fe_{2(1-x)}O_6$ ($C_3A_xF_{1-x}$)

- Cubic lattice parameters are changed with the replacement of Al^{3+} by Fe^{3+}. The range of variation of x from 0.035 to 0.1 is explained by the experimental conditions. The replacement of Al by Fe is accompanied by a change in position of bands in the IR absorption spectra and the appearance of a new band suggests the presence of FeO_4 groups.

3.5.3.2 $Ca_{12}Al_{14x}Fe_{14(1-x)}O_{33}$ ($C_{12}[A_xF_{1-x}]_7$)

- $C_{12}A_7$ takes some iron in solid solution that occurs by replacement of Al^{3+} by Fe^{3+} ($x = 0.89$). After substitution of Al by Fe, excess of Fe could replace calcium.

3.5.3.3 $CaAl_{2x}Fe_{2(1-x)}O_4$ (CA_xF_{1-x})

- In the case of $CaAl_2O_4$, Al has a tetrahedral coordination and occupies six different sites.
- Aluminium atoms exchangeable with iron can reach 11%.
- In $CaFe_2O_4$, iron atoms exchangeable with aluminium can reach 3%.
- $CaFe_2O_4$ crystallises in the orthorhombic system.
- In the crystal structure of CF, there are two distinct Fe positions, each surrounded by a distorted octahedron of oxide ions By Mössbauer spectroscopy, it is shown that this distribution of Al in CF causes distortion in the octahedral, leading to an increase in the quadrupole splitting at both the Fe sites. When more Al is added, a separate phase is formed and Fe also starts going to the distorted tetrahedral sites belonging to Al in CA and this increases with x (CA_xF_{1-x}).
- When x increases, more CA with Fe distribution is created and when $x = 0.8$, the concentration of the distorted octahedral sites of Fe in CF and the distorted tetrahedral sites of Fe in CA is quite comparable.

3.5.3.4 $Ca_4Al_{2x}Fe_{2(1-x)}O_{10}$ ($C_2A_xF_{(1-x)}$)

- The system $Ca_2(Al_xFe_{(1-x)})_2O_5$ constitutes a continuous solid solution from $x = 0$ up to 0.67 (table 3.10).
- The end member $Ca_2Fe_2O_5$ crystallises in the orthorhombic system and has the space group $Pnma$.
- From IR, it is shown that the isomorphous replacement Al–Fe is accompanied by changes in positions of bands and the appearance of a new band which is assigned to tetrahedral FeO_4 groups (BEN 1974, BEN 1976).

TAB. 3.10 – Crystallography characteristics of solid solution of $Ca_4Al_2Fe_2O_{10}$ and $CaAl_{12x}Fe_{12(1-x)}$.

Composition	$Ca_4Al_2Fe_2O_{10}$	$CaAl_{12x}Fe_{12(1-x)}$		
		$x = 1/6$	$x = 1/3$	$x = 2/3$
Chemical formula	$Ca_4Al_{2x}Fe_{2(1-x)}O_{10}$	$CaAl_2Fe_{10}O_{19}$	$CaAl_4Fe_8O_{19}$	$CaAl_8Fe_2O_{19}$
Oxide formula	$4CaO \cdot Al_2O_3 \cdot Fe_2O_3$	$CaO \cdot Al_2O_3 \cdot 5Fe_2O_3$	$CaO \cdot 2Al_2O_3 \cdot 4Fe_2O_3$	$CaO \cdot 4Al_2O_3 \cdot 2Fe_2O_3$
Cement formula	C_4AF	CAF_5	CA_2F_4	CA_4F_2
Mineral	Brownmillerite			
Lattice	Orthorhombic	Triclinic	Hexagonal	Hexagonal
Z	4			2
a (Å)	5.58	9.02	5.83	5.681
b (Å)		9.97		
c (Å)	14.5°	10.81	22.14	22 27
α		59°8		
β	5°34	73°40		
γ		66°10		
V (Å³)	432	764	651	622.56
d (g/cm³)	3.74	4.17	2.29	4.179
JCPDS N°	74 1346	46 38	49 1586	25 120
Ref	BER 1959	HAM 1969	KUL 1994	DYS 1972

- The solid solution is not isostructural throughout the entire composition range. A structural change, involving the relative symmetry of tetrahedral chains between each successive layer, takes place around $x = 0.25$ changing from $Ib2m$ Brownmillerite symmetry above $x = 0.25$ to the $Pnma$ symmetry below $x = 0.25$.
- The ferrite phase is ferromagnetic and the iron rich member presents magnetic properties.
- The cell dimension of $Ca_2Fe_2O_4$ increases during heating to 685 °C at which temperature the phase transition of $Pcmn$ to $Ibm2$ occurred.
- By Mössbauer spectroscopy, it is shown that for $x < 0.7$, the aluminium prefers the lattice sites with the weaker magnetic field and that there is no statistical distribution of the aluminium on the two possible lattice sites.
- In ferrite having a similar crystal structure, the magnetization of the sub lattice with the octahedron sites is usually greater than the magnetization of the sub lattice with the tetrahedron sites.
- From Mössbauer spectroscopy, isomer shift decreases with increasing x and thus the covalency between Fe^{3+} and O^{2-} ions increases with x value, which reflects the decreasing lattice volume and the increasing ionic character of $Al^{3+}O^{2-}$.
- The cation distribution between octahedral and tetrahedral site is as follows:
  - $x = 0$ $Ca_2[Fe](Fe)O_5$.
  - $x = 0.25$ $Ca_2[Al_{0.12}Fe_{0.88}]\,(Al_{0.38}Fe_{0.62})O_5$.
  - $x = 0.33$ $Ca_2[Al_{0.21}Fe_{0.79}]\,(Al_{0.45}Fe_{0.55})O_5$.
  - $x = 0.66$ $Ca_2[Al_{0.47}Fe_{0.63}]\,(Al_{0.87}Fe_{0.13})O_5$.
- The compositions $x = 0$–0.33 show thermal hysteresis on cooling the material in a magnetic field of 6300 Koersted.
- This hysteresis in the compositions x up to 0.25 having the space group $Pcmn$ is due to the correlation between the spin direction and the local anisotropy field. In $Pcmn$, the c direction is the only direction where the spin can align without producing ferromagnetic moment. However, hysteresis gap and preference of c direction are decreased with decreasing of paramagnetic Fe^{3+} ions. The hysteresis gap disappears at composition $x = 0.5$ and 0.67 with space group $Ibm2$ which results in more restriction on the spin direction.
- Quantum chemistry by analysing the energy of molecular orbits along with net charge population and covalent order in the $Ca_2Fe_{2-x}Al_xO_5$ system shows that when the ratio Al/Fe increases, Fermi energy falls gradually, net charge of Al is higher than that of Fe and covalent bond of Al–O is weaker than that Fe–O.
- The composition C_4AF is not a distinct compound. Al^{3+} with the smaller ionic radius (0.57 Å for Al and 0.64 Å for Fe) prefers the tetrahedral site but the preference decreases with an increase in x, thus Al and Fe are randomly distributed in tetrahedral and octahedral sites. For $x > 0.33$, Al atoms are distributed nearly equally between octahedral and tetrahedral sites while for $x = 0.5$, Al atoms substitute for about ¼ of the octahedral sites and ¾ of the tetrahedral sites. In the crystal structure of Brownmillerite ($x = 0.5$ Ca_2FeAlO_5), there are four Ca_2FeAlO_5 formula units in an orthorhombic cell with $a = 5.584$ Å, $b = 14.60$ Å and $c = 5.374$ Å.

- Its structure can be described as a $CaMO_{2.5}$ Perovskite structure with 1/6 of that anion sites are vacant.
- The oxygen vacancies order in chains along the 100 direction, leading to a structure composed of alternate layers of corner sharing MO^6 octahedra and chains of MO^4 tetrahedra stacked along 001 where M is a cation other than Ca.
- Calcium ion is surrounded by irregular array of seven oxygen ions with average Ca–O distance equal to 2.461 Å.
- The octahedral site shows a small distortion along the 001 direction: the metal oxygen bonds which lie within the octahedral layer plane 001 are slightly shorter than those which are almost perpendicular to the layer (along 001). Average distances FeAlO (octahedral) and AlFeO (tetrahedra) are 2.001 and 1.797 Å.
- By TEM and by studying calcium ferrite chemically extracted from industrial cement, it is shown that a little amount of Si (around 1%) can enter the structure of calcium aluminoferrite). The following formula has been found; $Ca_6Fe_{3.98}Al_{1.48}Si_{0.53}O_{15}$ in a laboratory clinker.
- According to EELS, the substitution by tetravalent cations into Brownmillerite is not accompanied by a significant reduction in the valence of the iron atoms. This suggests that charge compensations operate through the increase in oxygen content in the structure. Si^{4+} seems to play a dominant part in altering the Fe^{3+} spin state from the original high spin in the pure ferrite to the low spin state in the extracted ferrite phase.
- The substitution of Si^{4+} for Al^{3+} or Fe^{3+} on the tetrahedral lattice was regarded as a dominant mechanism to bring about their differences in Mössbauer superfine parameters.

3.5.3.5 $CaAl_{4x}Fe_{4(1-x)}O_7$, $C[A_xF_{1-x}]_2$

- CA_2 takes some iron in solid solution that occurs by replacement of Al^{3+} by Fe^{3+} ($x = 0.87$). The solid solution extends from CA_2 to 13% mole CF_2 in the range 1000/1320 °C. Data for the change in d-spacing for the reflexion (043) along the join CA_2–CF_2 show an increase up to 13% mole CF_2 and a plateau for higher concentration of Fe^{3+} (DAY 1965).

3.5.3.6 $CaAl_{12x}Fe_{12(1-x)}O_{19}$, $C[A_xF_{1-x}]_6$

- Calcium hexaaluminate forms solid solution with hypothetical calcium hexaferrite (CF_6).
- This solid solution extends from CA_6 to 52 mol % of CF_6 at 1330 °C. CA_6 can contain up to 60% Fe^{3+} ($x = 0.46$).
- The structure of $CaAl_4Fe_8O_{19}$ (CA_2F_4) is hexagonal with Magnetoplombite structure with $a = 5.83$ Å and $c = 22.14$ Å solid solution of CA_6 ($a = 5.56$ Å and $c = 21.98$ Å) and CF_6 ($a = 5.877$ Å and $c = 22.91$ Å).

- It is paramagnetic at room temperature attributed to the substitution of Al^{3+} replacing the magnetic ions occupying the sites in the fivefold position which contribute to the dominant magnetic moment to the material.
- Mössbauer spectra of $CaAl_{12-x}F_xO_{19}$ show Fe^{3+} in tetrahedral, trigonal, pyramidal and octahedral sites with a preference for occupation of the tetrahedral site.
- A triclinic form $CaAl_{12x}Fe_{12(1-x)}$ has also been detected for $x = 1/6$.

3.5.3.7 $Ca[Al_{2x}Fe_{2(1-x)}]_3O_{10}$, $C[A_xF_{(1-x)}]_3$ (T Phase or AFC)

- The range of compositions which at subsolidus crystallize to yield the T phase represented by the formula $Ca[Al_{2x}Fe_{2(1-x)}]_3$, lies along the join $CaO \cdot 3Al_2O_3$–$CaO \cdot 3Fe_2O_3$ between 44.5 and 81.5 mol %.
- The X-ray pattern can be indexed on the basis of a monoclinic pseudo cell defined by the values given in table 3.11.
- The unit cell varies as a function of the ratio A/F.
- The solid solutions $Ca[Al_{2x}Fe_{2(1-x)}]_3$ have a similar X-ray diffractogram.
- The T phase is composed of at least three homologues of a silicoferrite of calcium called SFCA each with similar but unique powder X-ray diffraction pattern.
- The three unique phases are:
  (1) $M_{14}O_{20}$, SFCA type $Ca_2Al_5Fe^{3+}{}_7O_{20}$ with 25 CaO, 31 Al_2O_3 and 44 Fe_2O_3 mol %.
  (2) $M_{34}O_{48}$; an intergrowth SFCA type 2, $Ca_{5.1}Fe^{3+}{}_{18.7}Al_{9.3}Fe^{2+}{}_{0.9}O_{48}$, with 27 CaO, 24 Al_2O_3 and 49 Fe_2O_3 mol %.
  (3) $M_{20}O_{28}$, SFCA type 1, $Ca_{3.18}Fe^{3+}{}_{14.66}Al_{1.34}Fe^{2+}{}_{0.82}O_{48}$ with 28 CaO, 6 Al_2O_3 and 66 Fe_2O_3 mol %.
- The limits of the T phases are $Ca_4Al_4Fe_{20}O_{40}$ ($CA_{0.5}F_{2.50}$) and $Ca_4Al_{10}Fe_{14}O_{40}$ ($C_2A_{1.25}F_{1.7}$).
- T phase is composed of a layer of octahedral walls with an intermediated width of five octahedral layers alternating with tetrahedral layers, in which the different components of this type of layers are found in SFCA and SFCA type 1 (themselves alternate resulting in a composite unit cell volume for the new structure and stoichiometry $M_{34}O_{48}$ (SFCA type 2).
- The structure of the T phase is modified when the temperature increases but differently with the ratio A/F. Three types of diffractogram are found (table 3.12).
- At temperatures from 1200 to 1300 °C, CaF_2 presents a diffractogram (SFCA type 1) similar to the one recorded in the literature (LIS 1967). At 1350 °C, the diffractogram is modified without any modification of the analysis found by SEM. The diffractogram of the sample $CA_{0.75}F_{2.25}$ is modified from type 1 to type 2 at 1300 °C and type 3 at 1350 °C. The diffractogram of the sample $CA_{1.25}F_{1.75}$ is not modified.

TAB. 3.11 – Crystallography characteristics of the T phase *versus* A/F.

	a (Å)	b (Å)	c (Å)	β	V (Å^3)
$CA_{1.25}F_{1.75}$	9.8854	3.4838	12.9862	95°540	445.14
CAF_2	9.8646	3.4741	12.9089	95°209	440.57
$CA_{0.75}F_{2.25}$	9.8358	3.4616	12.8747	95°563	436.29

TAB. 3.12 – Evolution of the XRD pattern of T phases *versus* temperature (LIS 1967).

T (°C)	$CaAl_{1.5}Fe_{4.5}O_{10}$	$CaAl_2Fe_4O_{10}$	$CaAl_{2.5}Fe_{3.5}O_{10}$
1200			
1250	Type 1	Type 1	Type 1
1300	Type 2	Type 1	Type 1
1350	Type 3	Type 2	Type 1
1400			Type 1 liquid

TAB. 3.13 – Main references concerning the CaO Fe$_2$O$_3$–Al$_2$O$_3$ phase diagram.

Hansen *et al.* (HAN 1928)	Lea *et al.* (LEA 1956)
McMurdie (MAC 1937)	Newkirk *et al.* (NEW 1958)
Tavasci (TAV 1936)	Majumdar (MAJ 1965)
Swayze (SWA 1946)	Lister *et al.* (LIS 1967)
Parker (PAR 1925)	Imlach *et al.* (IML 1971–2)
Brisi (BRI 1954)	Palomo *et al.* (PAL 1988)
Toropov (TOR 1955)	

- This modification of the structure can be due to the presence and the variation of the Fe^{2+} content that is significant between 1300 °C (0.21% FeO) and 1350 °C (0.8% FeO).
- It is noteworthy that the evolution of the structure is reversible between types 1 and 2.

3.5.4 Stability and Phase Diagrams

3.5.4.1 Solidus in Air

- Table 3.13 summarizes the main references concerning the CaO–Fe$_2$O$_3$–Al$_2$O$_3$ phase diagram.
- Subsolidus compatibility relations are illustrated by isothermal planes taken at 1170 °C (figure 3.29). This temperature is selected to be above the stability limit of CF$_2$ but below that of Fe$_2$O$_3$–Al$_2$O$_3$.
- The phase assemblages are described in table 3.14.

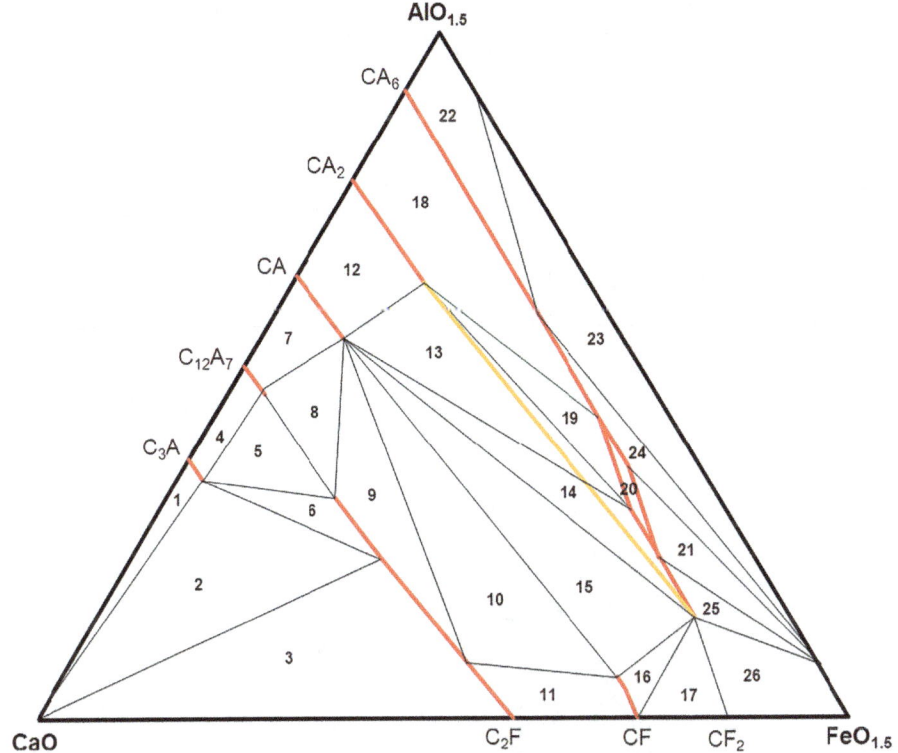

FIG. 3.29 – Subsolidus compatibility relations of the CaO–Fe$_2$O$_3$–Al$_2$O$_3$ phase diagram illustrated by isothermal planes taken at 1170 °C (Solid solutions are plotted in red).

3.5.4.2 Liquidus in Air

- Phase diagram of the CaO–Al$_2$O$_3$–Fe$_2$O$_3$ system at liquidus temperature has been constructed on the basis of data found in the literature. It is shown in figure 3.30.
- This system is not completely ternary due to the gradual increases in Fe^{2+} content close to the Fe metal corner and the occurrence of spinel phases. However, the quantity the Fe^{2+} content is low even at liquidus temperature.
- Phases coexisting with liquid include one phase CaO of constant composition and ranges of solid solutions C$_3$A, C$_{12}$A$_7$, CA, CA$_2$, CA$_6$, Al$_2$O$_3$–Fe$_2$O$_3$, Fe$_3$O$_4$, CF$_2$, CF, C$_2$F and a ternary phases (T).
- A ternary phase Ca[Al$_{2x}$Fe$_{2(1-2x)}$]$_3$O$_{10}$ and solid solutions Ca$_3$Al$_{2x}$Fe$_{2(1-x)}$O$_6$, Ca$_{12}$Al$_{14x}$Fe$_{14(1-x)}$O$_{33}$, CaAl$_{2x}$Fe$_{2(1-x)}$O$_4$, CaAl$_{4x}$Fe$_{4(1-x)}$O$_7$, CaAl$_{12x}$Fe$_{12(1-x)}$O$_{19}$, Ca$_4$Al$_{2x}$Fe$_{2(1-x)}$O$_{10}$ occur as primary phase in the CaO–Al$_2$O$_3$–Fe$_2$O$_3$ system.
- T phase, Ca[Al$_{2x}$Fe$_{2(1-x)}$]$_3$O$_{10}$, is stable from room temperature until its congruent melting temperature at 1470 °C.

TAB. 3.14 – Phase assemblages corresponding to figure 3.29.

(1)	C–C$_3$A–(C$_3$A)ss	(10)	CAss CF–C$_2$F	(19)	(CA$_2$)ss–(CA$_6$ss)–Tss
(2)	C–(C$_3$A)ss–C$_2$F	(11)	CF–C$_2$F	(20)	Tss
(3)	C–C$_2$F	(12)	CAss–(CA$_2$ss)	(21)	(CA$_6$)ss–Tss–F
(4)	(C$_3$A)ss)–(C$_{12}$A$_7$)ss	(13)	CAss–CA$_2$–Tss	(22)	(CA$_6$)ss–Ass
(5)	C$_3$A–C$_{12}$A$_7$–C$_2$F	(14)	CAss–Tss	(23)	(CA$_6$)ss–A–F
(6)	(C$_3$A)ss–C$_2$F	(15)	CAss–CF–Tss	(24)	(CA6)ss–F
(7)	CAss–(C$_{12}$A$_7$)ss	(16)	CF–Tss	(25)	Tss–F
(8)	(C$_{12}$A$_7$)ss–CAss–C$_2$F	(17)	CF–CF$_2$–Tss	(26)	CF$_2$–Tss–F
(9)	CAss–C$_2$F	(18)	(CA$_2$)ss–(CA$_6$)ss		

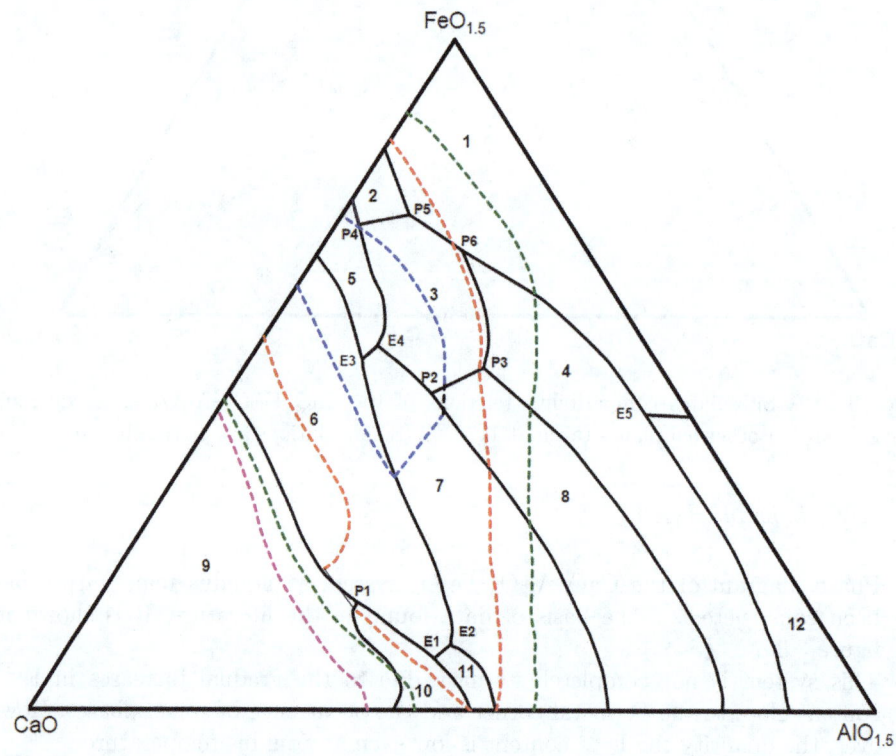

FIG. 3.30 – Liquidus of the CaO–AlO$_{1.5}$–FeO$_{1.5}$ system. Key: (1) Fe$_3$O$_4$, (2) Fe$_2$O$_3$, (3) T, (4) CA$_6$, (5) CF, (6) C$_2$F, (7) CA, (8) CA$_2$, (9) CaO, (10) C$_3$A, (11) C$_{12}$A$_7$, (12) Al$_2$O$_3$. Isotherms at 1600, 1500, 1400 and 1300 °C pink, green, red and blue, respectively.

- In air, the following assemblages are in equilibrium with liquid (invariants points in the CaO–Al$_2$O$_3$–Fe$_2$O$_3$ system – table 3.15).

TAB. 3.15 – Invariant points in the CaO–Al$_2$O$_3$–Fe$_2$O$_3$ system.

Eutectic	Phase assemblages	T°	Peritectic	Phase assemblages	T°
E1	CA–C$_{12}$A$_7$–C$_2$F	1336 °C	P1	C–C$_3$A–C$_2$F	1389 °C
E2	C$_{12}$A$_7$–C$_3$A–C$_2$F	1335 °C	P2	CA–CA$_2$–T	1265 °C
E3	CA–CF–T	1180 °C	P3	CA$_2$–CA$_6$–T	1404 °C
E4	C$_2$F–CA–CF	1190 °C	P4	CF–T–Fe$_2$O$_3$	1195 °C
E5	CF–T–CF$_2$	1175 °C	P5	T–Fe$_3$O$_4$–Fe$_2$O$_3$	1280 °C
				T–CA$_6$–Fe$_3$O$_4$	1435 °C

Note: T for Ca[Al$_{2x}$Fe$_{2(1-x)}$]$_3$.

3.5.5 Liquidus in Reducing Atmosphere

- Liquid surface has been studied at the partial oxygen pressure PO$_2$ = 1, 0.21, 10^{-5}, 10^{-8} atm.
- Liquid formation begins at lower temperatures if the atmosphere is reducing: this is a consequence of the formation of phases that are less rich in oxygen.
- The schematic flow diagram shows the phase present at a given temperature and pressure and also the nature of the invariant points. However, it does not give information of their bulk composition (flowchart given in figure 3.31).
- In a four component system, the equilibrium between 3 crystalline phases, one liquid phase and one gaseous phase represents a univariant equilibrium. The point of intersection generated by three adjacent primary phases is not an invariant point but a piercing point at which the true liquidus invariant curve cuts the chosen isobaric surface.
- Straight line represents the crystalline phase in equilibrium along quaternary univariant curve. Liquid invariant points are generated by the intersection of four univariant curves. The crystalline phases are identified in square boxes. The starting point of all quaternary invariant curves is the corresponding to ternary invariant point.
- Quaternary invariant points are shown in table 3.16.
- At PO$_2$ = 10^{-8} atm, primary phase volumes are of Spinel, CA$_2$, CA, C$_{12}$A$_7$, C$_3$A, C$_2$F, CaO. Thus, CA$_6$ and Al$_2$O$_3$ are not found as primary phases. Large intersection areas exist for spinel and CaO while areas for C$_3$A, CA, CA$_2$ and CF$_2$ are small.
- At PO$_2$ = 10^{-5} atm, intersection areas have been located for C$_2$F, T phase, Spinel, CA$_2$, CA, C$_{12}$A$_7$, C$_3$A, CaO and C$_4$fF$_4$. The primary phase of CF is not cut by the isobaric surface.
- The areas of the isobaric surface containing the primary fields CaO, C$_3$A, C$_{12}$A$_7$, CA, and C$_2$F are little affected by reducing pressure in the range from 1 to 10^{-8} atm.
- Between air and PO$_2$ = 10^{-5} atm, the primary phase area of Fe$_2$O$_3$, CF and C$_2$F disappear and are replaced by C$_4$fF$_4$. The primary phases CA$_2$, CA$_6$, and T are reduced.

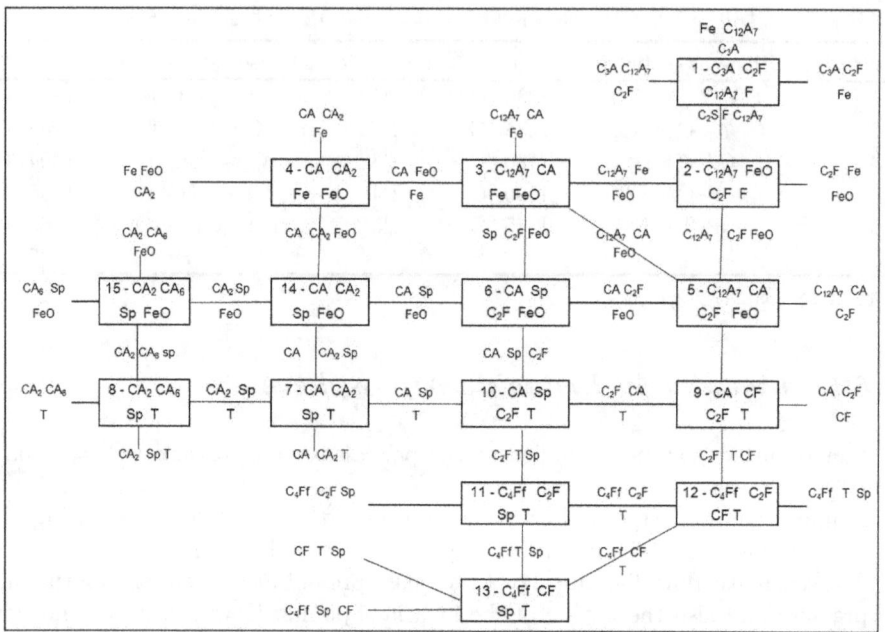

FIG. 3.31 – Flowchart of the CaO–Al$_2$O$_3$–FeO system. Note: T for Ca[Al$_{2x}$Fe$_{2(1-x)}$]$_3$, Sp for Spinel.

TAB. 3.16 – Invariant points in the CaO–Al$_2$O$_3$–FeO system.

(1) C$_3$A–C$_2$F–C$_{12}$A$_7$–C$_2$S–F	(9) CA–CF–C$_2$F–T
(2) C$_{12}$A$_7$–FeO–C2F–F	(10) CA–Sp–C$_2$F–T
(3) C$_{12}$A$_7$–CA–F–FeO	(11) CA–CA$_2$–Sp–T
(4) C$_{12}$A$_7$–CA$_2$–Fe–FeO	(12) CA$_2$–CA$_6$–Sp–T
(5) C$_{12}$A$_7$–CA–C$_2$F–FeO	(13) C$_4$Ff–C$_2$F–CF–T
(6) CA–Sp–C$_2$F–FeO	(14) C$_4$Ff–C$_2$F–Sp–T
(7) CA–CA$_2$–Sp–FeO	(15) C$_4$Ff–CF–Sp -T
(8) CA$_2$–CA$_6$–Sp–FeO	

Note: T for Ca[Al$_{2x}$Fe$_{2(1-x)}$]$_3$, Sp for Spinel.

- On reducing atmosphere still to 10^{-8} atm, the areas of CA$_2$, CA$_6$, and Al$_2$O$_3$ are further reduced whilst T and C$_4$fF$_4$ vanished.
- In oxygen, the melting of the most Fe rich T phase is congruent and the liquid has a FeO content of about 5%. In air, melting becomes incongruent giving CA$_6$ as primary phase while at PO$_2 = 10^{-5}$ atm, the primary phase is the Spinel.
- T phase has a primary phase area on the PO$_2 = 10^{-5.5}$ atm isobaric surface.

3.5.6 Model of the System

- The lack of thermodynamic data relative to the solid solutions prevents the development of model for the CaO–Al_2O_3–Fe_2O_3 system.

3.5.7 Formation and Synthesis

- C_4AF, C_6AF_2, and C_6A_2F are produced by sintering at 1350 °C.
- The best fit of the experimental kinetic data is obtained with the equation derived for diffusion-controlled reaction in sphere by Jander. From the oxide, the activation energy is 438 kJ/mole for C_3A and 297 kJ/mole for CF.
- The composition of ferrite depends on the cooling rate; the more liquid phase, the richer in alumina. By fast rate of cooling, ferrite crystallises with low iron content.

Chapter 4

Quaternary Chemical Systems

4.1 Introduction

- A quaternary system describes the phase relations *versus* temperature and pressure of all possible mixtures of four constituents.
- The information obtained by experiments or by thermodynamic modelling must be treated to be clear and useful for the applications. Among the different approaches, three methods are principally used:

  (a) Locating of the constituents and the data in a regular tetrahedron.
  (b) Use a perspective drawing.
  (c) Construction of a flowchart thermodynamic modelling.

- The details of the methods are developed as follows:

  ○ Principle of phase equilibrium in section 4.1.1.
  ○ Examples of phase equilibrium calculation in sections 4.2 and 4.3.

4.1.1 Principle of Phase Equilibrium in a Quaternary System

4.1.1.1 Regular Tetrahedron

- A quaternary system, such as the CaO–SiO_2–Al_2O_3–Fe_2O_3 system, may be represented by a regular tetrahedron, each apex of which represents 100% of one component.
- The four faces of the tetrahedron, showing the relations in a four components polythermal condensed system, are equilateral triangle representing the four limiting ternary systems.

- The knowledge of the phase relations existing in these ternary systems is of greater assistance in working with quaternary mixtures.
- Each of the limiting ternary systems is divided into areas limited by boundary curves.
- These boundary curves in the limiting ternary systems become curved surface within the tetrahedron.
- In the limiting ternary systems, three boundary curves intersect at a ternary invariant point at a particular temperature with 3 solid phases in equilibrium with only that particular liquid whose composition is represented by the ternary point.
- Each such ternary invariant points in the limiting ternary system becomes an univariant line (3 solid phases in equilibrium with a liquid whose composition lies on the line within the tetrahedron).
- Three curved surfaces (faces of primary phase volume) intersect within the tetrahedron to form such curved lines.
- Four such univariant lines intersect within the tetrahedron at a quaternary invariant point at a particular temperature with four solid phases in equilibrium with only that particular liquid having the composition represented by the point.
- The study of equilibrium relations in a condensed quaternary system consists in:
    (1) Delineating the portion of the tetrahedron occupied by each primary phase volume.
    (2) Determining the compositions and temperatures along the edges and at all apices of each primary phase volumes.
    (3) Determining the mutual relations within the tetrahedron of the many primary phase volumes involved.
    (4) Finding the location of sufficient numbers of isotherms to define adequately the liquidus and solidus within the system.
- For some quaternary systems of four oxides, it is possible to divide the system into several parts some of which are ternary or quaternary system within the larger fundamental quaternary system and thus describe it by portions.
- For other quaternary systems, it is convenient to select three of the four components and add progressively 10, 20 and 30% of the fourth component to get a series of planes through the tetrahedron parallel to one face taken as the base or a series of arbitrary plane through the tetrahedron.

4.1.1.2 Perspective of the Liquidus in the System CaO–SiO_2–Al_2O_3–Fe Oxides

- The liquidus relations are shown in a perspective drawing (see section 4.2).
- The data are showing a tetrahedron where one face has been cut away allowing a view of the interior (figure 4.1). The geometric or compositional relationships do however become clearer if they can be shown in such a perspective drawing.
- The example presented in figure 4.1 has been constructed with 4 constituents A, B, C and D.

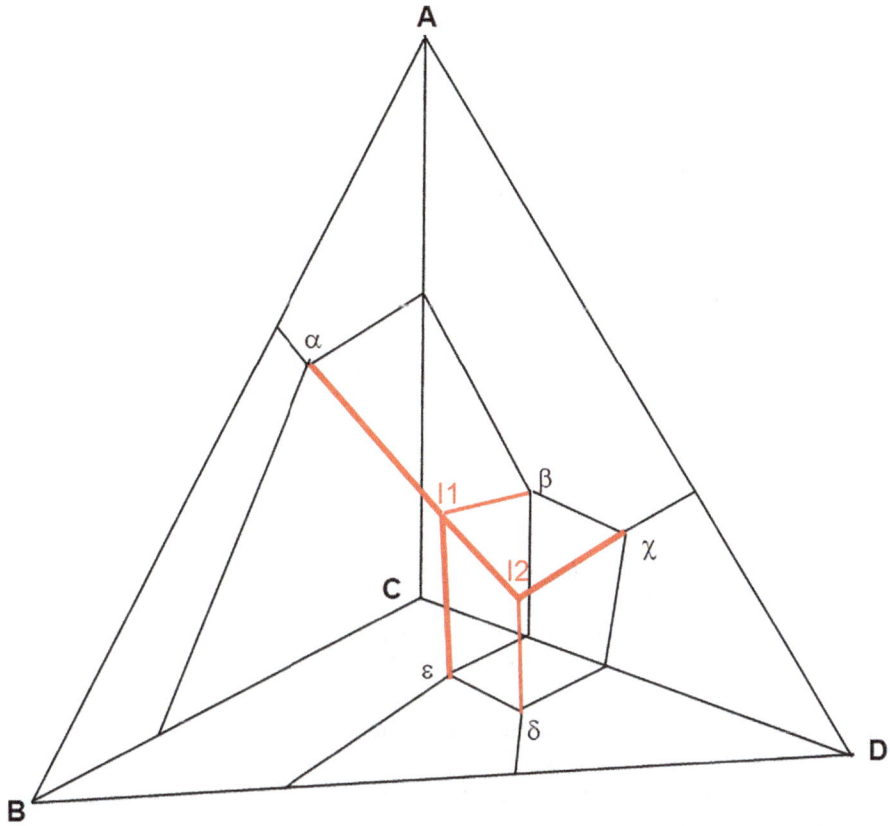

FIG. 4.1 – Method of representation of the liquidus.

- Four limiting ternary systems are ABC (left triangle), ACD (right triangle), BCD (base) and ABD (front, diagram is removed).
- One ternary invariant point is assumed in the system ABC (α), two invariant points are assumed in the systems ACD (respectively β, χ) and BCD (respectively δ, ε).
- In the quaternary system, the existence of 2 quaternary invariant points is assumed (I1 and I2).
- Univariant lines start from ternary invariant points until a quaternary point is reached (red lines within the tetrahedron).
- The representation is not strictly accurate in the geometric sense but care has been taken not to introduce any large distortion, although the space between curves has occasionally been widened or the position of intersection shifted very slightly for the sake of clarity. In the interior of tetrahedron, the primary phase volumes are delineated by a series of volume surfaces.

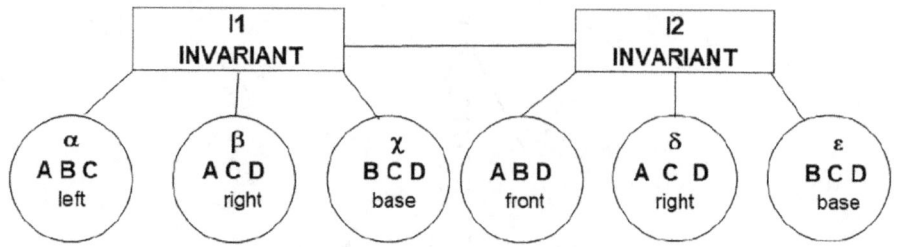

FIG. 4.2 – Flowchart of diagram corresponding to system presented in figure 4.1.

- If we assume that each quaternary invariant points, I1 I2, is a eutectic, the univariant curves which connect invariant points must have thermal maxima on them.

4.1.1.3 Construction of a Flowchart

- One method of depicting quaternary invariant point is by means of a schematic flowchart diagram that can show the phases present, temperature, pressure and existence of univariant points but they give no information concerning their bulk compositions. Such a diagram has been drawn for the present system and is shown in figure 4.2; the flowchart corresponds to the previous diagram.
- Rectangles are used to show invariant equilibria with liquid (or univariant if liquid is absent) in the quaternary system.
- Circles are used to show invariant equilibria with liquid in the limiting ternary diagrams; four lines start from invariant quaternary point to reach ternary invariant points or another quaternary invariant points.

4.1.1.4 Thermodynamic Modelling

4.1.1.4.1 Principle of Phase Equilibrium Calculation
- The chemical and mineralogical compositions are the key parameters for the control of the manufacture of many industrial products (refractories, cement, slags, ceramics, glass...).
- While the chemical composition can be easily measured and controlled, the mineralogical composition is also strongly dependent on the process.
- Obtaining the mineralogical compositions from the chemical composition and the parameters of the process is achieved experimentally or/and by thermodynamic modelling.
- From a set of experimental results obtained by keeping a mixture of constituents at a given temperature long enough to reach an equilibrium of reaction, it is possible to construct a phase diagram. These diagrams can be calculated from thermodynamics which allow us to better understand how to control the

manufacturing process which consequently helps us to optimize the properties of the final product.
- Phase diagrams present the thermodynamic properties of a system in a condensed manner. That is to say, they are a graphical illustration of the phase rule (Gibbs) and this method finds its limit in the numerous experiments required, especially for multi-component systems.
- To go further and overcome this difficulty, thermodynamic modelling was developed, as a representation of thermodynamic data, and amongst other benefits these calculations can show up deficiencies in the thermodynamic data and optimize experimental research.
- To date, there has been very little use of phase diagrams by the cement industry, especially at high temperatures. Tentative calculations have been done for Portland cement clinkering, but nothing exists for calcium aluminate cements. Two reasons for this can be put forward; firstly, the accuracy of the data has been insufficient to avoid the need for experimentation, secondly, the difficulty to handle phase diagrams for multi-component systems. The limited precision can be due to the method itself (experimental or modelling) and to non-equilibrium.

4.1.1.4.2 Example of Phase Equilibrium Calculation

- The chemical equilibrium of a system is determined by minimization of its Gibbs energy with respect to the amount and composition of the phases present.
- Three steps are necessary; the collection and use of thermodynamic data (enthalpy, entropy, specific heat...), a model to evaluate the free energy and software to optimize minimisation. Many publications exist describing models (i.e. polymeric or sublattice model, etc.) and the software (Thermo-calc, Parrot, Dictra...).
- The variation of the free energy is given by the expression:

$$\Delta G = \Sigma x_i \, G_i^0 + \Sigma \, x_i \, \mathrm{Ln} \, x_i + \Delta G^{ex}.$$

The first term is the sum of unary data. The reference state is the weighted sum of the enthalpies of the constituent elements at $278.15\,°C$ (CaO, $AlO_{1.5}$, SiO_2, $FeO_{1.5}$).
The second term is the configurational entropy to the Gibbs energy of mixing.
The third term is the excess free energy of mixing.

- In this book, the following assumptions are made:
  ○ Only equilibrium between condensed phases is considered.
  ○ The oxide liquid is modelled as a solution of non-ideally interacting species.
  ○ The formalism of Redlich and Kister (RED 1948) is used to determine ΔG^{ex}.

- With Redlich Kister coefficients for free energy, it is possible to calculate the full equilibrium phase diagram. It is necessary to adjust the entropy or free energy function values for solid calcium aluminate to match the known liquidus curves.
- The thermodynamic data used to compute the phase diagrams come from the JANAF database, Thermodata or from an estimation calculated from existing phase diagrams.

TAB. 4.1 – Estimation of the free energy of formation of the main compounds found in the CaO–SiO$_2$–Al$_2$O$_3$–Iron oxides systems as a function of temperature (J mol^{-1}).

	A	B	D	Reference
CA$_6$	−1340.699	2.6		ALL 1981
CA$_2$	−3200	5.524		ALL 1981
CA	−6040	6.11434		ALL 1981
C$_{12}$A$_7$	−413	9.067		ALL 1981
C$_3$A	−3400	5.607		ALL 1981
C$_2$F	164 803	220.005	0.00331	MCH 1974
CF	−6000	4.2		KUB 1967
CF$_2$	−3200	3.25		KUB 1967
CS	−45 531	0.10967	1.7382	Thermodata
C$_3$S$_2$	−47 940	−2.2002	3.0218	Thermodata
C$_2$S	−42 108.5	0.98141	3.2308	KIN 1951
C$_3$S	−32 407	−0.7503	3.7142	KIN 1951
A$_3$S$_2$	2074.80	3.44969	−0.1582	Thermodata
FA	−71 665	−10.068		Thermodata
TC(A$_{1,5}$ F$_{1,5}$)$_6$	1 106 615.875	674.44623		Estimated from the authors
C$_2$AS	−20 677	4.0156	−0.66084	RIB 1980
CAS$_2$	−22 459.8	−1.1277		RIB 1980

Key: Estimation is made using a simplified equation; $A - BT - DT^2/2$ resulting from the complete equation; $A - BT + CT(1 - LnT) - DT^2/2 - E/2T - FT^3/6$.

- Table 4.1 shows the free energy of formation of the main compounds found in the CaO–SiO$_2$–Al$_2$O$_3$–Iron oxides systems.

4.2 CaO–SiO$_2$–Al$_2$O$_3$–Fe$_2$O$_3$ System in Air

4.2.1 Quaternary Constituents of the CaO–SiO$_2$–Al$_2$O$_3$–Fe Oxides System

- The following quaternary compounds and solid solutions are identified in the CaO–SiO$_2$–Al$_2$O$_3$–Fe$_2$O$_3$ system.
- The quaternary constituents are: Ca[Al$_x$Fe$^{3+}_{(1-x)}$]$_2$SiO$_6$ (Fassaite, Ca Tschermak's type pyroxene), [Ca$_x$Fe$^{2+}_{(1-x)}$]$_3$[Al$_y$Fe$^{3+}_{(1-y)}$][(SiO$_4$)]$_3$ (garnet family), Ca$_{20}$Al$_{32-2x}$Fe$_x$Si$_x$O$_{68}$ (also called Pleochroite) (KAP 1980), a silico ferrite of calcium and aluminium (called SFCA).
- A solid solution of silica in the ternary phase Ca[Al$_x$Fe$_{(3-x)}$]$_2$O$_{10}$, a solid solution of iron oxide in Gehlenite Ca$_2$(Al$_x$Fe$_{(1-x)}$)SiO$_7$ and at little extend iron oxide in Anorthite belong to the system CaO–Al$_2$O$_3$–SiO$_2$–Fe oxides.
- A phase with a composition 2CaO·4(Fe^{3+}Al)$_2$O$_3$·SiO$_2$ has been found experimentally (HIJ 1968). It is called 'X'. It exhibits the same XRD pattern as CaFe$_4$O$_7$, Ca$_{7.2}$Fe$_{0.8}$Fe$^{3+}_{30}$O$_{53}$ Ca$_3$Fe$_{15}$O$_{25}$.

Quaternary Chemical Systems

- The other phases encountered in the system between 700 and 1400 °C are Pseudo-wollastonite, Wollastonite, ferri-Fassaite solid solution (ss), Hematite, Anorthite (ss), beta Corundum, Hibonite (CA_6) and liquid.
- The positions of the quaternary compounds and the solid solutions in the diagram of the CaO–SiO_2–Al_2O_3–Fe_2O_3 system are represented in a schematic way in figure 4.3.

4.2.2 Mineralogy

4.2.2.1 Esseneite

- Esseneite has the composition $Ca[Al_x Fe^{3+}_{(1-x)}]_2SiO_6$ and for $x = 0.5$ i.e. $CaFe^{3+}AlSiO_6$.
- It belongs to the calcium Tschermack's type pyroxene.
- As a mineral, it occurs in fused sedimentary rocks associated with naturally combusted coal seams near Gillete (Wyoming). It is named for Eric Essene, American Geologist.
- Fassaite solid solution ranges from about FTS_{78} ($CaFe_2SiO_6$ ferri-Tschermak's molecule), $CATS_{22}$ ($CaAl_2SiO_6$ aluminium Tschermack's molecule) to FTS_{48} $CATS_{52}$.
- Esseneite ($x = 0.5$ FTS_{50} $CATS_{50}$ or $CaFe^{3+}AlSiO_6$) crystallizes in the monoclinic system in the $C2/c$ space group with the pyroxene structure. It is stable at low pressure under high oxygen fugacity (GHO 1989).
- The structure of $CaFe^{3+}AlSiO_6$ is intermediated between that of Diopside and calcium Tschermak's pyroxene $CaAl^{3+}Al^{3+}SiO_6$. It forms a complete solid solution with Diopside below 1250 °C at 1 atm (COS 1995/AKA 1985).
- The topology of the $C2/c$ type pyroxene structure is determined by the way in which the three types of polyhedra are connected (M2 polyhedron, M1 octahedron and T tetrahedron). The calcium occupies the polyhedral site M2. The octahedral M1 site is occupied by $0.82Fe^{3+}$ and $0.18Al^{3+}$. The tetrahedral site (T) is occupied by Si^{4+} and Al^{3+}.
- There is no Fe^{3+} in the tetrahedral site of the mineral Esseneite as opposite as the experimentally synthesized compound $CaFe^{3+}AlSiO_6$. The site preference of Fe^{3+} and Al^{3+} can be deduced for Fassaite solid solution, Fe^{3+} and Al^{3+} might enter the four- and six-fold coordination structure as well.

4.2.2.2 Garnet

- Garnet has the composition $[Ca_xFe^{2+}_{(1-x)}]_3 [Al_yFe^{3+}_{(1-y)}] [(SiO_4)]_3$. Garnets are nesosilicate having the general formula $X_3Y_2(SiO_4)$. The X site is occupied by divalent Fe^{2+} and Y site by trivalent Fe^{3+} in an octahedral/tetrahedral network with SiO_4^{4-} occupying the tetrahedra. Andradite ($Ca_3Fe^{3+}_2Si_3O_{12}$) forms a complete solution solid with Grossularite up to 866 °C.

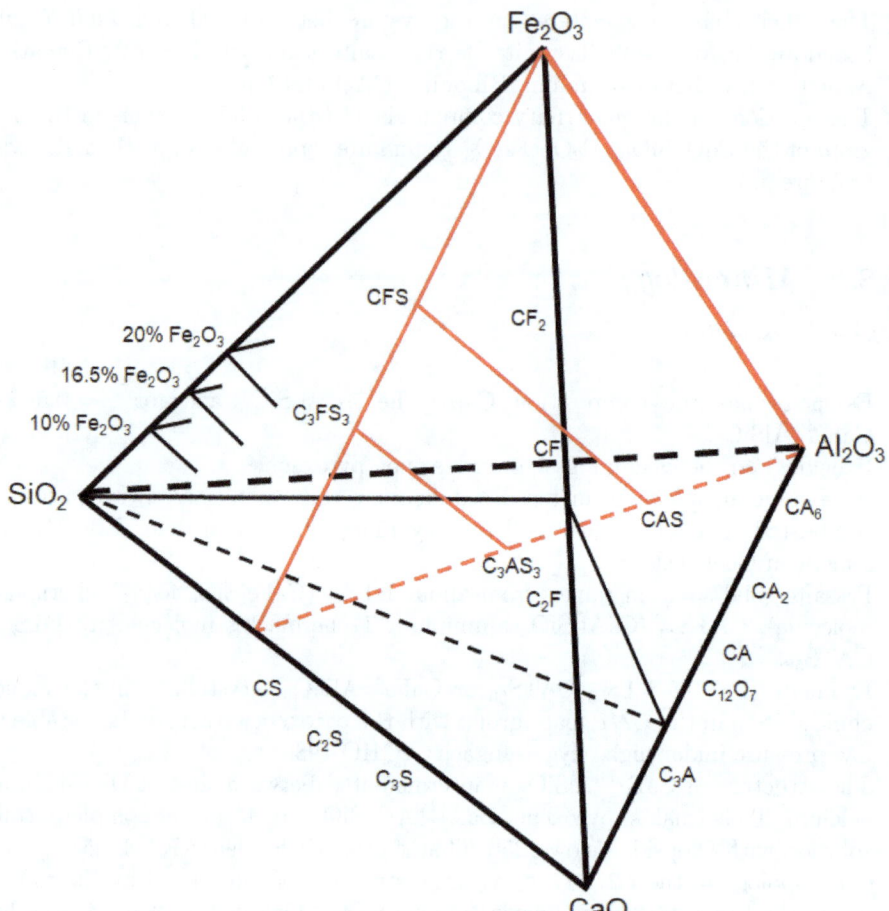

FIG. 4.3 – CaO–SiO$_2$–Al$_2$O$_3$–Fe oxides system and position of the partial systems studied.

- Garnet groups include a group of minerals which are found in many places. They are used in interpreting the genesis of many igneous, the temperature time history of the rocks in which they grew.

4.2.2.3 Q Phase

- Q phase has the composition $Ca_{20}[Al_xFe^{3+}_{(1-x)}]_{26}Fe^{2+}_3Si_3O_{68}$. Q phase is also called Pleochroite or fibres. It is not found as mineral but occurs in high alumina industrial cements.
- Pleochroite has been recognized to be analogous with the phase Q, its magnesian equivalent.
- $Ca_{20}Al_{32-2x}Mg_xSi_xO_{68}$ with x varying from 2.5 to 3.5. Iron occurs as both Fe^{2+} and Fe^{3+} and some of the Fe^{3+} substitute Al^{3+}. The partition of structural iron

between its two principal oxidation states Fe^{2+} and Fe^{3+} cannot be measured directly. Mixed layer intergrowths between Gehlenite and Pleochroite and similarity of structure have been revealed by TEM.

4.2.2.4 Iron Gehlenite

- Iron Gehlenite has the composition $Ca_2Al_xFe_{(1-x)}SiO_7$. It belongs to the Melilite family. Iron Gehlenite ($x = 0$) is found in industrial slags but rarely recognized in nature. Melilite (from the Greek '*Meli*' meaning honey and '*Lithos*' meaning stone) is found close to igneous rocks poor in silica and alkalis.
- In iron Gehlenite, iron replaces aluminium, forming a limited solid solution. The degree of substitution is not well defined but likely to be between $x = 0.67$ and $x = 0.5$.
- The following compositions have been proposed: $Ca_{1.96}Al_{1.35}Fe_{0.67}Si_{1.03}O_7$ and $Ca_{1.99}Al_{1.8}Fe_{0.2}Si_{1.0}O_7$ (PAT 2002).
- A solid solution of Gehlenite of formula $Ca_{2.03}Al_{0.97}Fe_{0.29}Si_{0.788}O_7$ has been obtained in a laboratory clinker produced by fusion. It is silica deficient and contains 14.7 mol % Fe_2O_3 (SCH 1983). The hypothetical end member ($Ca_2(Al_xFe_{(1-x)})_2SiO_7$ $x = 0$) decomposes into α'-C_2S and Hematite below 1315 °C. Addition of Fe_2O_3 to Gehlenite gives a compound structurally related to SFCA and iron Gehlenite according to the reaction:

$0.63Fe_2O_3 + Ca_2Al_2SiO_7 \rightarrow 0.29Ca_{1.26}Al_{1.33}Fe_{3.51}Si_{0.46}O_{10}$ (SFCA) + $0.14 Ca_{4.26}Al_{3.38}Fe_{0.5}Si_{0.19}O_{14}$ (IG).

4.2.2.5 SFCA

- SFCA has the composition $Ca_{3.1}Al_{1.5}Fe_{8.3}Si_{1.1}O_{20}$. SFCA is not found as a mineral but is formed in fired lime-flux and it is the main compound in iron sinters.
- A number of synthetic phases have been prepared which have the Aenigmatite structure type SFCA described (HAM 1989). It exists with a wide solid solution of all cations. It explains the numerous chemical formulas proposed for this compound: $Ca_2(Fe^{3+}Al)_8SiO_{14}$ (HUC 1971, 1974, 1978), $Ca_5(Fe,Al)_{18}Si_2O_{36}$ (INO 1982), $Ca_{2.3}Mg_{0.8}Al_{1.5}Fe_{8.3}Si_{1.1}O_{20}$ (HAM 1989).
- SFCA lies in the plane $C(Al,F)_3$–C_4S_3. The compositional range is between 7 and 12 wt % C_4S_3 component and follows the coupled substitution mechanism; $2Fe^{3+} \rightarrow Si^{4+} + Ca^{2+}$.
- The substitution involving Ca^{2+} and Si^{4+} for $2M^{3+}$ (parallel to the pseudo binary CF_3/CA_3) is more limited having a maximum range between 3 and 11 wt % C_4S_3 component. This maximum occurs at 1240 °C for bulk composition containing around 10 wt % Al_2O_3.
- SFCA crystallises in the triclinic system (table 4.2).

TAB. 4.2 – Crystallography characteristics of the minerals belonging to the of the CaO–SiO$_2$–Al$_2$O$_3$–Fe oxides system.

Chemical formula	Ca$_{20}$Al$_{26}$Fe$_3$Si$_3$O$_{68}$	CaFe$_{0.6}$Al$_{1.4}$SiO$_6$
Oxide formula	20CaO·13Al$_2$O$_3$·3FeO·3SiO$_2$	CaO·0.6FeO·0.71Al$_2$O$_3$·SiO$_2$
Cement formula	C$_{20}$A$_{13}$f$_3$S$_3$	Cf$_{0.3}$A$_{0.71}$S
Mineral	Pleochroite	Esseneite
Lattice	Orthorhombic	Monoclinic
Ref	KAP 1980	DYS 1972

Chemical formula	Ca$_{2.45}$Fe$^{3+}_{9.04}$ Al$_{1.74}$ Fe$^{2+}_{0.16}$ Si$_{0.60}$ O$_{20}$
Oxide formula	2.45CaO·9.04Fe$_2$O$_3$·1.74Al$_2$O$_3$·0.16FeO·0.6SiO$_2$
Cement formula	C$_{2.45}$F$_{9.04}$A$_{1.74}$f$_{0.16}$Si$_{0.6}$
Mineral	SFCA
Lattice	Triclinic
Ref	MUM 1988, 1998, 2003

- A formula of the solid solution can be written x(Ca$_4$Si$_3$)(1−x)[CaAl$_6$(1−y)Fe$_6$O$_{20}$) with x between 0.09 and 0.15 and y between 0.183 and 0.55.
- SFCA can be divided into two main types on the basis of composition and morphology. The first is a low Fe form (SFCA) and the second one is a high Fe low Si form SFCA I.
- SFCA I has a characteristic platy morphology, although it may sometimes appear needle like or acicular in cross section.
- SFCA exhibits a prismatic form and its morphology has often been referred to as columnar, blocky or lath.
- SFCA, SFCA I and SFCA II have an homologous structure of type M$_{14+6n}$O$_{20+8n}$:
  - $n = 0$ SFCA Ca$_2$(Fe,Al)$^{3+}_{12}$O$_{20}$ Ca$_{2.45}$Fe$^{3+}_{9.04}$Al$_{1.74}$Fe$^{2+}_{0.16}$Si$_{0.6}$O$_{20}$.
  - $n = 1$ SFCA I Ca$_2$(Fe,Al)$^{3+}_{16}$Fe$^{2+}_{2}$O$_{28}$Ca$_{3.18}$Fe$^{3+}_{14.66}$Al$_{1.34}$Fe$^{2+}_{0.82}$O$_{28}$.
  - $n = 2$ SFCA II Ca$_2$(Fe,Al)$^{3+}_{20}$Fe$^{2+}_{4}$O$_{36}$Ca$_{5.1}$Fe$^{3+}_{18.7}$Al$_{9.3}$Fe$^{2+}_{0.9}$O$_{48}$.
- An untwined single crystal of SFCA, grown using a flux melted technique, has been used for studying its structure.
- SFCA has a layer structure with metal atoms occupying octahedral and tetrahedral sites within the layer-framework of oxygen atoms.
- Six of the nine octahedral sites are occupied by Fe^{3+} (M1, M2, M3, M7 and M11) and the three last by Ca^{2+} (one of these is substituted by Fe^{3+}).
- Of the six tetrahedral sites (M4, M5, M6, M8, M9 and M12) only four are occupied by Fe^{3+}. These two having shorter bonding interaction are considered to be locations in which extensive substitution of Si and Al occurs. Equal amounts of Ca^{2+} and Si^{4+} are involved in a coupled substitution for (Fe, Al)$^{3+}$ (DAW 1983, INO 1982, HAM 1989).

4.2.2.6 Spinel

- It has the composition $(FeO, Mg)O \cdot (Fe^{3+}Al)_2O_3$ (NAV 1982).
- It is a mineral of metamorphic origin found in limestone and dolomite marble and low silica and alkali igneous rocks.
- It is composed of an approximately cubic close packed array of anions with divalent cations occupying the tetrahedral sites and trivalent cations occupying octahedral interstitial sites. 1/8 of the tetrahedral holes are filled with M^{2+} cations and ½ of the octahedral holes are filled with X^{3+} cations. The cation to anion ratio is ¾.
- The normal distribution is $A^{2+}(B^{3+}{}_2)O_4$ and the inverse distribution $B(AB)O_4$. The ions (B) occupy the octahedral sites.
- It is convenient to characterize the cation distribution by specifying the fraction of tetrahedral sites (x) occupied by the b cations; $A_{(1-x)}B_x(A_xB_{(2-x)})O_4$:
  - $x = 0$ for normal Spinel $A(B_2)O_4$.
  - $x = 1$ for inverse Spinel $B(AB)O_4$.
  - $x = 2/3$ for random Spinels.

4.2.2.7 $Ca\,[Al_x\,Fe_{(3-x)}]_2\,O_{10}$ (T Phase)

- At 1350 °C, the addition of 5% of SiO_2 (0.30 mol %) to $CaAl_2Fe_2O_{10}$ leads to SFCA type compound and Hematite according to the following reaction:

  $0.7CaAl_2Fe_2O_{10} + 0.3\,SiO_2 \rightarrow 0.7CaAl_{0.9}Fe_{1.7}Si_{0.45}O_{10} + 0.3Fe_2O_{10}$.

- When more silica is added (0.56% mole), another phase crystallises (solid solution of Anorthite) according to the reaction:

  $0.44CaAl_2Fe_2O_{10} + 0.56SiO_2 \rightarrow 0.2CaAl_{1.3}Fe_{1.2}Si_{0.67}O_{10}$
  $+ 0.24CaAl_{0.81}Fe_{0.16}Si_{1.98}O_8 + 0.61Fe_2O_3$.

- The quantity of silica present in the structure of $CaAl_2Fe_2O_{10}$ increases (0.45, 0.58 and 0.67 mol %) when more is added, rates of 0.3, 0.46 and 0.56 mol % respectively, while the XRD diagram is similar as this obtained at 1200 °C with pure $CaAl_2Fe_2O_{10}$.
- The close agreement of the cell values of the T phase $Ca_2Al_4Fe_8O_{20}$ with half values of the lattice parameters might well be taken as evidence that SFCA structure type exists to the $CaO-Al_2O_3-Fe_2O_3$ system (MUM 1988, OLS 1948). The tetrahedral Si found in SFCA could be replaced by additional Al resulting in the formula $Ca_2Al_3{}^T(Al^O,Fe)_8O_{20}$, in approximate agreement with the ternary phase proposed for the T phase (T = tetrahedron, O = Octahedron).

4.2.2.8 Phase X

- Phase X has the composition $2CaO \cdot 4(Fe^{3+}Al)_2O_3 \cdot SiO_2$. This phase has been found experimentally at the laboratory scale. Its formula is $(2CaO)_4\,(Fe^{3+},$

TAB. 4.3 – Main references relevant to the $CaO-SiO_2-Al_2O_3-Fe_2O_3$ system in air.

Experimental works	Experimental works	Modelling works
Hansen et al. (HAN 1928)	Lea (LEA 1956)	Sorrentino (SOR 2008)
Mc Murdie (1937) (MAC 1937)	Newkirk (NEW 1958)	
Tavasci (1936) (TAV 1936)	Majumdar (MAJ 1965)	
Swayze (SWA 1946)	Lister et al. (LIS 1967)	
Parker (PAR 1952)	Imlach et al. (IML 1971–2)	
Brisi (BRI 1954)	Palomo et al. (PAL 1988)	
Toropov (1955) (TOR 1955)		

Al)$_2O_3SiO_2$) (HIJ 1968) close to ($CaFe^{3+}_4O_7$) or ($Ca_{7.2}Fe^{2+}_{0.8}Fe^{3+}_{30}O_{53}$) or ($Ca_3Fe^{3+}_{15}O_{25}$).

- Phase X phase exhibits similar powder X ray diffraction pattern. Its existence is not confirmed.

4.2.3 Mineralogy Structure Stability

4.2.3.1 Introduction

- Table 4.3 summarizes the main references relevant to the $CaO-SiO_2-Al_2O_3-Fe_2O_3$ system in air.
- The quaternary system is limited by 4 ternary systems: $CaO-SiO_2-Al_2O_3$, $CaO-SiO_2-Fe_2O_3$, $CaO-Al_2O_3-Fe_2O_3$ and $SiO_2-Al_2O_3-Fe_2O_3$.
- Table 4.4 shows the main phases found in the $CaO-SiO_2-Al_2O_3-Fe_2O_3$ system identified by XRD and SEM with the most probable range of solid solution at 1250 °C.

4.2.3.2 Solidus in the $CaO-SiO_2-Al_2O_3-Fe_2O_3$ System in Air

- The identification of the four phase assemblages was carried out by a trial and error approach. Using a model (figure 4.3), a theoretical assemblage of four phases is assumed.
- Each assemblage formed by 25 wgt % of each phase is experimentally determined (SEM and XRD).

If a discrepancy occurs between the assumption and the result of the experiments, the assemblage is rejected.

- In a quaternary system such as $CaO-SiO_2-Al_2O_3-Fe_2O_3$, four phases constitute a univariant phase assemblage (invariant if liquid is present). At equilibrium, the number of phases can be lower but not higher than four.
- Invariant points in the ternary system are linked to quaternary invariant equilibria by univariant lines. Four univariant equilibria will converge on each quaternary point. Not all univariant equilibria which radiate from the quaternary points necessarily reach a ternary system.

TAB. 4.4 – Main compounds and their solid solutions.

C_3Sss	C_2Sss	C_3S_2ss	$CSss$	C_3Ass	$C_{12}A_7ss$
$C_3A_{0.01}F_{0.01}S_{0.98}$	$C_2A_{0.01}F_{0.01}S_{0.98}$	$C_3A_{0.1}S_{1.9}$	$CA_{0.01}F_{0.01}S_{0.98}$	$C_3A_{0.95}F_{0.05}$	$C_{12}A_{6.56}F_{0.35}S_{0.09}$
CA	CA_2	$(CA_6)_1$	$(CA_6)_2$	$(CA_6)_3$	Fss
$CA_{0.9}F_{0I1}$	$CA_{1.75}F_{0I25}$	$CA_{4.4}F_{1.6}$	$CA_{3.4}F_{2.6}$	$CA_{2.9}F_{3.1}$	$F_{0.9}A_{0.1}$
Ass	$F1$	$F2$	$F3$	$(C_2AS)ss$	$(CAS_2)ss$
$A_{0.9}F_{0I1}$	C_6A_2F	C_4AF	C_6AF_2	$C_2A_{0.9}F_{0.1}S$	$CA_{0.9}F_{0.1}S_2$
$T1$	$T2$	$T3$	SFC	ESS	
$CA_{0.76}F_{2.23}$	$CA_{1.14}F_{1.86}$	$CA_{1.62}F_{1.38}$	$0.5(C_4S_3)0.5F_3$	$CA_{0.5}F_{0.5}S$	

- 38 assemblages of four phases (volume) have been found (table 4.5) and based on these results, a flowchart has been constructed (figure 4.4).
- The assemblages of 4 phases are represented in rectangles while circles are used to show univariant equilibria (liquid absent). Different colours for circles are used to differentiate the four ternary systems; $CaO-SiO_2-Al_2O_3$, $CaO-Fe_2O_3-SiO_2$, $CaO-Al_2O_3-Fe_2O_3$ and $Al_2O_3-Fe_2O_3-SiO_2$.
- Figure 4.5 represents the section at 50% of Fe_2O_3 1200 °C in air (without liquid) found theoretically and experimentally. The figure 4.6 represents an enlarged part of the centre of the tetrahedron.
- In table 4.5, the white cells represent the assemblages of the phases found in the section at 50% Fe_2O_3 while the grey cells represent the assemblages present in the system $CaO-SiO_2-Al_2O_3-Fe_2O_3$ but not in the section at 50% of Fe_2O_3.

4.2.3.3 Liquidus in the $CaO-SiO_2-Al_2O_3-Fe_2O_3$ System

- The quaternary $CaO-SiO_2-Al_2O_3-Fe_2O_3$ diagram has been studied in air through different systems or joins; $CaSiO_3-Al_2O_3-Fe_2O_3$, $Ca_3Fe_2^{3+}Si_3O_{12}$ (Andradite) − $Ca_3Al_2^{3+}Si_3O_{12}$ (Grossularite$_2^{3+}SiO_6$ (FTS) − $CaAl_2^{3+}SiO_6$ (CATS), four sections including three at constant percentage of Fe_2O_3 (10, 16.5 and 20 wgt % Fe_2O_3) and the fourth taken through the $Ca_2Al_2O_5-Ca_2Fe_2O_5-SiO_2$ section.
- The content of ferrous iron is in general sufficiently low that the results can be presented in a good first approximation as sections through the quaternary system at constant Fe_2O_3 content.
- Figure 4.3 is a general view of the position of these systems in the whole $CaO-SiO_2-Al_2O_3-Fe_2O_3$ diagram.

4.2.3.4 Sections and Joins

4.2.3.4.1 $CaSiO_3-Al_2O_3-Fe_2O_3$ Join (Huck 1974)
- The $CaSiO_3-Al_2O_3-Fe_2O_3$ join is bound by the following limiting systems: $Al_2O_3-Fe_2O_3$, $CaSiO_3-Al_2O_3$ and $CaSiO_3-Fe_2O_3$.
- Hematite $(Fe_2O_3)_{ss}$, Corundum $(Al_2O_3)_{ss}$, Hibonite (CA_6), Gehlenite $(Ca_2Al_2SiO_7)_{ss}$, Fassaite $(CaAl_2SiO_6)$, Pseudowollastonite $(CaSiO_3)$ and Andradite $(Ca_3Fe^{3+}_2Si_3O_{12})$ are found as primary phases in this join. They are written with their mineral name.
- A new phase called X has been found, but not confirmed.
- In the $Al_2O_3-Fe_2O_3$ system, $Al^{3+}Fe^{3+}O_3$ become stable above 1318 °C and coexist with Hematite or Corundum.
- The solubility of Hematite with Corundum is 15% by weight at 1250 °C and the solubility of Corundum with Hematite is about 18% by weight at the same temperature.

TAB. 4.5 – List of 38 four phase assemblages.

(1)	C–C$_3$S–C$_3$Ass–Ferrite	(20)	C$_2$ASss–CAS$_2$ss–Fe$_2$O$_3$–T
(2)	C$_3$S–C$_2$Sss–C$_3$Ass–Ferrite	(21)	C$_2$ASss–T–CF$_2$–CF
(3)	C$_2$Sss–C$_3$Ass–C$_{12}$A$_7$ss–Ferrite	(22)	CAss–CA$_2$ss–C$_2$ASss–T
(4)	C$_2$Sss–CAss–C$_{12}$A$_7$ss–Ferrite	(23)	CS–C$_2$ASss–CAS$_2$ss–ESS
(5)	C–C$_3$S–C$_2$F–Ferrite	(24)	C$_2$ASss–CAS$_2$ss–ESS–Fe$_2$O$_3$
(6)	C$_3$S–C$_2$Sss–C$_2$F–Ferrite	(25)	C$_2$ASss–CF$_2$–T–Fe$_2$O$_3$
(7)	C$_2$Sss–C$_2$ASss–C$_2$F–Ferrite	(26)	CA$_2$ss–CA$_6$ss–C$_2$ASss–T
(8)	C$_2$Sss–CAss–C$_2$ASss–Ferrite	(27)	CS–CAS$_2$ss–ESS–Fe$_2$O$_3$
(9)	C$_2$Sss–C$_2$ASss–C$_2$F–CF	(28)	CAS$_2$ss–SiO$_2$–Fe$_2$O$_3$–Fe$_2$O$_3$ss
(10)	C$_2$ASss–CAss–C$_2$F–Ferrite	(29)	C$_2$ASss–CAS$_2$ss–CA$_6$ss–T
(11)	C$_2$Sss–C$_3$S$_2$–C$_2$ASss–Fe$_2$O$_3$	(30)	C$_2$ASss–CAS$_2$ss–CA$_6$–CA$_6$ss
(12)	C$_2$Sss–C$_2$ASss–Fe$_2$O$_3$–SFC	(31)	CA$_2$ss–CA$_6$–CA$_6$ss–C$_2$AS
(13)	C$_2$Sss–C$_2$ASss–SFC–CF	(32)	CS–CAS$_2$ss–Fe$_2$O$_3$–SiO$_2$
(14)	CAss–C$_2$ASss–C$_2$F–CF	(33)	CAS$_2$ss–A$_3$S$_2$–Fe$_2$O$_3$ss–SiO$_2$
(15)	CS–C$_3$S$_2$–C$_2$ASss–Fe$_2$O$_3$	(34)	CAS$_2$ss–T–Fe$_2$O$_3$–Fe$_2$O$_3$ss
(16)	C$_2$ASss–SFC–CF$_2$–Fe$_2$O$_3$	(35)	CA$_6$ss–CAS$_2$ss–T–Fe$_2$O$_3$ss
(17)	C$_2$ASss–SFC–CF–CF$_2$	(36)	CA$_6$–CA$_6$ss–CAS$_2$ss–Al$_2$O$_3$
(18)	CAss–C$_2$ASss–CF–T	(37)	CAS$_2$ss–A$_3$S$_2$–Al$_2$O$_3$–Fe$_2$O$_3$ss
(19)	CS–C$_2$ASss–ESS–Fe$_2$O$_3$	(38)	CA$_6$ss–CAS$_2$ss–Al$_2$O$_3$–Fe$_2$O$_3$ss

Key: For simplification the names Ferrite and T represent, respectively, the solid solutions of CF$_2$ and T. The phase assemblages corresponding to the grey background are not found in the section at 50% Fe$_2$O$_3$.

- In the CaSiO$_3$–Al$_2$O$_3$ system, the Gehlenite, Anorthite and Anorthite–CA$_6$ joins intersect at one atmosphere. It divides the join into:
  (a) Wollastonite–Anorthite–Gehlenite.
  (b) Anorthite–Gehlenite–CA$_6$.
  (c) Anorthite–CA$_6$–Corundum.
- In the CaSiO$_3$–Fe$_2$O$_3$ system, liquid plus Hematite and Pseudowollastonite are stable phases between 1260 and 1137 °C.
- Gehlenite and Anorthite participate in phase assemblages of the Al-rich portion on the CaSiO$_3$–Al$_2$O$_3$–Fe$_2$O$_3$ join. As example Gehlenite plus Anorthite plus Pseudowollastonite or Wollastonite or CA$_6$ (CaO–Al$_2$O$_3$–SiO$_2$ system).
- Pseudowollastonite and Wollastonite crystallize in the calcium rich portion of the CaSiO$_3$–Al$_2$O$_3$–Fe$_2$O$_3$ join.
- Solid solutions of ferri-Fassaite are stable as phase in the central portion of the CaSiO$_3$–Al$_2$O$_3$–Fe$_2$O$_3$ system. It coexists in mono and polyphase assemblages of the CaSiO$_3$–Al$_2$O$_3$–Fe$_2$O$_3$ system.
- The assemblages involving Hematite solid solution are:
  (a) Hematite ss + Garnet ss + Fassaite at 1050 °C.
  (b) Hematite ss + Pseudowollastonite + Fassaite at 1200 °C.

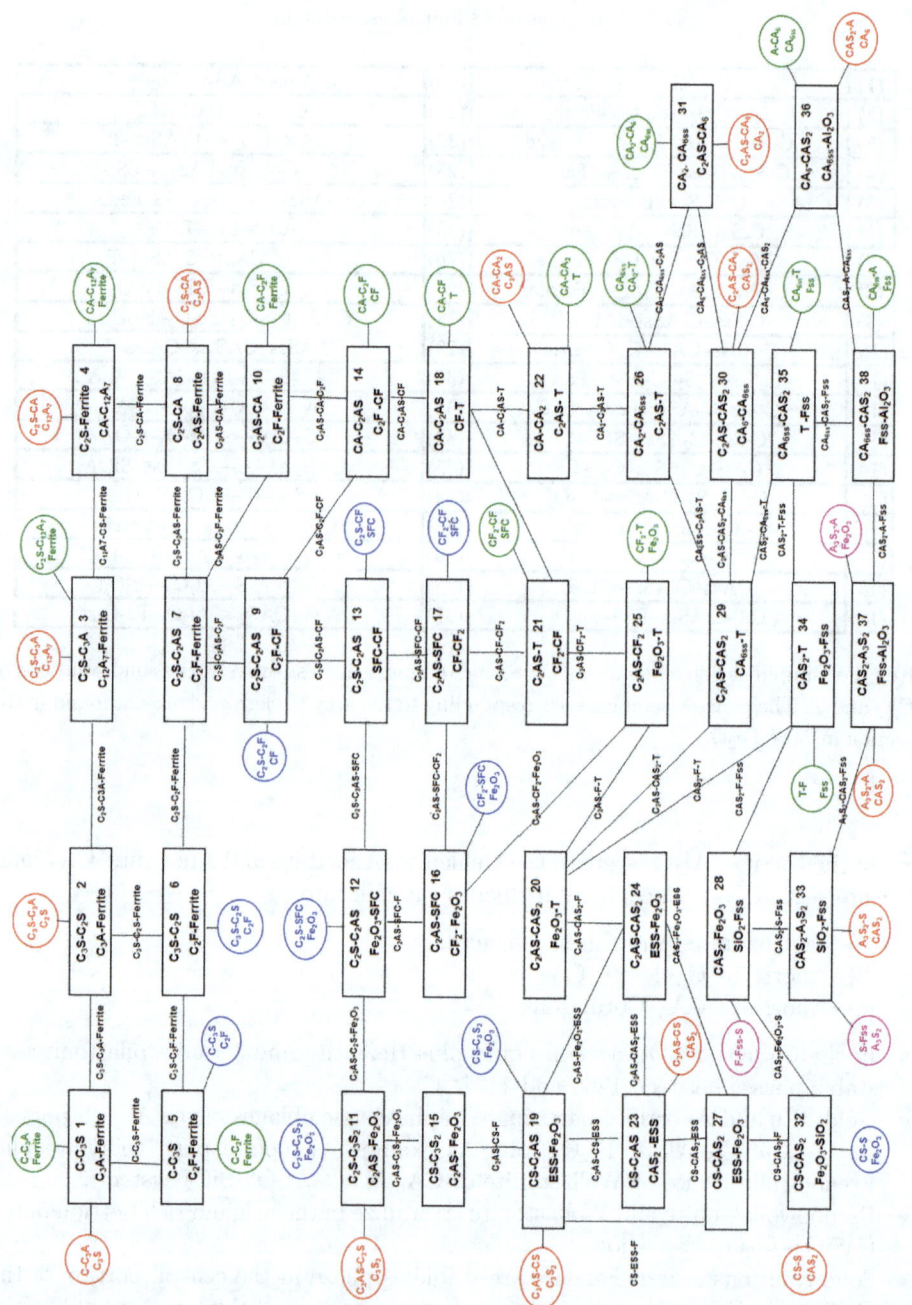

FIG. 4.4 – Flowchart of the CaO–SiO$_2$–Al$_2$O$_3$–Fe oxides system. Key: The number refers to table 4.5 and main compounds to table 4.4. The assemblages of 4 phases are represented in rectangles while circles are used to show univariant equilibria (liquid absent) in the following ternary systems; CaO–SiO$_2$–Al$_2$O$_3$, CaO–Fe$_2$O$_3$–SiO$_2$, CaO–Al$_2$O$_3$–Fe$_2$O$_3$ and Al$_2$O$_3$–Fe$_2$O$_3$–SiO$_2$.

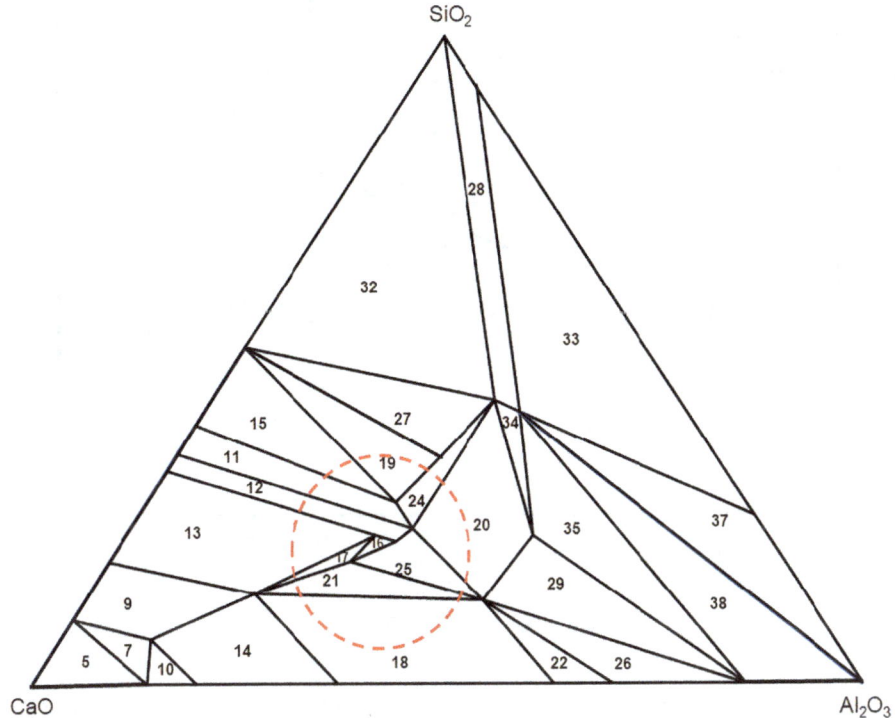

FIG. 4.5 – Section of solidus à 50% Fe_2O_3. Key: The numbers refer to the volume table 4.5. Circle is zoomed in figure 4.6.

(c) Hematite ss + Fassaite + liquid at 1240 °C.

- In the $CaSiO_3$–Al_2O_3–Fe_2O_3 portion, liquid crystallises by eutectic reaction of Anorthite + Fassaite + Pseudowollastonite   and   Anorthite + Gehlenite + Pseudowollastonite + Fassaite at 1200 °C.
- The phase assemblage Hematite + Garnet is stable below 1137–1178 °C close to the $CaSiO_3$–Fe_2O_3 join.
- Hematite plus Garnet and Hematite plus Garnet plus Fassaite decompose to Hematite + Pseudowollastonite and Hematite plus Pseudowollastonite + Fassaite at high temperatures.
- The following quaternary invariant equilibria are found (excluding those involving the phase X, whose presence is not confirmed):

    (a) Anorthite + Hematite + Pseudowollastonite + Fassaite (1203 °C).
    (b) Anorthite + Gehlenite + Pseudowollastonite + Fassaite below 1203 and 1202 °C.

4.2.3.4.2 $(Ca_3Al_2Si_3O_{12})$–$(Ca_3Fe^{3+}{}_2Si_3O_{12})$ Join (Grossularite–Andradite Join)

- Grossularite is a stable phase within the Gehlenite–Anorthite–Wollastonite portion, up to at least 855 °C and one atmosphere.

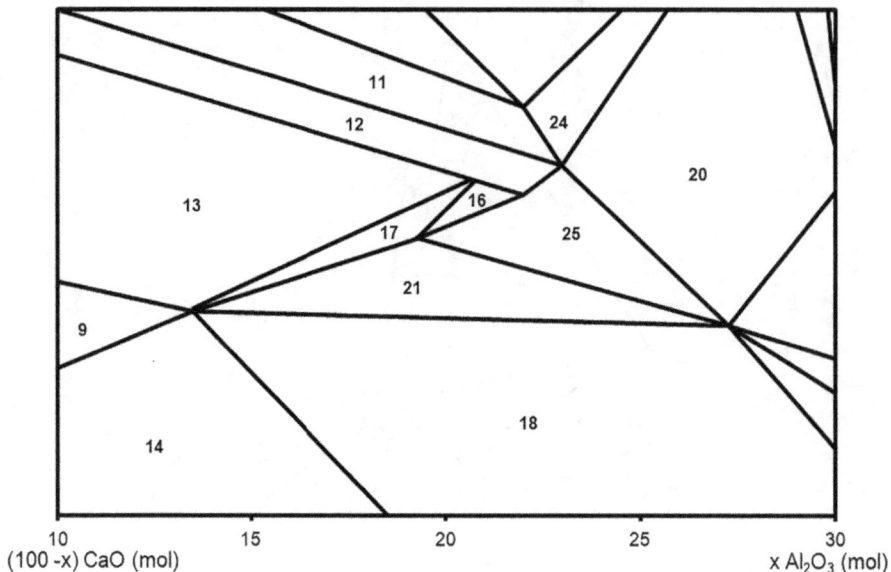

FIG. 4.6 – Zoom part of the section of solidus à 50% Fe_2O_3. Key: The number refers to table 4.5.

- Above 855 °C, Grossularite breaks down to Gehlenite plus Wollastonite plus Anorthite.
- Andradite is a stable phase up to 1137 °C and coexists with Al_2O_3 and Fe_2O_3 above this temperature, it decomposes to Hematite and Pseudowollastonite.
- Andradite forms a complete solid solution with Grossularite below 800 °C. The incorporation of Grossularite raised its thermal stability with a maximum at 1178 °C at a composition formed by 98% Andradite and 2% Grossularite. Andradite breaks down to Pseudowollastonite, Hematite and Fassaite.
- At a composition of Andradite (53.01) and Grossularite (46.99) at 935 °C, the reaction of Garnet giving Wollastonite and Fassaite intersects the ($Ca_3Fe^{3+}{}_2Si_3O_{12}$)–($Ca_3Al_2Si_3O_{12}$) join. Thus, Garnets of the composition beyond Andradite (53.01) and Grossularite (46.99) decompose to a mix of Garnet, Wollastonite, Gehlenite and Anorthite below 935 °C whereas the Wollastonite and Fassaite above 935 °C.
- Wollastonite inverts to Pseudowollastonite at 1135 °C along the ($Ca_3Al_2Si_3O_{12}$)–($Ca_3Fe^{3+}{}_2Si_3O_{12}$) join.
- The assemblage Pseudowollastonite, Fassaite and Hematite starts to melt at 1203 °C.
- The temperature of 1225 °C is identical with the eutectic at which Pseudowollastonite + Fassaite + liquid are stable on the Pseudowollastonite–Fassaite join.
- The Pseudowollastonite–Fassaite join divides the Andradite-Grossularite join into an iron and aluminium rich portion. The iron rich portion is strictly ternary in behaviour whereas a quaternary treatment is required in the aluminium rich part.

TAB. 4.6 – Assemblages of 2, 3 and 4 phases.

2 Phases	3 Phases	4 Phases
Pwo + L	Hem + Pwo + L	Geh + Pwo + CAS_2 + L
C_2AS + L	Pwo + Fas + L	Wo + C_2AS + An + Fas
Hem + Pwo	Hem + Pwo + Fas	Wo + C_2AS + An + Gar
Pwo + Fas	Gar + Pwo + Fas	Pwo + C_2AS + An + Fas
	Gar + Wo + Fas	
	Fas + Geh + Pwo	
	Pwo + Geh + An	
	Hem + Pwo + Gar	

- In the later part, Pseudowollastonite, Gehlenite, Anorthite and Fassaite react with formation of liquid at 1202 °C (quaternary invariant equilibrium with four solids and a liquid). The liquid composition is not on the $CaSiO_3$–Al_2O_3–Fe_2O_3 join, but in the $CaSiO_3$–Fe_2O_3–SiO_2 subsystem. The temperature of 1208 °C represents a piercing point.
- At higher temperature, the join and Grossularite intersect the liquidus volume of Pseudowollastonite on the iron rich part and of Gehlenite on the aluminium rich portion of the $CaSiO_2$–Al_2O_3–Fe_2O_3 plane, respectively.
- Table 4.6 presents the compatibilities of 2, 3 and 4 phases.

4.2.3.4.3 Stability Relations on the $(CaFe^{3+}_2SiO_6)$–$(CaAl_2SiO_6)$ or FTS–CATS Join

- The maximum extension of the Fassaite ss ranges from about $FTS_{78}CATS_{22}$ to $FTS_{48}CATS_{52}$ between 1178 and 1203 °C.
- $FTS_{54}CATS_{46}$ melts incongruently to Hematite + Anorthite + liquid at 1294 °C.
- Hematite is the primary phase above 1203 °C in the iron rich part of the join to 1294 °C.
- It is replaced by Magnetite from 1370 and 1385 °C.
- Table 4.7 represents the compatibilities of 2, 3 and 4 phases.

4.2.3.4.4 $Ca_2Al_2O_5$–$Ca_2Fe_2O_5$–SiO_2 Join

- Studies in this section have been carried out with SiO_2 contents ranging up to 15% (figure 4.7).
- The section is not a true ternary system because there is no compound at one of the end-member composition $Ca_2Al_2O_5$ at least not at pressure below several kbars. Nevertheless, the section is close to a ternary system.
- The liquidus section shows primary phase fields of Ferrite and Melilite (Gehlenite solid solution). The Melilite extends towards $Ca_2Al_2SiO_7$ end-member and the Ferrite towards $Ca_2Al_2O_5$ (cross–hatched lines in figure 4.7).

TAB. 4.7 – Assemblages of 2, 3 and 4 phases.

2 Phases	3 Phases	4 Phases
Hem + L	An + CA_6 + L,	Fatss + Geh + An + CA_6
CA_6 + L	Hem + Pwo + L	
An + L	Hem + Fat + L	
Hem + Pwo	Hem + Pwo + Fat	
Hem + Gar	Hem + Pwo + Gar	
	Hem + Fat + Gar	
	Hem + An + L	

- Primary fields of $Ca_{12}Al_{14}O_{23}$ ($C_{12}A_7$), $CaAl_2O_4$ (CA) and Ca_2SiO_4 (C_2S) appear. The boundary between the fields of Anorthite and Melilite is suggested by a dashed line.
- Many of the phases which are present in this section at liquidus temperature have compositions which do not lie in the selected section; C_2S, CA and $C_{12}A_7$ for example.
- The section contains very nearly the compositions at two of the quaternary invariant points within the CaO–SiO_2–Al_2O_3–Fe_2O_3 system.
- The addition of SiO_2 progressively changes liquidus temperatures and primary phases of compositions having a $CaO/(Al_2O_3 + Fe_2O_3)$ ratio equal to 2.
- Compositions which lie very close to the C_2A–C_2F edge have either Ferrite solid solution or $C_{12}A_7$ as their primary phases.
- If the primary phase is Ferrite, its iron content will be greater than the liquid and preferentially incorporated in the Ferrite ss. $C_{12}A_7$ contains only 2–4% $C_{12}F_7$.
- The addition of more silica to these compositions causes a rapid change in liquidus relationships.
- At high C_2F contents, more than 4–5% SiO_2 causes the primary phase to change from Ferrite to C_2S.
- However, at low iron content, the addition of SiO_2 suppresses the initial formation of $C_{12}A_7$ but the primary phase replacing it, is CA rather than a calcium silicate.
- Temperature falls along the $C_{12}A_7$–CA boundary curve toward a piercing point I4 at which $C_{12}A_7$, CA and a Ferrite solid solution nearly saturated in Al_2O_3 coexist with liquid (table 4.8).
- At slightly higher C_2F contents, the fields of CA and C_2S do not touch except at a single point whose composition and temperature are close to the invariant point I5 (C_2S + Melilite + Ferrite + CA at T° 1290 °C).
- In the region of these invariant points, addition of SiO_2 causes a slight drop in liquidus temperatures upon moving away from the Ferrite edge but then the liquidus temperature rises again.

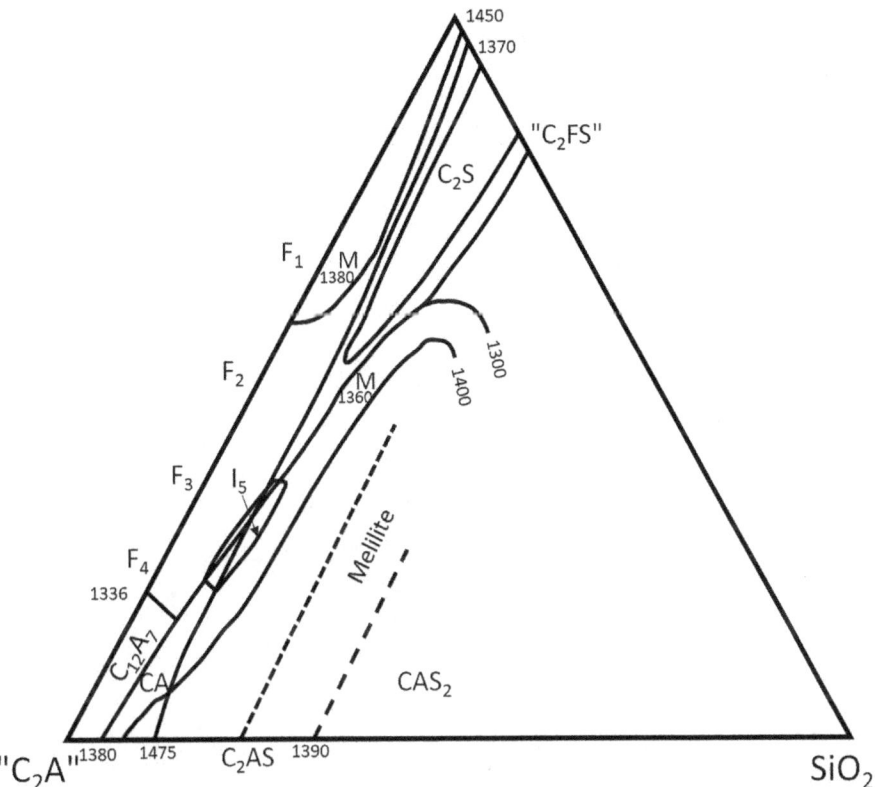

FIG. 4.7 – Liquidus of the C_2A–SiO_2–Fe_2O_3 diagram (SOR 1973). Temperature in °C.

4.2.3.4.5 Section at 10% Fe_2O_3

- Figure 4.8 shows a portion of the 10% isoplethal section superimposed in projection on the CaO–SiO_2–Al_2O face of the quaternary system.
- The dashed lines represent primary phase fields on the ternary face whereas those on the 10% Fe_2O_3 plane are shown by heavier solid lines. Arrows indicate directions of falling temperatures, while liquidus isotherms at the 10% Fe_2O_3 sections are shown by light continuous contours.
- Primary phase fields of CaO; the calcium aluminates (C_3A, $C_{12}A_7$, CA and CA_2) and the calcium silicates (C_3S and C_2S). The iron bearing solid solutions of Gehlenite appear in this section.
- In addition, the 10% section also contains a small primary field of the Ferrite phase. Its appearance indicates the relatively insoluble nature of the Ferrite phase in the CaO–SiO_2–Al_2O_3–Fe_2O_3 melts. The intrusion of the Ferrite solid solutions into this section results in the diminution of other primary phase regions notably those of $C_{12}A_7$ and CA.
- The Ferrite phase does not extend to the CaO–Al_2O_3–Fe_2O_3 face and this is supported by the fact that the 10% Fe_2O_3 isoplethal does not intersect the Ferrite field located in this system (DAY 1965).

TAB. 4.8 – Characteristics of the piercing point in the C_2A–C_2F–SiO_2 section.

Piercing points	Phases	CaO	SiO_2	Al_2O_3	Fe_2O_3	T (°C)
I4 (figure 4.8)	$C_{12}A_7$ + Ferrite + CA	49.9	5.2	36.4	8.5	1320
I5	C_2S + Melilite + Ferrite + CA	44.5	6	33	16.5	1290

Composition in weight %.

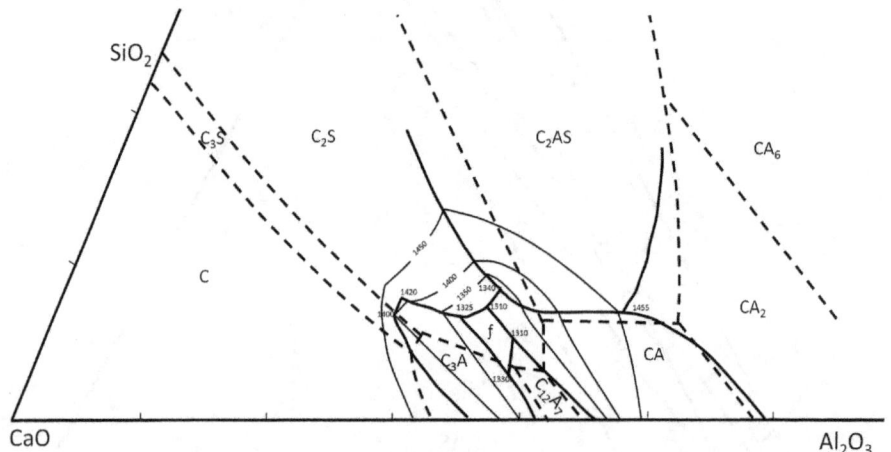

FIG. 4.8 – Liquidus section at 10% Fe_2O_3 (SOR 1973). Temperature in °C.

- Eight piercing points, at which curves of univariant equilibria pass through the section have been located:

    (1) C + C_3S + C_3A (1400 °C).
    (2) C_3S + C_3A + C_2S (1420 °C).
    (3) Ferrite + C_3A + C_2S (1325 °C).
    (4) Ferrite + C_3A + $C_{12}A_7$ (1330 °C).
    (5) Ferrite + CA + C_2S (1310 °C).
    (6) Ferrite + CA + $C_{12}A_7$ (1310 °C).
    (7) CA + C_2S + Melilite (1340 °C).
    (8) CA + CA_2 + Melilite (1455 °C).

4.2.3.4.6 Section at 16.5% Fe_2O_3

- Figure 4.9 shows a portion of the 16.5% isoplethal section superimposed in projection on the CaO–Al_2O_3–SiO_2 face of the quaternary system.
- Primary phase fields of CaO, C_3A, C_3S, C_2S, C_2AS and Ferrite appear. This section contains several of the quaternary invariant points. At points I1 and I5, four crystalline phases and liquid coexist:

    I1: C + C_3A + C_3S + Ferrite and liquid (1341 °C).
    I5: Ferrite + C_2AS + C_2S + CA and liquid at I5 (1290 °C).

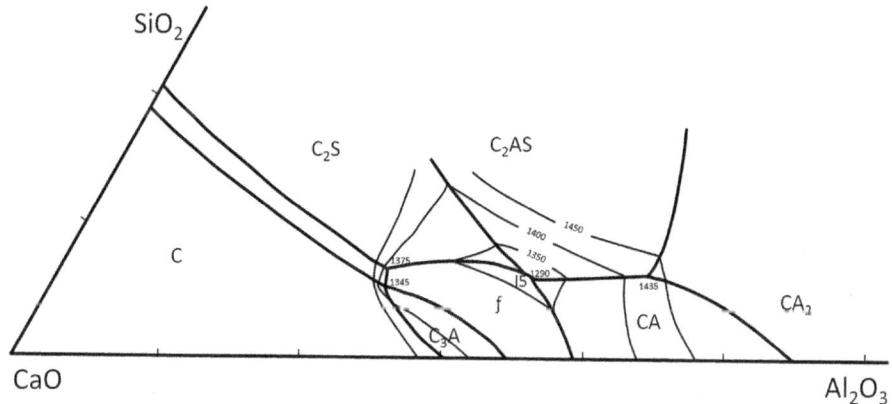

FIG. 4.9 – Liquidus section at 16.5% Fe_2O_3 (SOR 1973). Temperature in °C.

TAB. 4.9 – Invariant points in the alumina rich part of the CaO–SiO_2–Al_2O_3–Fe_2O_3 diagram.

	Phases	T (°C)	CaO	Al_2O_3	SiO_2	Fe_2O_3
I1	Ferrite + C_3A + C_3S + C	1341	55	22.7	5.8	16.5
I2	Ferrite + C_3A + C_3S + C_2S	1340	54.5	23.5	5.5	16.5
I3	Ferrite + C_3A + $C_{12}A_7$ + C_2S	1290	51.2	55.8	5	8
I4	Ferrite + CA + $C_{12}A_7$ + C_2S	1300	48.5	37	5.4	9
I5	Ferrite + CA + C_2S + C_2AS	1290	44.5	33	6	16.5

Composition in weight %.

- Five invariant points have been detected in this part of the CaO–SiO_2–Al_2O_3–Fe_2O_3 phase diagram (table 4.9).

4.2.3.4.7 Section at 20% Fe_2O_3

- Figure 4.10 shows a portion of the 20% isoplethal section superimposed in projection on the CaO–SiO_2–Al_2O_3 face of the quaternary system.
- The intersections of primary phase volumes with the sections are shown by heavy lines and isotherms by lighter lines. On this section, the field of C_3A is much reduced in extent and the field of $C_{12}A_7$ has vanished, while the Ferrite phase field has expanded.
- Six piercing points have been located:

  (1) C + C_3A + Ferrite (1360 °C).
  (2) C + C_3S + Ferrite (1400 °C).
  (3) C_3S + C_2S + Ferrite (1420 °C).
  (4) C_2S + C_2AS + Ferrite (1300 °C).

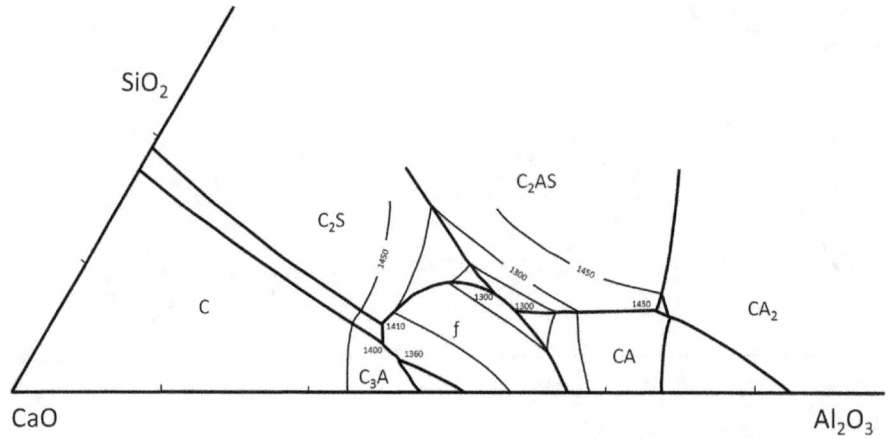

FIG. 4.10 – Liquidus section at 20 °C Fe_2O_3 (SOR 1973). Temperature in °C.

(5) CA + C_2AS + Ferrite (1300 °C).
(6) CA_2 + C_2AS + Ferrite (1430 °C).

4.2.3.4.8 Perspective of the Liquidus in the CaO–SiO_2–Al_2O_3–Fe_2O_3 System

- From the data available in the literature and with the aid of the three liquidus sections at constant percentage of Fe_2O_3 and the fourth taken through the section $Ca_2Al_2O_5$–$Ca_2Fe_2O_5$–SiO_2, a relative complete picture can be built up of phase relations in the quaternary system.
- The liquidus relations are shown in a perspective drawing figure 4.11.
- Because the drawing is necessarily two dimensional, an attempt has been made to show mainly those invariant surfaces which lie at high lime contents and also those primary phase volumes which are adjacent to that of the monocalcium aluminate. The monocalcium aluminate volume is shown towards the left-hand corner of the figure and it is assumed that the melting of pure CA is just congruent.
- The lower portion of the composition tetrahedral is the CaO–SiO_2–Al_2O_3 face. Therefore, the phases which appear on the base are calcium silicates, calcium aluminates and aluminosilicates of these phases.
- The fields of CaO, C_3S, C_2S, $C_{12}A_7$, CA, C_2AS, CA_2 and C_3S_2 are shown. The left-hand face of the projection is the CaO–Al_2O_3–Fe_2O_3 face and it also contains the calcium aluminate and in addition the calcium ferrite. The C_2F composition is shown and from this point a range of solid solution extends toward 'C_2A'.
- This latter composition is not shown on the diagram but lies with very slightly higher lime contents than the $C_{12}A_7$ composition which is shown.
- The face on the right-hand side of the projection is CaO–SiO_2–Fe_2O_3 face. Over the range of compositions studied, it does not contain any ternary phases so that only the fields of the calcium silicates appear.

Quaternary Chemical Systems

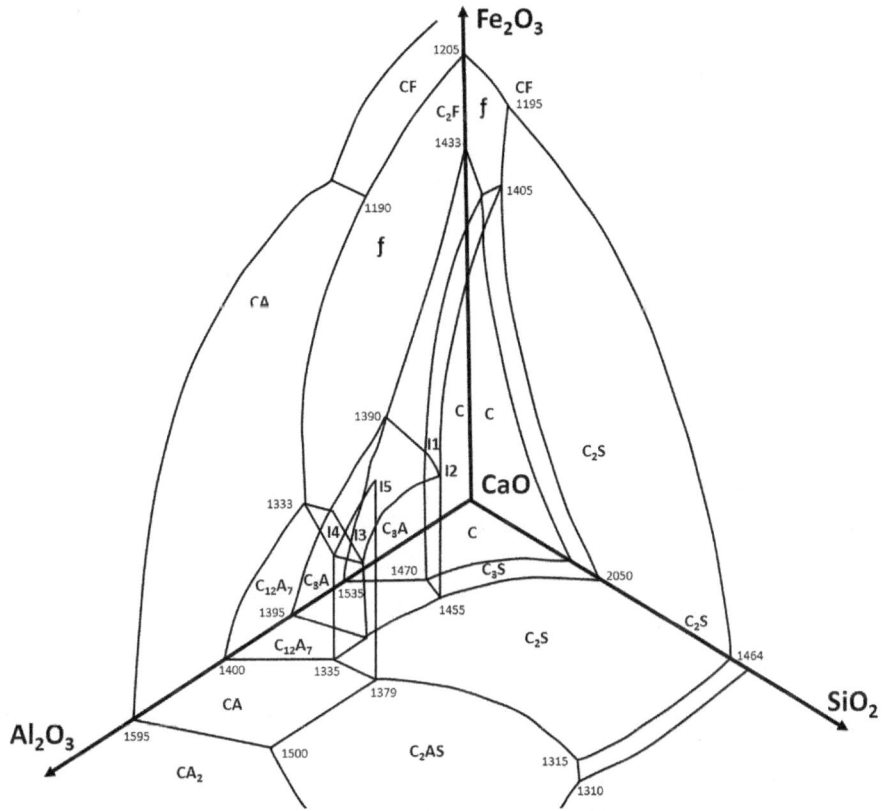

FIG. 4.11 – Perspective of the liquidus in the CaO–SiO_2–Al_2O_3–Fe_2O_3 system (SOR 1973). Temperature in °C.

- The viewpoint of the projection is selected to give the appearance of looking through the Al_2O_3–SiO_2–Fe_2O_3 face of the system towards the CaO apex which is the point further away from the viewer. The details of the Al_2O_3–SiO_2–Fe_2O_3 face are shown in chapter ternary A–S–F system.
- The representation shown here is not strictly accurate in the geometric sense but care has been taken not to introduce any large distortion, although the space between curves has occasionally been widened or the position of intersection shifted very slightly for the sake of clarity.
- In the interior of tetrahedron, the primary phase volumes are delineated by a series of volume surfaces. There are no quaternary phases that are phases which contain essential quantity of all four components.
- The primary phase volumes are outlined by heavy lines with arrow showing the direction of falling temperature. Because there are no quaternary phases, each volume terminates against one, two or three of the ternary faces. Thus, CaO

appears on three faces; CaO–SiO$_2$–Fe$_2$O$_3$, CaO–Al$_2$O$_3$–SiO$_2$ and CaO–Al$_2$O$_3$–Fe$_2$O$_3$.
- Binary phase or a solid solution based on one of the binary phases such as C$_2$S or C$_2$F, respectively, appears on two faces.
- Ternary phase, such as Gehlenite, appears on one face.
- Therefore, some of surfaces bonding the quaternary volumes appear in one or more limiting ternary systems and these intercepts are known from data in the literature.
- The remainder of the surface bounding the quaternary phase volumes must be located experimentally.
- Some of the quaternary surfaces bonding phase volumes start from ternary systems. Others have no extension to ternary faces. This is one reason why quaternary phase relationships must be studied.
- Another reason is that the points of intersection of quaternary surface cannot be predicted. These points represent a definite chemical composition. The intersection of four surfaces generates such a point. Five of these have been located and the phase presence as well as the temperature is shown in the upper right and corner of figure 4.11 as well as in table 4.9.
- For instance at I3, a quaternary invariant point at which a ferrite solid solution, C$_2$S, CA and C$_{12}$A$_7$ are in equilibrium with liquid surface bounding primary phase volumes of each phases meet.
- It therefore follows that at each invariant point, four invariant curves must meet. Each univariant curve represents three phases plus liquid. Combinations of the four crystalline phases taken in sets of three must occur.
- A schematic diagram has been constructed in addition to perspective diagrams in order to represent these complex equilibria (figure 4.11).
- Each of the quaternary invariant points I1–I5 is a eutectic. Therefore, the univariant curves which connect invariant points must have thermal maxima on them.
- For example, points I3 and I4 are connected by univariant curve along which C$_{12}$A$_7$, Ferrite and C$_2$S are in equilibrium with liquid. The thermal maximum on this univariant curve is shown by two arrows, one pointing away from the maximum towards I4, the other towards I3.
- In this instance, the maximum lies approximately midway along the length of the univariant curve. The temperature of the thermal maximum is not known precisely, but it cannot be much greater than 20/30 °C above the temperature of I3 and I4 since these are 1290 and 1300 °C, respectively. The temperature of the maximum is unlikely to be higher than 1320–1330 °C and may be as low as 1300–1310 °C.
- The method by which these maxima can be located is shown more clearly with the aid of special drawings designed to show the relationship between the invariant points and the subsolidus compatibility volume.

Quaternary Chemical Systems

- The existence of thermal maxima along so many of the quaternary univariant curves divides the quaternary system into a number of sub quaternary systems. Thus, each of the phase assemblages present at an invariant point will behave as a separated quaternary system with respect to its crystallisation or melting under certain conditions.

4.2.4 Formation and Synthesis

- The $CaFe^{3+}AlSiO_6$ pyroxene has been synthesised at 20 kbar and 1375 °C (48 h) using a piston cylinder type apparatus (GHO 1989).
- Pleochroite was found to be the primary phase for silica rich compositions having a FeO content greater than 5 wt %. Fe^{2+} and Fe^{3+} are both present in Pleochroite.

4.3 $CaO–Al_2O_3–SiO_2–FeO$ System

4.3.1 Introduction

- The quaternary $CaO–SiO_2–Al_2O_3–FeO$ system represented by a regular tetrahedron whose apex represents 100% of one component is limited by the following ternary systems; $CaO–SiO_2–FeO$, $CaO–SiO_2–Al_2O_3$, $SiO_2–Al_2O_3–FeO$ and $CaO–Al_2O_3–FeO$.
- These ternary limiting systems have been studied in the preceding chapters.
- In order to explore the interior of the $CaO–Al_2O_3–SiO_2–FeO$ tetrahedron, a selection of subsystems (quaternary, ternary or binary) has been studied. For convenience, any lines or planes in the tetrahedron which is studied are designed with FeO at one apex.
- Figure 4.12 shows the position of the sub-systems joins described in the next paragraphs.

4.3.2 Ternary Systems Located Within the CaO, Al_2O_3, SiO_2, FeO Tetrahedron

- Many authors do not differentiate Fayalite from Olivine as it is the end member. Here the pure $CaO–SiO_2–Al_2O_3–FeO$ chemical system is considered. So in the next paragraphs, Olivine is considered but it would be more precise to use Fayalite as there is no Magnesium.

4.3.2.1 SiO_2–Anorthite–FeO Join (Red Plane in Figure 4.12)

- A schematic representation of this join is given in figure 4.13.
- Seven primary phase volumes within the tetrahedron are as follows (with liquid); Anorthite, Hercynite ($FeOAl_2O_3$), Tridymite, two immiscible liquids plus

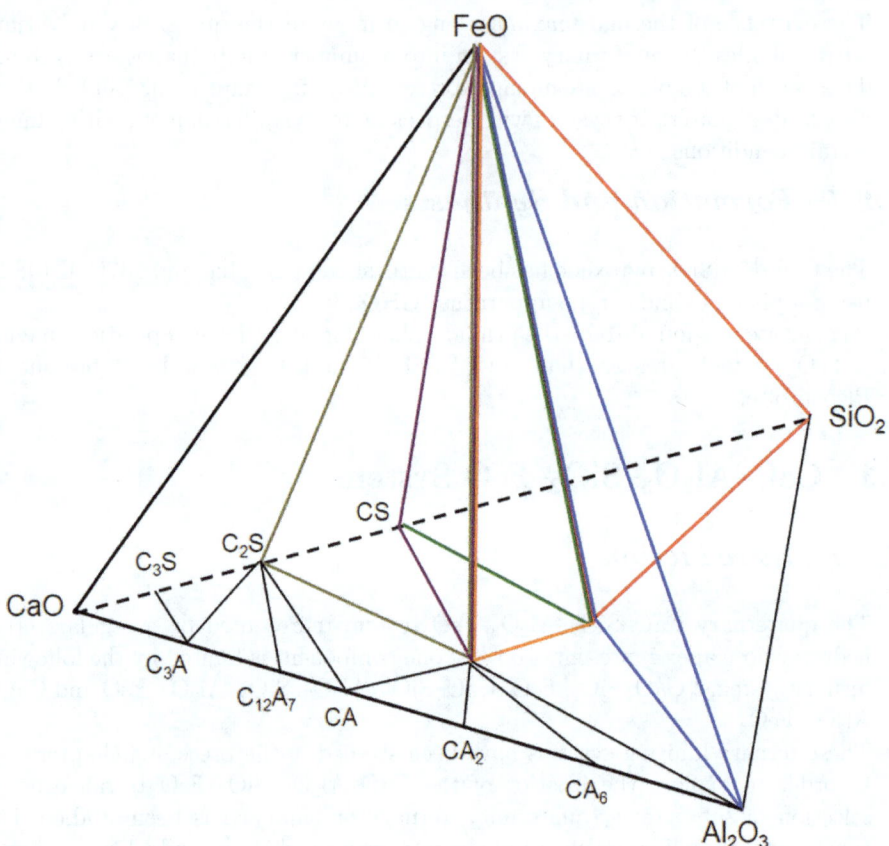

FIG. 4.12 – Position of the sub-system joins described in the next paragraphs.

Cristobalite, Olivine, Wüstite and Cristobalite. Cristobalite area is not represented in figure 4.13.

- The points a, b and c are piercing points of three univariant lines within the plane SiO_2–Anorthite–FeO figure 4.13 and table 4.10). Along each of these univariant lines, three crystalline univariant phases are in equilibrium with liquid whose compositions lie on the line.
- The phase assemblage along the three univariant lines within the tetrahedron are as follows:

  (1) Tridymite + Anorthite + Olivine + Liquid.
  (2) Anorthite + Olivine + Hercynite + Liquid.
  (3) Olivine + Hercynite + Wüstite + Liquid.

- Each univariant line (except one which joins two invariant points in different faces of the tetrahedron) leads to at least one quaternary invariant point.
- Four univariant lines meet at each quaternary invariant point.

Quaternary Chemical Systems 175

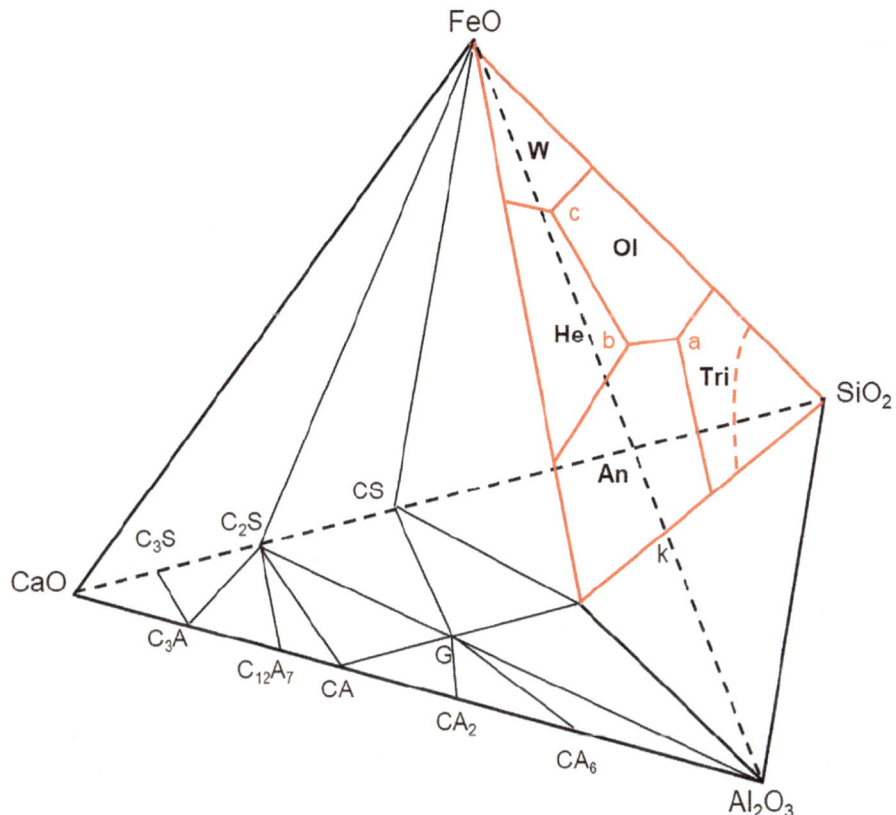

FIG. 4.13 – Red plane is a schematic representation of the SiO$_2$–Anorthite–FeO join. Key: An for Anorthite, He for Hercynite, Ol for Olivine, Tri for Tridymite and W for Wüstite.

4.3.2.2 CaSiO$_3$–Anorthite–FeO Join (Green Plane in Figure 4.12)

- A schematic representation of this join is given in figure 4.14.
- The following seven primary phase volumes within the tetrahedron are as follows with liquid; Anorthite, Hercynite, Wüstite, Melilite, Olivine, Pseudowollastonite (α-CaSiO$_3$) and Wollastonite (β-Wollastonite).
- The Anorthite–FeO side forms one side of the SiO$_2$–Anorthite–FeO, Anorthite–Al$_2$O$_3$–FeO and Anorthite–Gehlenite–FeO joins.
- The points d, e, f, g, h, i and j, are piercing points of seven univariant lines within the tetrahedron in the CaSiO$_3$–Anorthite–FeO plane (figure 4.14 and table 4.10).
- Along each of these seven curved univariant lines, three crystalline phases are in equilibrium with liquid whose compositions lie on one of the lines.

TAB. 4.10 – Composition of the piercing points.

Piercing points	a	b	c	d	e	f	g	h	i	j	k	l	m	n	o	p
Corundum											x					
Mullite																
Anorthite	x	x		x	x						x		x			
Hercynite		x	x					x	x		x	x	x			
Tridymite	x															
Olivine						x	x	x	x	x				x	x	
Fayalite	x	x	x													
Wollastonite				x	x	x	x				x					
Pseudowollastonite					x		x							x		
Melilite				x		x			x	x	x	x	x	x	x	x
Wüstite			x					x			x				x	x
C$_2$S																x

- The phase assemblage along these seven univariant lines are as follows:

  (1) Wollastonite + Anorthite + Melilite + liquid (d lies on this line).
  (2) Pseudowollastonite + Wollastonite + Anorthite + liquid (e lies on this line).
  (3) Wollastonite + Olivine + Melilite + liquid (f lies on this line).
  (4) Pseudowollastonite + Wollastonite + Olivine + liquid (g lies on this line).
  (5) Olivine + Wüstite + Hercynite + liquid (h lies on this line). The SiO$_2$–Anorthite–FeO join cuts this line and the ternary eutectic, Fayalite + Wüstite + Hercynite + liquid in the limiting FeO–Al$_2$O$_3$–SiO$_2$ system, is one terminus of this line.
  (6) Olivine + Hercynite + Melilite + liquid (i lies on this line).
  (7) Anorthite + Hercynite + Melilite + liquid (j lies on this line).

4.3.2.3 Anorthite–Al$_2$O$_3$–FeO Join (Blue Plane on Figure 4.12)

- A schematic representation of this join is given in figure 4.15.
- Four primary phase volumes are shown within the tetrahedron (in presence of liquid); Corundum (α-Al$_2$O$_3$), Anorthite, Hercynite and Wüstite.
- The Anorthite–FeO side also forms one side of the SiO$_2$–Anorthite–FeO, CaSiO$_2$–Anorthite–FeO and Gehlenite–Anorthite–FeO joins.
- Point k is the piercing point of a univariant line within the tetrahedron in the Anorthite–Al$_2$O$_3$–FeO plane (figure 4.15 and table 4.10).
- Along this univariant line, three crystalline solid phases are in equilibrium with liquid whose composition lies on this curved line.
- The point k is a piercing point (figure 4.15 and table 4.10).

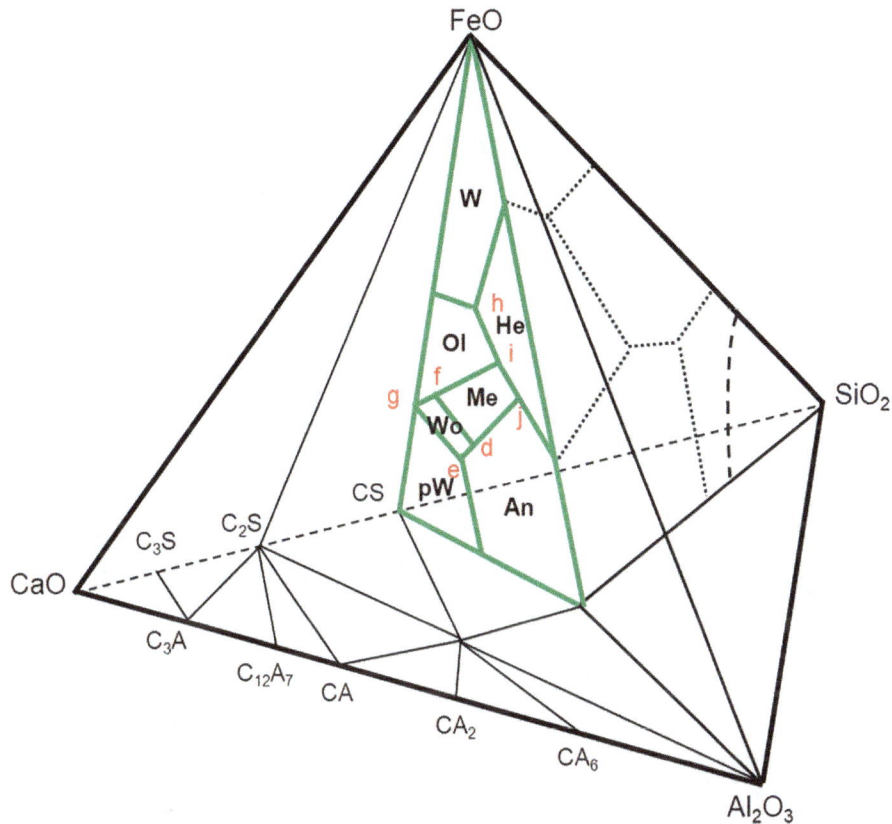

FIG. 4.14 – Green plane is a schematic representation of the CaSiO$_3$–Anorthite–FeO join. Key: An for Anorthite, He for Hercynite, W for Wüstite, Ol for Olivine, Me for Melilite, Wo for Wollastonite and pW for Pseudowollastonite.

- The phase assemblage along the univariant line is Anorthite + Corundum + Hercynite + liquid. All compositions which lie within the area Anorthite–Al$_2$O$_3$–Hercynite are completely crystalline below 1393 °C.

4.3.2.4 Gehlenite–Anorthite–FeO Join (Orange Plane on Figure 4.12)

- The Gehlenite–Anorthite–FeO join is shown in figure 4.16.
- This join cuts the following four primary phase volumes within the tetrahedron (with liquid): Anorthite, Gehlenite, Hercynite and Wüstite.
- Points l and m are piercing points of two univariant lines within the tetrahedron in the Gehlenite–Anorthite–FeO plane (figure 4.16 and table 4.10).
- Along each of these four curved univariant lines, three crystalline phases are in equilibrium with liquid whose composition lies on one of the lines.

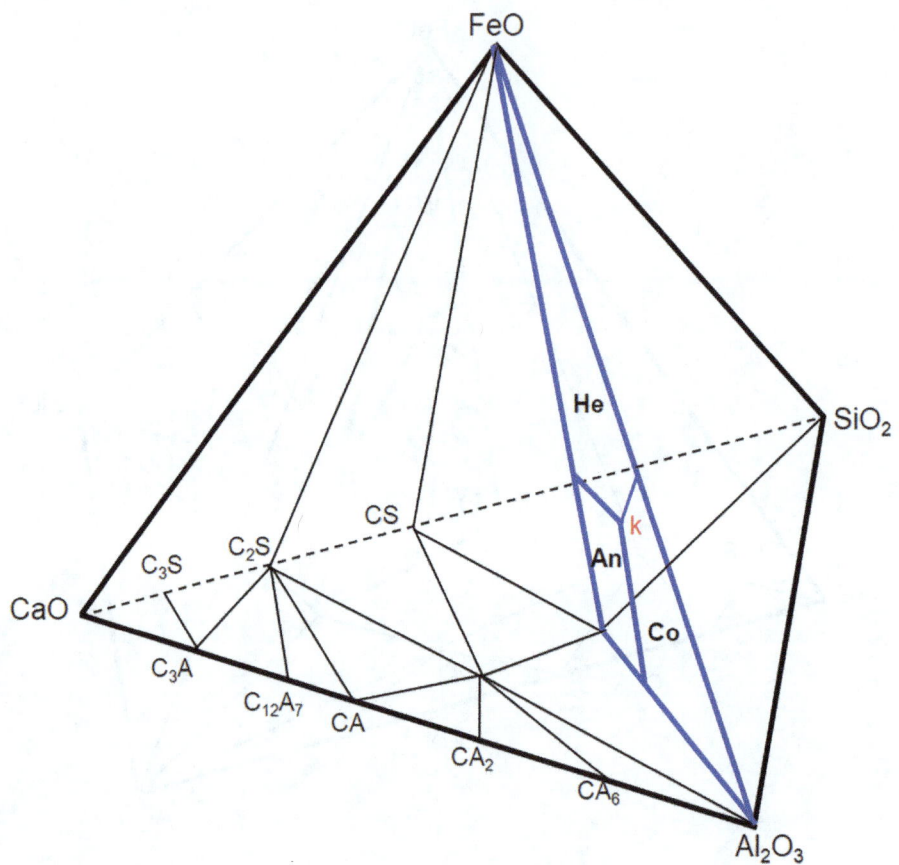

FIG. 4.15 – Blue plane is a schematic representation of the Anorthite, Al_2O_3–FeO join. Key: An for Anorthite, He for Hercynite and Co for Corundum.

- The phase assemblages along these two univariant lines are as follows:

  (1) Melilite + Anorthite + Hercynite + liquid (m lies on this line).
  (2) Melilite + Hercynite + Wüstite + liquid (l lies on this line).

4.3.2.5 $CaSiO_3$–Gehlenite–FeO Join (Pink Plane on Figure 4.12)

- It forms one face of interior of the $CaO·SiO_2 - CaO·SiO_2 - 2CaO·Al_2O_3·SiO_2 -$ FeO tetrahedron (figure 4.17).
- It is bounded by three binary systems; $CaO·SiO_2 - 2CaO·Al_2O_3·SiO_2$ (eutectic at 1318 °C 63.3% CS and 36.7%C_2AS), $2CaO·Al_2O_3·SiO_2$ – FeO, and $CaO·SiO_2$ – FeO.
- This join cuts the following four primary phase volumes within the tetrahedron; pseudowollastonite + liquid, Melilite + liquid, Olivine + liquid and Wüstite + liquid.

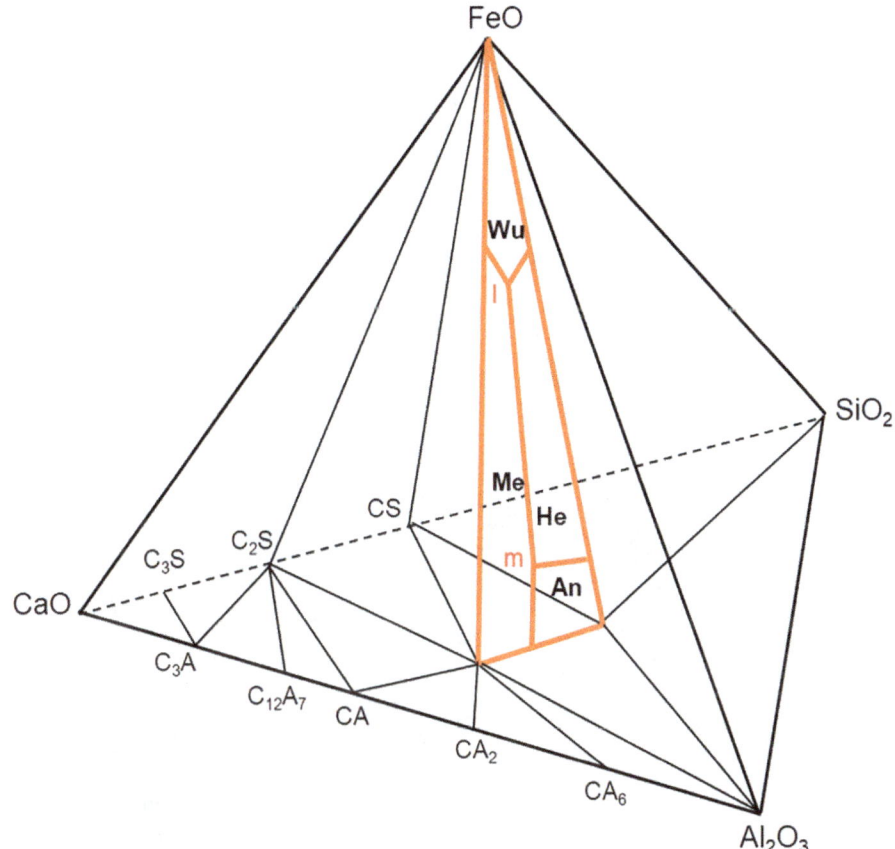

FIG. 4.16 – Orange plane is a schematic representation of the Gehlenite–Anorthite–FeO join. Key: An for Anorthite, He for Hercynite and Me for Melilite.

- Points n and o are piercing points of two univariant lines within the tetrahedron in the CaSiO$_3$–Gehlenite–FeO plane.
- Along each of these two curved univariant lines, three crystalline phases are in equilibrium with liquids having a composition that lies on one of the lines.
- The phase assemblages along these two univariant lines are as follows:

  (1) Pseudowollastoniute + Melilite + Olivine + liquid (n lies on this line).
  (2) Melilite + Olivine + Wüstite + liquid (o lies on this line).

4.3.2.6 $2CaO \cdot SiO_2$–Gehlenite–FeO Join (Brown Plane on Figure 4.12)

- It is a triangular plane bounded by three binary systems; $2CaO \cdot SiO_2$–Gehlenite, $2CaO \cdot SiO_2$–FeO and Gehlenite–FeO. It forms one face of interior of tetrahedron (figure 4.18).

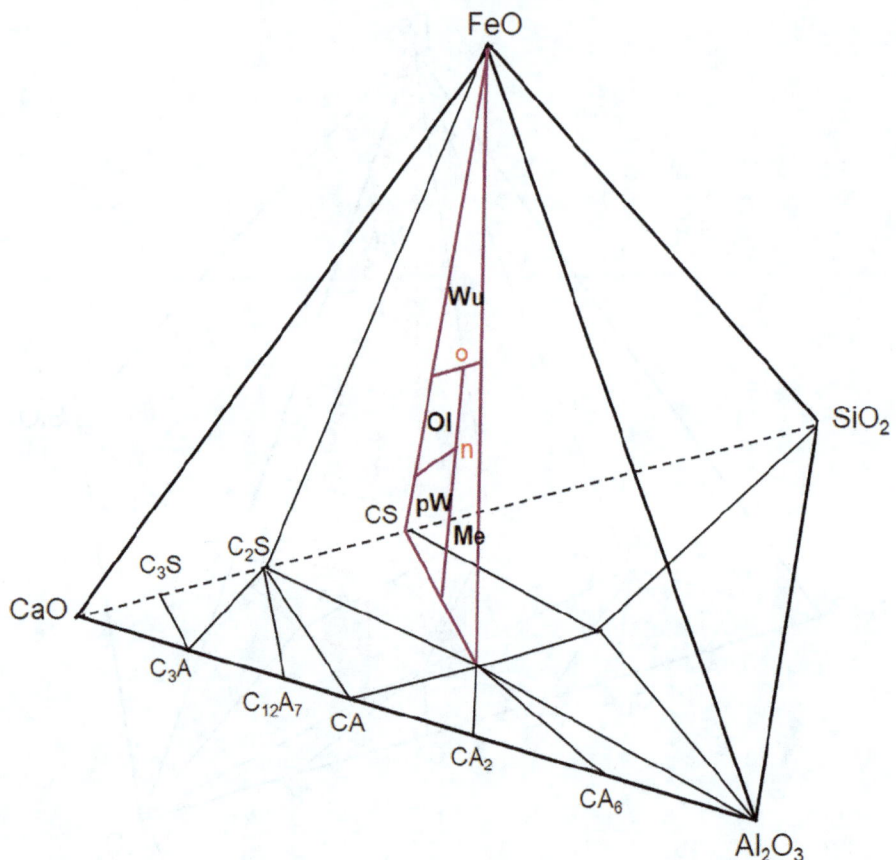

FIG. 4.17 – Brown plane is a schematic representation of the CaSiO$_3$–Gehlenite–FeO join. Key: W for Wüstite, Ol for Olivine, Me for Melilite and pW for Pseudowollastonite.

- The 2CaO·SiO$_2$–Gehlenite binary system is a join in the CaO–Al$_2$O$_3$–SiO$_2$ system with a eutectic at 1545 °C at the composition 37% 2CaO·SiO$_2$ and 63% 2CaO·Al$_2$O$_3$·SiO$_2$.
- The 2CaO·SiO$_2$–FeO binary system is a pseudo binary system in the CaO–SiO$_2$–FeO system with a eutectic at 1285 °C at a composition 26% CaO and 74% FeO.
- The Gehlenite–FeO system shows a minimum at 1273 °C at 18% of Gehlenite.
- The 2CaO·SiO$_2$–Gehlenite–FeO ternary system includes a ternary eutectic at 1250 °C. At this temperature, C$_2$S, Gehlenite and Wüstite are in equilibrium with a liquid of composition 21.3% C$_2$S, 21.3% C$_2$AS and 57.4% FeO.
- The 2CaO·SiO$_2$–Gehlenite–FeO ternary system forms one face of the interior 2CaO·SiO$_2$–CaO·SiO$_2$–Gehlenite–FeO tetrahedron. A second face corresponds to CaSiO$_3$–Gehlenite–FeO system previously described.

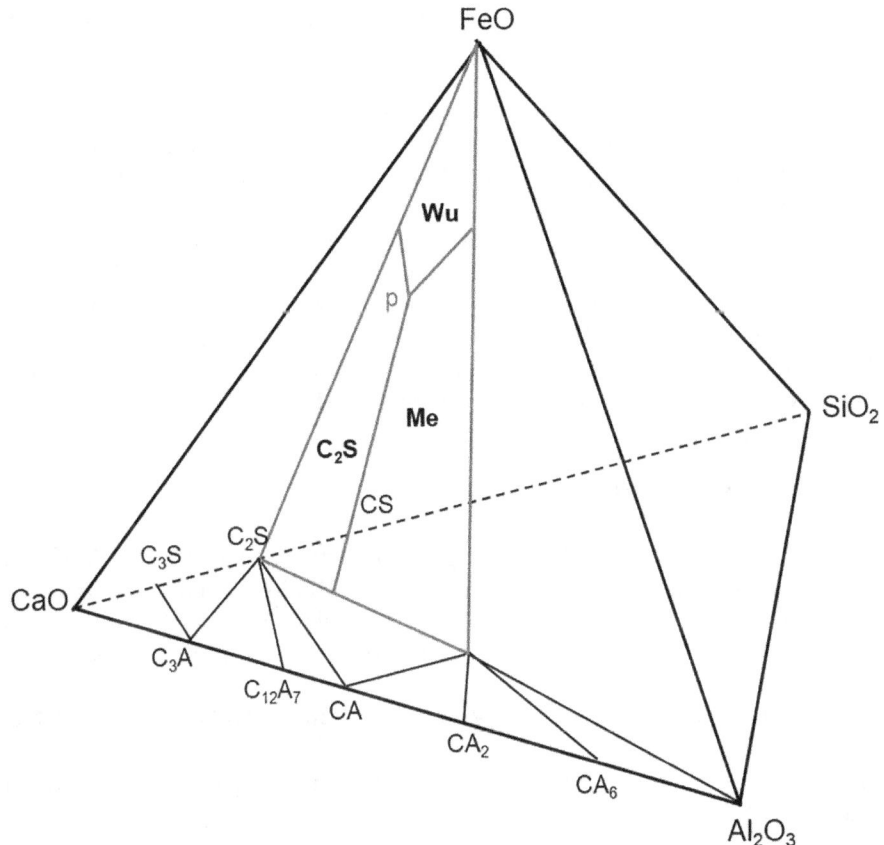

FIG. 4.18 – Black plane is a schematic representation of the 2CaO·SiO$_2$–Gehlenite–FeO join. Key: W for Wüstite and Me for Melilite.

- The point p is a piercing point (figure 4.18 and table 4.10).
- The phase assemblage along the univariant line is as follows:

  (1) Wüstite + Melilite + C$_2$S + liquid (p lies on this line).

4.3.3 Binary Systems

- The upper part of the diagram represents the silica rich compositions (planes A/B/C/D).
- The D/E/F planes represent the low silica compositions.
- The Anorthite–FeO lines are common to the A/B/C planes.
- The Gehlenite–FeO lines are common to the D/E/F planes.
- Line I–K represents the binary Gehlenite–Iron Åkermanite.

- A straight line joins two Melilites; $Ca_2FeSi_2O_7$ (iron Åkermanite) and $Ca_2Al_2SiO_7$ (Gehlenite).
- Gehlenite melts congruently at 1590 °C. Iron Åkermanite is stable only to 775 °C, above which $CaFeSiO_4$ and $CaSiO_3$ are the stable phases.
- Gehlenite and Åkermanite have the same basic structure in which pairs of SiO_2 tetrahedra are linked at one corner but in Gehlenite half of the SiO_4 groups are replaced by AlO_4 tetrahedra.
- The phases encountered in the different systems are Garnet solid solutions (Gar ss), Pseudowollastonite (PWo), Wollastonite (Wo), Ferrifassaite solid solutions (Fat ss), Hematite solid solutions (Hem ss), Magnetite solid solutions (mt ss), Gehlenite solid solutions (Geh ss), Ferrigelhenite, Anorthite (an), Corundum, CA_6 (or Hibonite).
- Pleochroite was found to be the primary phase for silica rich compositions having a FeO content greater than 5 wt % (SOU 1991–1 and SOU 1991–2); Fe^{2+} and Fe^{3+} are both present in Pleochroite.

4.3.4 Crystallized Solids

- Seven crystalline solids of fixed compositions are found in the portion of the quaternary system studied, namely: Corundum, Mullite, Tridymite, Cristobalite, Pseudowollastonite and Anorthite. The crystalline phases Wüstite, Olivine, Wollastonite and Melilite are not of fixed compositions but are solid solutions.

4.3.5 Univariant Lines and Quaternary Invariant Points

- In the figures 4.13–4.18, three curves meet at the following piercing points (table 4.10):
    - a, b and c in figure 4.13.
    - d, e, f, g, h, i and j in figure 4.14.
    - k in figure 4.15.
    - l and m in figure 4.16.
    - n and o in figure 4.17.
    - p in figure 4.18.
- Each of these points lies on a quaternary invariant line within the tetrahedron (figure 4.19).
- Along each of these lines, three crystalline phases are in equilibrium with liquid whose composition lies on the line.
- The part of flowchart of the CaO–SiO_2–Al_2O_3–FeO system corresponding to the sub-systems presented previously summarises the 12 invariant points (rectangles in figure 4.19). Moreover the ternary systems are represented by brackets.
- Figure 4.19 has been constructed to show the probable relations between univariant lines, ternary invariant points in the limiting systems and between the eleven quaternary invariant points found in the system.
- Each univariant line within the tetrahedron may connect:

Quaternary Chemical Systems

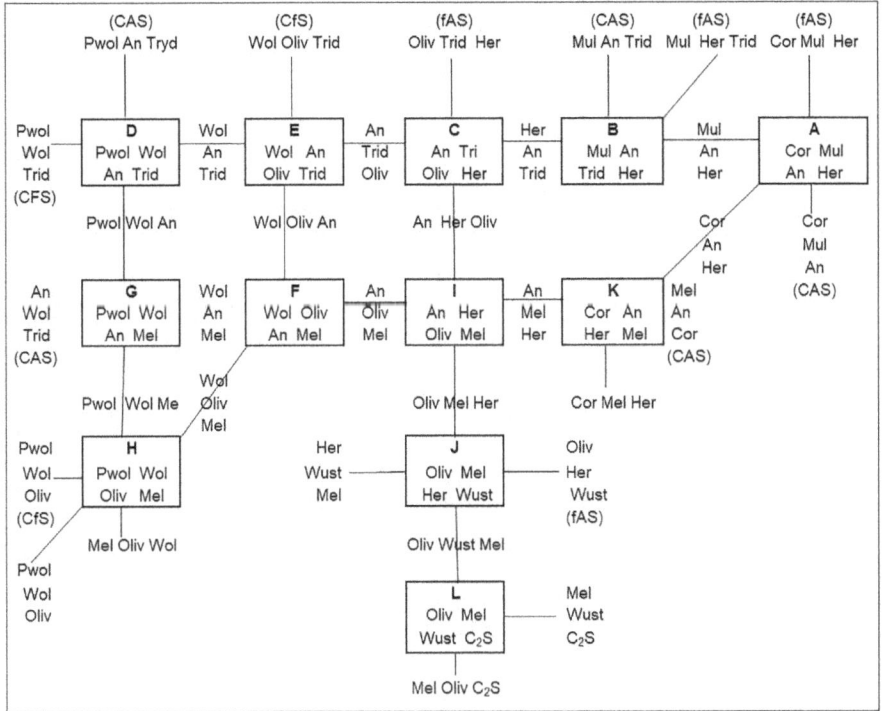

FIG. 4.19 – Flowchart of the CaO–SiO$_2$–Al$_2$O$_3$–FeO system corresponding to the sub-systems presented previously. Key: Invariant points are in the rectangles and letters correspond to the name of the invariant points. Wust for Wüstite, Oliv for Olivine, Mel for Melilite, Pwol for Pseudowollastonite, Wol for wollastonite, An for Anorthite, Herc for Hercynite, Cor for Corundum and Tri for Tridymite.

(1) In the two ternary invariant points which lie in different faces of the tetrahedron.
(2) Ternary invariant points from one of the four phases of the tetrahedron and a quaternary invariant point within the tetrahedron.
(3) Two quaternary points within the tetrahedron and the 18 univariant lines.

4.3.6 2CaO·SiO$_2$–CaO·SiO$_2$–Gehlenite–FeO Quaternary System

- The 2CaO·SiO$_2$–CaO·SiO$_2$–Gehlenite–FeO system is represented by an interior tetrahedron. The front face (2CaO·SiO$_2$–Gehlenite–FeO) has been removed for clarity.
- The system is described by a perspective drawing (figure 4.20) and a flowchart (figure 4.21).

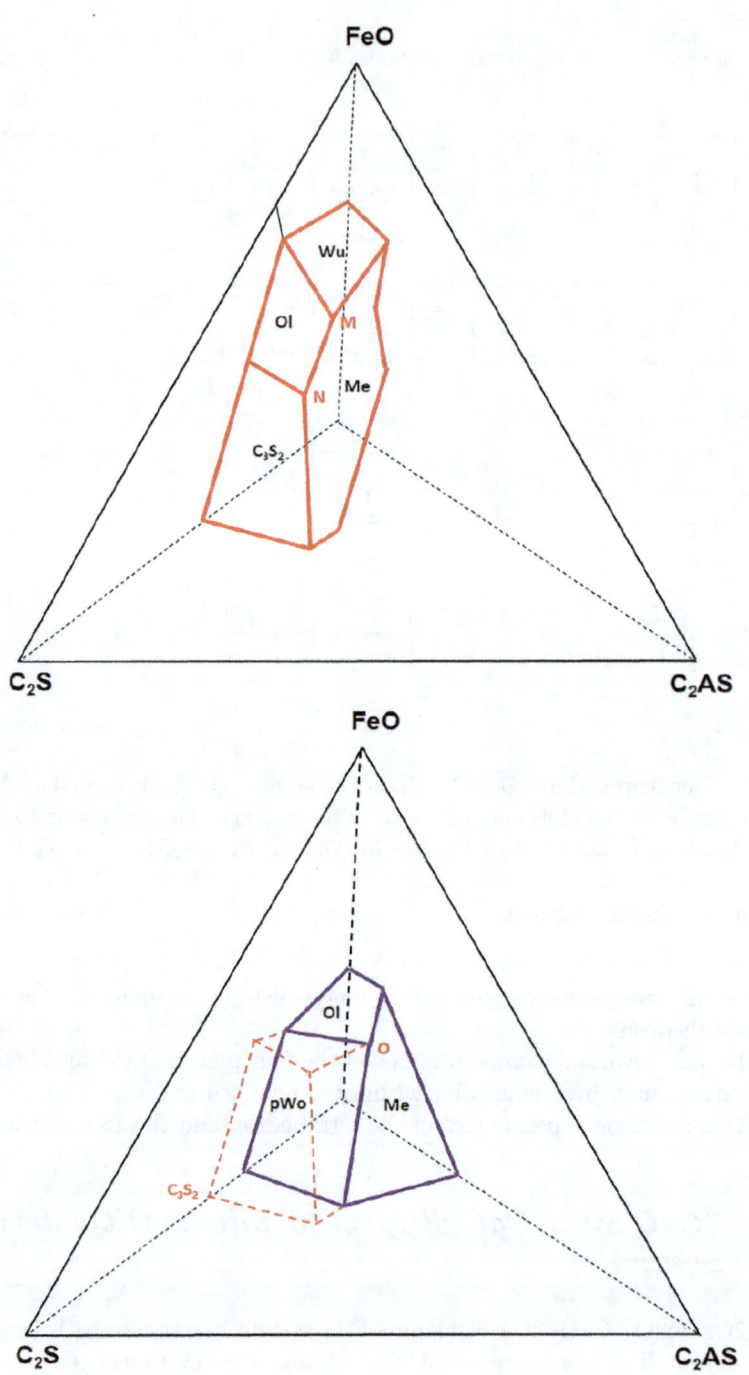

FIG. 4.20 – Perspective drawing of 2CaO·SiO$_2$–CaO·SiO$_2$–Gehlenite–FeO system and position of invariant points. Key: Wu for Wüstite, Ol for Olivine, Me for Melilite and pWo for Pseudowollastonite.

Quaternary Chemical Systems

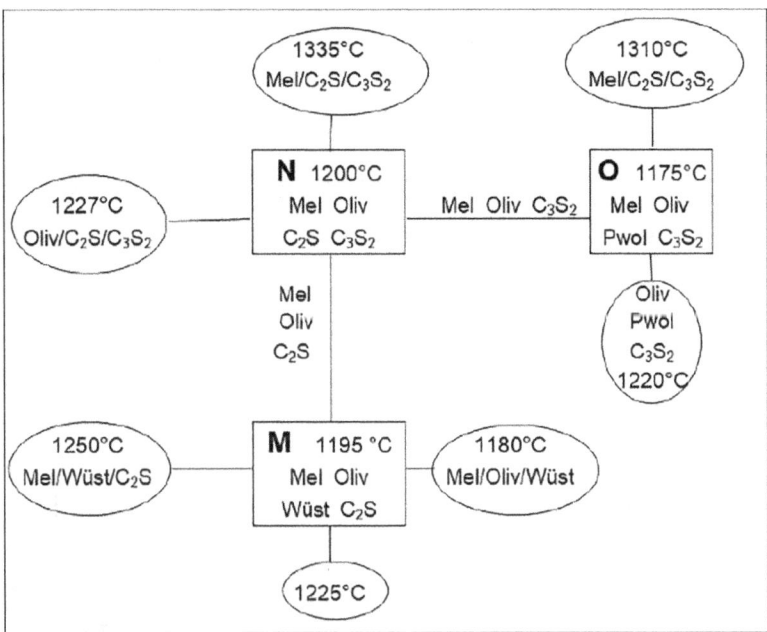

FIG. 4.21 – Relationship of univariant lines and invariant points in the $2CaO \cdot SiO_2$–$CaO \cdot SiO_2$–Gehlenite–FeO system.

- Each face of the tetrahedron contains one or more points at which three crystalline phases and liquid are in equilibrium; points on the liquidus surface at which three boundaries' curves meet.
- From each of the eight points on the faces, a quaternary univariant line extends into the tetrahedron.
- From two univariant lines which do not intersect the faces of the tetrahedron three crystalline and liquid are in equilibrium.
- Where the composition of the crystalline phase is variable which is especially true for Melilite, the crystalline phase changes composition along with the liquid.
- Just as in ternary systems diagrams, where three univariant lines intersect at an invariant point, so in a quaternary invariant point four invariant lines meet at a quaternary invariant point.
- The ten univariant lines of the tetrahedron necessitate the existence of three quaternary invariant points as shown schematically in the flowchart (figure 4.21 and table 4.11).
- At each of these points, 4 crystalline phases are in equilibrium with liquid whose composition is that at these points (table 4.11).
- Other ternary and quaternary invariant points exist, in addition to those shown, where two of the phases are C_2S and C_2AS.

TAB. 4.11 – Composition of the invariant points.

		C_2S	CS	C_2AS	FeO	CaO	Al_2O_3	SiO_2	T (°C)
M	C_2S + Ol + Me + Wü	5	41	15	55	32	6	27	1195
N	C_2S + C_3S_2 + Ol + Mel	10	51	15	24	37,5	5,5	33	1200
O	C_3S_2 + CS + Ol + Mel	5	61	12	22	32,5	4,5	36	1175

Composition in weight %.
Key: Wü for Wüstite, Ol for Olivine and Me for Melilite.

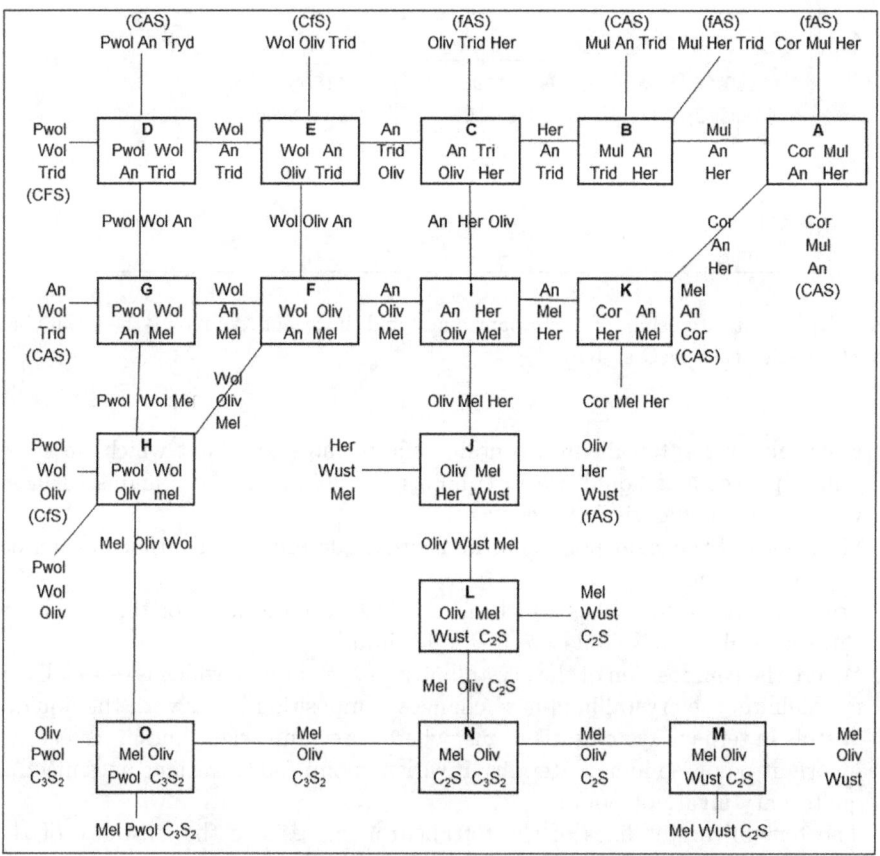

FIG. 4.22 – Complete flowchart of the CaO–SiO$_2$–Al$_2$O$_3$–FeO system. Key: Invariant points are in the rectangles and letters correspond to the name of the invariant points. Wust for Wüstite, Oliv for Olivine, Mel for Melilite, Pwol for Pseudowollastonite, Wol for wollastonite, An for Anorthite, Herc for Hercynite, Cor for Corundum and Tri for Tridymite.

- The composition and temperature of the invariant points can be estimated if we know the temperature and the composition of each invariant or piercing point on the faces of the tetrahedron and the direction of temperature drop along the quaternary univariant line.
- Combining the flow chart of the $2CaO \cdot SiO_2$–$CaO \cdot SiO_2$–Gehlenite–FeO system (figure 4.21) and the previously reported flow chart of the CaO–SiO_2–Al_2O_3–FeO system (figure 4.19) enables us to have a more complete flow chart of this complex system (figure 4.22).

Chapter 5

Quinary Chemical Systems

5.1 Introduction – Presentation of Quinary Data

- It is possible to show phase relationships in a five components system using geometrical methods when one of the independent variables is held constant or is forced to follow a set pattern.
- In the case of iron oxide system when either the oxygen pressure is kept constant, or the equilibria is investigated in contact with metallic iron, the geometrical treatment is analogous to those used in representing phase relations in a quaternary system.

5.2 CaO–SiO_2–Al_2O_3–FeO–Fe_2O_3 System

5.2.1 Stability and Phase Diagrams

- Figure 5.1 shows a section of the quinary system at constant total iron content 15% and constant ratio Fe^{2+}/Fe total.
- The oxygen isobars necessary to give the fixed reduction ratio have been superimposed.
- The oxygen isobars move across the liquidus surface. They are controlled mainly by the liquid composition.
- The influence of the temperature on the position of the isobar is slight relative to the importance of the chemical effect caused by changing the liquid position. The direction of the isobar indicates that SiO_2 has the most important effect on the reduction ratio.

DOI: 10.1051/978-2-7598-2480-9.c005
© Science Press, EDP Sciences, 2020

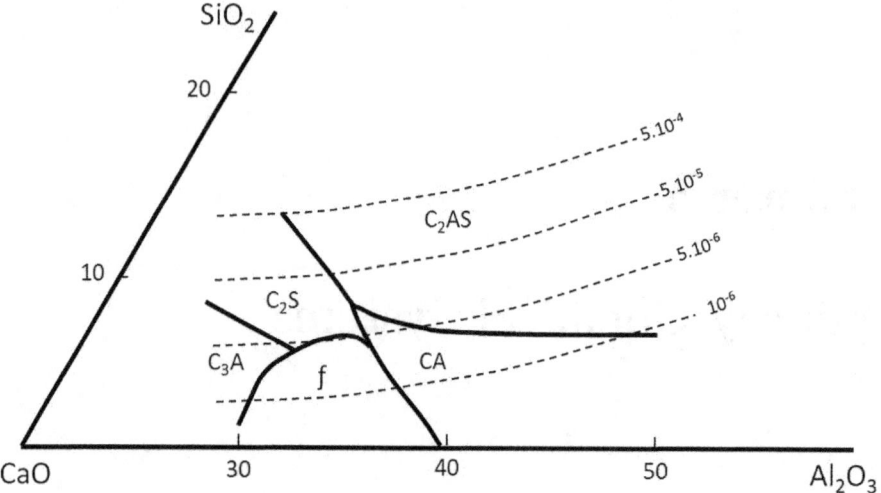

FIG. 5.1 – Section containing 15% iron oxide of the CASfF system (SOR 1973). Note: The oxygen pressure is adjusted so that the FeO content equals 5% and Fe_2O_3 10% by weight.

- Primary phase regions of; C_3A, C_2S, ferrite solid solutions, CA and Melilite have been found. These fields give rise to phase assemblages containing:

  (a) Melilite, C_2S, CA and liquid.
  (b) CA, ferrite solid solution, C_2S and liquid.
  (c) C_3A, C_2S, ferrite solid solution and liquid.

- Pleochroite does occur in area having a reduction of 0.3/0.4, but never as a primary phase. Compositions having Pleochroite as secondary crystallisation have Melilite as primary crystallisation.
- Experimental data show that eight of the six phase assemblages (including liquid), are defined. These are shown in the following flowchart of the quinary system (figure 5.2).
- This diagram shows the phases present, temperature pressure and the nature of invariant points but does not give information concerning their bulk compositions.
- Circle are used to show invariant equilibria with liquid (or univariant if liquid is absent) in one of the quaternary systems.
- Invariant points in these systems are linked to quinary invariant equilibria by univariant lines.
- The quinary equilibria are distinguished by being placed in rectangles.
- Univariant equilibria which radiated from the quinary points do not necessarily reach a quaternary system. Only those that have been located experimentally or entirely quinary are shown.
- Since five univariant equilibria will converge on each quinary point, it can be seen that all possible invariant equilibria are not located. The others which are presumed to exist can easily be listed if desired by tacking all combination four at a time of the five phases present at the quinary points.

Quinary Chemical Systems

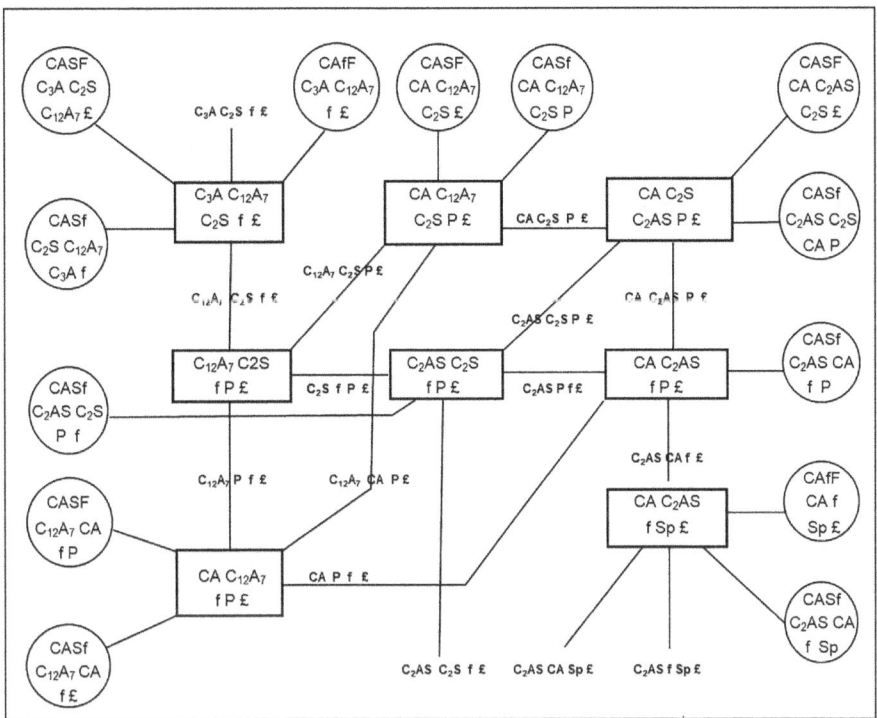

FIG. 5.2 – Flow diagram of the invariant and univariant equilibria in the quinary CASfF system. Key: £ = Calcium alumino ferrite, f = Wüstite and Sp = Spinel.

- The following can be observed on the nature of the phases present at these points:
  (1) Ferrite is always present as a ferric iron containing phases. All eight quinary points contains ferrite. The ferrite phase does not contain SiO_2 or FeO.
  (2) Pleochroite and C_3A are incompatible.
  (3) Pleochroite occurs in seven of the eight phase volumes.
  (4) The FeO containing phase could be Wüstite, Pleochroite or Spinel or mixture of these. As long as C_2S and C_2AS are present the ferrous iron occurs as Pleochroite. If CA and Melilite are present but C_2S absent, FeO will occur as both FeO and as Pleochroite.
  (5) These possible phase combinations have been developed for the maximum possible number of phases, but the figure also shows a number of four phase combinations. In these four phase combinations, at least one of the phases will be variable in composition. All the four phase combinations include the ferrite phases.
  (6) Gehlenite and $C_{12}A_7$ are incompatible.

5.2.2 Conclusions

- Phase diagram of the CASfF system is constructed from a set of experimental results obtained by keeping a mixture of components at a given temperature and pressure during a time long enough to obtain an equilibrium of the reactions.
- Our knowledge of the system is not sufficient to be able to follow in detail the sequence of the reactions but it is sufficiently detailed to describe the final crystallisation products.

References Part I

ABB 1975 – Abbatista F., Burdese A., Maja M. (1975) Le diagramme d'équilibre du système CaO-FeOx, *Rev. Int. Hautes Temp.* **12**, 337.
ADA 1980 – Adamkovicova K., Kosa L., Proks I. (1980) The heat of fusion of $CaSiO_3$, *Silik.* **24**, 193.
AGG 1972 – Aggarwal P.S., Gard J.A., Glasser F.P., Biggar G.M. (1972) Synthesis and properties of dicalcium aluminate $2CaO\text{-}Al_2O_3$, *Cem. Concr. Res.* **2**, 291.
AGR 1960 – Agrell S.O., Smith J.V. (1960) Cell dimension, solid solution, polymorphism and identification of mullite and sillimanite, *J. Am. Ceram. Soc.* **43**, 69.
AGR 1961 – Agrell S.O., Gay P. (1961) Kilchoanite, a polymorph of rankinite, *Nature* **189**, 743.
AKA 1985 – Akasaka M., Ohashi H. (1985) 57Fe-Mössbauer Study of Synthetic Fe^{3+} Melilites, *Phys. Chem. Miner.* **12**, 13.
AKS 1975 – Aksay I.A., Pask J.A. (1975) Stable and metastable equilibria in the system SiO_2-Al_2O_3, *J. Am. Ceram. Soc.* **58**, 507.
ALI 1989 – Ali M.M., Raina S.R., Singh V.K. (1989) Kinetics and diffusion studies in CA_2 formation, *Cem. Conc. Res.* **19**, 45.
ALL 1955 – Allen W.C., Snow R.B. (1955) The orthosilicate-iron oxide portion of the system $CaO\text{-}FeO\text{-}SiO_2$, *J. Am. Ceram. Soc.* **38**, 264.
ALL 1981 – Allibert M., Chatillon C., Jacob K.T., Lourtau R. (1981) Mass spectrometry and electrochemical studies of thermodynamic properties of liquid and solid phases in the system $CaO\text{-}Al_2O_3$, *J. Am. Ceram. Soc.* **64**, 307.
ALO 1978 – Alota S., Borgranni C.B., Geli J. (1978) Investigations on binary and ternary calcium ferrite, *VIIe journée int. Sidérurgie, Versailles* 23.
ANG 1990 – Angel R.J., Carpenter M., Finger L. (1990) Structural variation associated with compositional variation and order disorder behaviour in anorthite rich feldspars, *Amer. Miner.* **75**, 150.
ANG 1996 – Angel R.J., Ross N.L., Seifert F., Fliervoet T.F. (1996) Structural characterization of pentacoordinated silicon in a calcium silicate, *Nature* **384**, 441.
APP 1972 – Appendino P. (1972) Research on the alumina lime ferric oxide system, *il cemento* **2**, 69.
ARA 1959 – Aramaki S., Roy R. (1959) Revised equilibrium diagram for the system Al_2O_3-SiO_2, *Nature* **184**, 631.
ARA 1988 – Arakcheeva A., Karpinsky V. (1988) Polytipic relation in the structure of the group of hexagonal ferrite Ca ferrite of the structure type $Ca_2F_{16}O_{25}$, *Sov. Phys. Cristallographia* **33**, 378.
ARA 1990 – Arakcheeva A.V., Karpinsky O.G. (1990) Crystal structure of hexagonal $Ca_2F_{22}O_{33}$, *Sov. Phys. Cryst.* **35**, 650.
ARU 1957 – Aruja E. (1957) The unit cell of orthorhombic pentacalcium trialuminate, C_5A_3, *Acta Cryst.* **10**, 337.
ASA 1968 – Asada M., Omori Y., Sabongi K. (1968) Study on the properties of calcium ferrites in the Fe-Ca-O system, *Trans. Iron Steel Inst. Japan* **8**, 245.
ATL 1958 – Atlas L.M., Sumida W.K. (1958) Solidus, subsolidus and subdissociation phase equilibria in the system Fe-Al-O, *J. Amer. Ceram. Soc.* **41**, 150.
AUB 1970 – Aubry J., Berthet A., Duchène R., Etienne H., Evrard O., Jeannot F., Gleitzer C., Offroy C., Perrot P. (1970) Stabilisation du protoxyde de fer par formation de solution solides, *Ann. Chim.* **5**, 299.
BAI 1986 – Baisanov S.O. (1986) Phase equilibrium in the pseudo wollatonite-anorthite system, *Russ. J. Inorg. Chem.* **10**, 1553.

BAL 1970 – Baldock P.J., Parker A., Sladdin I. (1970) X ray powder diffraction data for calcium monoaluminate and calcium dialuminate, *J. Appl. Crystallogr.* **3**, 188.
BAR 2013 – Bartl H., Scheller T. (2013) A propos de la structure de $C_{12}A_7$, *Neues Jahrbuch für mineralogie* **12**, 547.
BAR 1980 – Barnes P., Fentiman C.H., Jeffery J.W. (1980) Structurally related dicalcium silicate phases, *Acta Cryst.* **A36**, 353.
BAR 1993 – Barry T.I., Dinsdale A.T., Gisby J.A. (1993) Predictive thermochemistry and phase equilibria of slags, *JOM* **45**, 32.
BEN 1974 – Bensted J., Varma S.P. (1974) Some applications of infrared and raman spectroscopy in cement chemistry. Part 3 - hydration of Portland cement and its constituents, *Cem. Technol.* **5**, 440.
BEN 1974 – Bensted J., Varma S.P. (1974) Some application of infra red and raman spectroscopy in cement chemistry, *Cem. Technol.* **5**, 440.
BEN 1976 – Bensted J. (1976) The ferrite phase an infra red spectroscopy study, *Il Cemento* **1**, 45.
BEN 1979 – Bensted J. (1979) γ-C_2S and its hydraulicity, *Cem. Concr. Res.* **8**, 73.
BER 1959 – Bertaut E.F., Blum P., Sagnières A. (1959) Structure du Ferrite Bicalcique et de la Brownmillerite, *Acta Cryst.* **12**, 149.
BER 1971 – Berggren J. (1971) Refinement of the crystal structure of dicalcium ferrite, *Acta Chem. Scand.* **25**, 3616.
BER 1984 – Berman R.G., Brown T.H. (1984) A thermodynamic model for multicomponent melts with application to the system CaO-Al_2O_3-SiO_2, *Geochim. Cosmochim. Acta* **48**, 661.
BER 1988 – Bergman B., Song C.H. (1988) Decomposition of wüstite in the system Ca-Fe-O, *J. Amer. Ceram. Soc.* **71**, 121.
BER 1989 – Bergman B., Song C.H. (1989) Microstructural study of phase equilibria in the Fe-Ca-O system involving $CaFe_5O_7$, *J. Am. Ceram. Soc.* **72**, 1364.
BIG 1967 – Bigaré M., Guinier A., Maziéres C., Regourd M., Yannaquis N., Eysel W., Hahn T., Woermann E. (1967) Polymorphism of tricalcium silicate and its solid solutions, *J. Amer. Ceram. Soc.* **50**, 609.
BJO 1984 – Björkman B. (1984) A solid state EMF study of the system CaO-$Ca_2Fe_2O_5$-feO-Fe, *Scand. J. Metall.* **13**, 193.
BJO 1985 – Björkman B. (1985) An assessment of the system Fe-O-SiO_2 using a structure based model for the liquid silicate, *Calphad* **9**, 271.
BLA 1969 – Black P. (1969) Rankinite and kilchoanite from Tokatoka, New Zealand, *Mineral. Mag.* **37**, 517.
BOH 1991 – Bohlen S.R., Montana A., Kerrick D.M. (1991) Precise determination of the equilibria kyanite $<->$ sillimanite and kyanite $<->$ andalusite and revised triple point for Al_2SiO_5 polymorphs, *Am. Mineral.* **76**, 677.
BOW 1924 – Bowen N.L., Greig I.W. (1924) The system Al_2O_3.SiO_2, *J. Am. Ceram. Soc.* **7**, 238.
BOW 1932 – Bowen N.L., Schairer J.F. (1932) The system FeO-SiO_2, *Am. J. Sci.* **24**, 177.
BOW 1935 – Bowen N.L. (1935) Ferrosilite as a natural mineral, *Am. J. Sci.* **30**, 481.
BOW 1975 – Bowman A.F. (1975) *An investigation of Al_2SiO_5 phase equilibrium utilizing the SEM*, M.Sc. thesis Oregon.
BRA 1960 – Braun P.B., Kwestroo W. (1960) On some calcium-iron-oxygen compounds, *Philips Res. Rep.* **15**, 394.
BRI 1954 – Brisi C. (1954) System CaO-Al_2O_3-Fe_2O_3, *Annali di Chimica* **4**, 500.
BRI 1959 – Brindley G.W., Nakahira M. (1959) The kaolinite mullite reactions series I. Survey of outstanding problems, *J. Am. Ceram. Soc.* **42**, 311.
BRI 1989 – Brian B.M.C., Coop M. (1989) Natural silica, silicon chemical, *Ind. Miner.* March 1989, 43.
BRO 1958 – Brownell W.E. (1958) Subsolidus relations between mullite and iron oxide, *J. Am. Ceram. Soc.* **41**, 226.
BRO 1971 – Brown G.C., Fyfe W.S. (1971) Kyanite-andalusite equilibrium, *Contribut. Mineral. Petrol.* **33**, 227.
BUD 1953 – Budnikov P.P., Tresvyatski S.G., Kushokovski V.I. (1953) Making the phase diagram of the system Al_2O_3-SiO_2 more precise, *Dokl. Akad. Nauk SSSR.* **93**, 281.

BUI 1968 – Buist D.S. (1968) A study of calcium hexaluminate, *Min. Mag.* **36**, 676.
BUR 1952–1 – Burdese A. (1952–1) Reduction equilibria of the system CaO-Fe_2O_3, *Metall. Ital.* **44**, 343.
BUR 1952–2 – Burdese A. (1952–2) Sul sistema CaO-FeO-Fe_2O_3, *Ric. Sci.* **24**, 782.
BUR 1952–2 – Burdese A., Cirilli V. (1952–2) Ricerche sul sistema CaO-Fe_2O_3-Fe_3O_4, *Ric. Sci.* **22**, 1564.
BUR 1961 – Burnham C.W., Buerger M.J. (1961) Refinement of the crystal structure of andalusite, *Zeitschrift für Kristallographie - Cryst. Mater.* **115**, 269.
BUR 1963–1 – Burnham C. (1963–1) Refinement of the crystal structure of sillimanite, *Zeitschrift für Kristallographie - Cryst. Mater.* **118**, 127.
BUR 1963–2 – Burnham C. (1963–2) Refinement of the crystal structure of kyanite, *Zeitschrift für Kristallographie - Cryst. Mater.* **118**, 337.
BYK 1981 – Byker H.J., Craig R.E.R., Eliezer I., Eliezer N., Howald R.A., Visvanadham P. (1981) Thermodynamic treatment of the CaO-SiO_2 system, *Calphad* **5**, 217.
CAM 1972 – Cameron M., Sueno S., Papike J.J. (1972) High temperature crystal chemistry of acmite, hedenbergite and spodumene join, *Geol. Soc. Amer. Sci. Abstr. Programs* **4**, 466.
CAM 1973 – Cameron M., Sueno S., Prewitt C.T., Papike J.J. (1973) High temperature crystal chemistry of acmite, diopside, hedenbergite, jadeite, spodumène and ureyite, *Amer. Mineral.* **58**, 594.
CAN 1986 – Canha R.H., Segadaes A.M. (1986) Compatibility relationships in the system CaO-SiO_2-Fe_2O_3-FeO, *Journal de Physique Colloques* **47**, 461.
CAR 1931 – Carlson E.T. (1931) Decomposition of C_3S in T° range 1000–1300 °C, *J. Res. Natl. Bur. Stand.* **7**, 893.
CHA 1972 – Chatterjee A.K., Zhmoidin G. (1972) The phase equilibria diagram of the system CaO-Al_2O_3-CaF_2, *J. Mater. Sci.* **7**, 93.
CHA 1982 – Chang D.R., Howald R.A., Roy B.N. (1982) Updating binary system files: CaO-$AlO_{1.5}$, *Calphad* **6**, 83.
CHE 1962 – Chessin H., Turkdogan E.T. (1962) A crystallographic investigation of Calcium Diferrite, *J. Am. Ceram. Soc.* **45**, 597.
CHE 1989 – Chesnokov B.V., Lotova E.V., Pavlyutchenko V.S., Usova L.V., Bushmakin A.F., Nishanbaev T.P. (1989) Svyatoslavite, $CaAl_2Si_2O_8$ (orthorombic). a new mineral, Zapiski RMO (Proceedings of the Russian Mineralogical Society), In Russian. Vol 118, 1989, pp. 111–114.
CHE 1989 – Chesley J.A., Burnet G. (1989) A two-stage reaction sequence for C_3S formation, *Cem. Conc. Res.* **19**, 837.
CHO 2002 – Choi H.J., Lee J.G. (2002) Synthesis of mullite whiskers, *J. Am. Ceram. Soc.* **85**, 481.
CHU 1988 – Chung W.S., Murayama T., Ono Y. (1988) Direct measurement of reduction equilibria of calcium ferrite with CO-CO_2 gas mixture by EMF method using solid oxide electrolyte in the temperature range 1273 to 1373°K, *Trans. Iron Steel Inst. Jpn.* **28**, B-4.
CIN 1998 – Cinibulk M.K. (1998) Effect of precursors and dopants on the synthesis and grain growth of calcium hexaluminate, *J. Am. Ceram. Soc.* **81**, 3157.
CIR 1952–1 – Cirilli V., Burdese A. (1952–1) The calcium oxide-wüstite system, *Proc. Int. Symp. React. Solids Gothenburg Sweden* 867.
CIR 1952–2 – Cirilli V., Burdese A. (1952–2) Riduzione del ferrito monocalcio a differenti temperature, *Metallurgia* **44**, 371.
COC 1993 – Cochet G., Sorrentino F. (1993) Limestone filled cement: properties and uses, *Mineral admixtures in cement and concrete*, (S.L. Sarkar, S.N. Ghosh, Eds). Vol 4, pp. 266–295.
COL 1970 – Colville A.A. (1970) The crystal structure of $Ca_2Fe_2O_5$ and its relation to the nuclear electric field gradient at the iron sites, *Acta Cryst.* **26**, 1469.
COS 1995 – da Costa E., Coheur J.P., Vanderheyden B., Munnix R. (1995) Slag formation in the adhering layer of granules and its reaction with nuclei in iron ore sintering, *ISIJ Int.* **35**, 138.
COT 1992 – Coté B., Massiot D., Poe B., Mc Millan P., Tautelle F., Coutures J.P. (1992) Liquids and glasses structural differences in the CaO-Al_2O_3 system, as evidenced by ^{27}Al NMR spectroscopy, *Journal de physique, Colloque C2 supplément au journal de physique III* **2**, 223.

CRI 1991 – Criado E., Aza S.D. (1991) Phase relationship in the subsystem $2CaO.Al_2O_3.SiO_2$ - $CaO.Al_2O_3.2SiO_2$ - $CaO.6Al_2O_3$, Proceedings conference UNITECR 91 Aachen (Germany), September 1991, pp. 566–573.
CRO 1939 – Crook W.J. (1939) The series iron oxides-lime, *J. Am Ceram. Soc.* **22**, 313.
CUN 1998 – Tas A.C. (1998) Chemical preparation of the binary compounds in the calcia-alumina system by self-propagating combustion synthesis, *J. Am. Ceram. Soc.* **81**, 2853.
CZA 1981 – Czaya R. (1981) Refinement of the structure of gamma-Ca_2SiO_4, *Acta Cryst.* **B27**, 848.
DAN 1984 – Daněk V. (1984) Calculation of liquidus surface of $CaSiO_3$ in the CaO-FeO-SiO_2 and CaO-Fe_2O_3-SiO_2 systems, *Chem. Zvesti* **38**, 379.
DAR 1948 – Darken L.S. (1948) Melting points of iron oxides on silica; phase equilibria in the system Fe-Si-O as a function of gas composition and temperature, *J. Am. Chem. Soc.* **70**, 2046.
DAV 1952 – Davis G.L., Tuttle O.F. (1952) Two new crystalline phases of the anorthite composition, $CaO·Al_2O_3·2SiO_2$, *Am. J. Sci.* Bowen Volume Publication No **1200**, 107.
DAV 1972 – Davis R.F., Pask J.A. (1972) Diffusion and reaction studies in the system Al_2O_3-SiO_2, *J. Am. Ceram. Soc.* **55**, 525.
DAW 1983 – Dawson P.R., Ostwald J., Hayes K.M. (1983) Calcium ferrite in the new low silica sintering technology, *BHP Tech. Bull.* **27**, 47.
DAY 1906 – Day A.L., Shepherd E.S., Wright F.E. (1906) The lime silica series of minerals, *Am. J. Sci. (4th series)* **22**, 265.
DAY 1965 – Dayal R.R. (1965) The system CaO-Al_2O_3-Fe_2O_3, PhD Thesis, Aberdeen University.
DAY 1980 – Day H.W., Kumin H.J. (1980) Thermodynamic analysis of the aluminium silicate triple point, *Am. J. Sc.* **280**, 265.
DEC 1957 – Decker B.F., Kasper J.S. (1957) The structure of calcium ferrite, *Acta Cryst.* **10**, 332.
DEC 1998 – de Capitani C., Kirschen M. (1998) A generalized multicomponent excess function with applicattion to immiscible liquids in the system CaO-SiO_2-TiO_2, *Geochimica et Cosmochimica Acta* **62**, 3753.
DEE 1962 – Deer W.A., Howie R.A., Zussman J. (1962). *An Introduction to the Rock-Forming Minerals*. Longmans, London, 528 pages.
DIN 1990 – Ding J.P., Li D.Y., Fu P.Q (1990) X-ray powder structural analysis of the spinel polymorph of Fe_2SiO_4, *Powder Diffr.* **5**, 221.
DOR 1979 – Dörner P., Gaukler L.J., Krieg H., Lukas H.L., Petzov G., Weiss J. (1979) On the calculation and representation of multicomponent systems, *Calphad* **3**, 241.
DRA 1970 – Dragoi I., Teoreanu I., Flueras G.H. (1970) Uber einige eingenschaften der schmelzen im teilsystem CaO-$2CaO.SiO_2$-$12CaO7.Al_2O_3$ in abhangigkeit vom des thermishen gleichgewichtsverhalthalnissen, Réf. ATILH n°07,754 (in German), *Buletinul Stiintific Si Tehnic Al Institutului Politehnic Timisoara, Seria Chimie, Roum.* **15**, 229.
DUF 1969 – Dufour M.C., Perrot P. (1969) Utilisation des réactions de transport chimique dans l'évaluation des enthalpies libres: applications aux oxydes mixtes de Fe III et de calcium, *Rev. Chimie Minérale* **6**, 427.
DYS 1972 – Dyson D.J., Juckes L.M. (1972) A silica-deficient pyroxene in iron ore sinters, *Mineral. Mag.* **38**, 872.
ELI 1981 – Eliezer I., Eliezer N., Howald R.A., Viswanadham P. (1981) Thermodynamic properties of calcium aluminates, *J. Phys. Chem.* **85**, 2835.
ERI 1993 – Eriksson G., Pelton A.D. (1993) Critical evaluation and optimization of the thermodynamic properties and phase diagrams of the CaO-Al_2O_3, Al_2O_3-SiO_2 and CaO-Al_2O_3-SiO_2 systems, *Metall. Mater. Trans. B* **24B**, 807.
EVA 1965 – Evans B.W. (1965) Application of a reaction rate method to the breakdown equilibria of muscovite and muscovite plus quartz, *Am. J. Sci.* **263**, 647.
EVR 1977 – Evrard O., Jeannot F., Malaman B., Tannieres N., Aubry J. (1977) Détermination sur monocristal de la maille cristalline de CaFe5O7, *C.R. Acad. Sci. Ser. C* **284**, 445.
EYS 1970 – Eysel W., Hahn T. (1970) Polymorphism and solid solution of Ca_2GeO_4 and Ca_2SiO_4, *Z. Kristallogr.* **131**, 322.
FEI 1986 – Fei Y., Saxena S.K. (1986) A thermodynamical data base for phase equilibria in the system Fe-Mg-Si-O at high pressure and temperature, *Phys. Chem. Miner.* **13**, 311.

FIL 1949 – Filonenko N.E., Lavrov I.V. (1949) Calcium hexaaluminate in the system $CaO-Al_2O_3$-SiO_2, *Doklady Akademii Nauk. SSSR.* **66**, 673.
FIL 1950 – Filonenko N.E., Lavrov I.V. (1950) Investigations of equilibrium conditions in the Al_2O_3 corner of the rernary system $CaO-Al_2O_3$-SiO_2, *J. Appl. Chem. USSR* **23**, 1105.
FIL 1953 – Filonenko N.E., Lavrov I.V. (1953) Fusion of mullite, *Doklady Akademii Nauk. SSSR* **89**, 141.
FIS 1956 – Fischer W.A., Hoffman A. (1956) Das Zustandschausbild Eisenoxydul-Aluminiumoxyd, (in German), *Archiv für das Eisenhüttenwesen* **27**, 343.
FIT 1991 – Fitz Gerald J.D., Ringwood A.E. (1991) High pressure rhomboedral perovskite phase $Ca_2AlSiO_{5.5}$, *Phys. Chem. Miner.* **18**, 40.
FOL 1997 – Folco L., Mellini M. (1997) Crystal chemistry of meteoritic kirschsteinite, *Eur. J. Mineral.* **9**, 969.
FUM 1966 – Fumo D.A., Morelli M.R., Segadães A.M. (1966) Combustion synthesis of calcium aluminates, *Mat. Res. Bull.* **31**, 1243.
GEN 1963 – Gentile A.L., Foster W.R. (1963) Calcium hexaaluminate and its stability relations in the system $CaO-Al_2O_3$-SiO_2, *J. Am. Ceram. Soc.* **46**, 74.
GHO 1979 – Ghosh S.N., Paul A.K., Handoo S.K. (1979) Decomposition of $CaCO_3$ and formation of C_3S, *J. Mater. Sci.* **14**, 1011.
GHO 1989 – Ghose S., Okamura F.P., Ohashi H. (1989) The crystal structure of $CaF_3 + SiAlO_6$ and the crystal chemistry of Fe^{3+}, Al^{3+} substitution in calcium Tschermak's pyroxene, *Contribut. Mineral. Petrol.* **92**, 530.
GIB 1994 – Gibb T.C., Herod A.J., Munro D.C., Peng N. (1994) Synthesis under high pressure and characterisation by Mössbauer spectroscopy of non-stoichiometric $Ca_2Fe_2O_{5.12}$, *J. Mater. Chem.* **4**, 1451.
GIT 1970 – Gitzen W.H. (1970) *Alumina as a ceramic material.* American Ceramic Society.
GOE 1980 – Goel R.P., Kellog H.H., Rrain J.L. (1980) Mathematical description of the thermodynamic Properties of the systems Fe-O and Fe-O-SiO_2, *Metall. Trans. B* **11**, 107.
GOG 1961 – Goggi G. (1961) Les équilibres solides liquides dans le système ternaire $CaO-2CaO.SiO_2$-$5CaO.3Al_2O_3$, *Silic. Indus.* **26**, 17.
GOK 1991 – Weinberg A.A., Goktas M.C. (1991) Preparation and crystallization of sol–gel calcia-alumina compositions, *J. Am. Ceram. Soc.* **74**, 1066.
GOL 1975 – Golovastikov R., Matveeva R.G., Belov N.V. (1975) Crystal structure of the tricalcium silicate $3CaO.SiO_2 = C_3S$, *Sov. Phys. Crystallogr.* **20**, 441.
GOO 1970 – Goodwin D.W., Lindop A.J. (1970) The crystal structure of $CaO.2(Al_2O_3)$, *Acta Cryst.* **B26**, 1230.
GOR 1951 – Goria C., Burdese A. (1951) Lime alumina system X ray studies of mono crystal CA_2 produced from aluminous cement, *Ric. sci.* **21**, 1613.
GRA 1969 – Grant R.W. (1969) Nuclear electric field gradient at the iron sites in $Ca_2Fe_2O_5$ and Ca_2FeAlO_5, *J. Chem. Phys.* **51**, 1156.
GRE 1927-1 – Greig J.W. (1927-1) Immiscibility in silicates melts, *Am. J. Sci. 5th Ser.* **13**, 133.
GRE 1927-2 – Greig J.W. (1927-2) On liquid immiscibility in the system $FeO-Fe_2O_3-Al_2O_3-SiO_2$, *Am. J. Sci.* **14**, 473.
GRE 1973 – Grenier J.C., Menil F., Pouchard M., Hagenmuller P. (1973) Sur quelques phases dérivées du ferrite bicalcique $Ca_2Fe_2O_3$, *C.R. Acad. Sci.* **277 C**, 647.
GRZ 1986 – Grzeszczyk S.T. (1986) Solid solutions of C_3A-Na_2O in $C_{12}A_7$, *Cem. Concr. Res.* **16**, 798.
GUG 1982 – Guglielmi M., Principi G. (1982) Gel-glass transformation in the $SiO_2-Fe_2O_3$ system, *J. Non-Cryst. Solids* **48**, 161.
GUI 1968 – Guinier A., Regourd M. (1968) Structure on Portland cement mineral, *5th Int. Congr. Chem. Cem. Tokyo* **1**, 1.
GUL 1994 – Gülgün M.A., Popoola O.O., Kriven W.M. (1994) Chemical synthesis and characterization of calcium aluminate powder, *J. Am. Ceram. Soc.* **77**, 531.
GUO 1988 – Guo Q., Wang S. (1988) *Sci. Sintering Ser. B* **31**, 1515.

GUR 1950 – Gurry R.W., Darken L.S. (1950) The composition of CaO-FeO-Fe$_2$O$_3$ and MnO-FeO-Fe$_2$O$_3$ melts at several oxygen pressures in vicinity of 1600, *J. Am. Chem. Soc.* **72**, 3906.

GUS 1974 – Gustafson W.I. (1974) The stability of andradite, hedenbergite and related minerals in the system Ca-Fe-Si-O-H, *J. Petrol.* **15**, 455.

HAG 1986 – Hageman V.B., van den Berg G.J.K., Janssen H.J., Oonk H.A.J. (1986) A reinvistigation of liquid immiscibility in the SiO$_2$-CaO system, *Phys. Chem. Glasses.* **27**, 100.

HAH 1968 – Hahn T., Eyzel W. Woermann E. (1968) Crystal chemistry of C3S solid solutions, 5th Int Congress Cement Chemistry Tokyo 1968 supplementary paper vol I, 1–55, pp. 61–66.

HAL 1991 – Hallstedt B. (1991) Assessment of the CaO-Al$_2$O$_3$ System, *J. Am. Ceram. Soc.* **73**, 15.

HAM 1989 – Hamilton J.D.G., Hoskins B.F., Mumme W.G., Borbidge W.E., Montague M.A. (1989) The crystal structure and crystal chemistry of Ca$_{2.3}$Mg$_{0.8}$Al$_{1.5}$Fe$_{8.3}$O$_{20}$ (SFCA): solid solution limits and selected phase relationship of SFCA in the SiO$_2$-Fe$_2$O$_3$-CaO(-Al$_2$O$_3$) system, *Neues. Jb. Miner. Abh.* **161**, 1.

HAN 1928 – Hansen W.C., Brownmiller L.T., Bogue R.H. (1928) Studies on the system calcium oxide -alumina-ferric oxide, *J. Am. Chem. Soc.* **50**, 396.

HAN 1987 – Hanic F., Kamarad J., Stracelsky J., Kapralik I. (1987) The p–T diagram of Ca$_2$SiO$_4$, *Br. Ceram. Trans. J.* **86**, 194.

HAR 1984 – Hara S., Taraki A., Ogine K. (1984), *2nd Int. symp on metallurgical slags and fluxes* (H. A. Fine, D.R. Gaskell, Eds). TMS AIME Warrendale PA, pp. 441-451.

HAR 1990 – Hart L.D., Lense E. (1990) *Alumina chemicals: science and technology handbook*. Wiley, 617 page.

HAY 1937 – Hay R., White J., Caufield T.H. (1937) The ternary system FeO-Al$_2$O$_3$-SiO$_2$, *J. Soc. Glass Tech.* **21**, 270.

HAY 1940 – Hay R., White J. (1940) Slags system, *J. West Scot. Iron b Steel Inst.* **88**.

HAY 1966 – Hays J.F. (1966) Stability and properties of synthetic pyroxene CaAl$_2$SiO$_5$, *Amer. Min.* 1524.

HEN 2006 – Henao H.M., Ohno H., Itagaki K. (2006) Effect of Al$_2$O$_3$ or MgO addition on liquidus of FeOx corner in FeOxSiO$_2$-CaO slag at 1250 and 1300 °C. *Proc SOHN Int Symp at San Diego, California, TMS, Warrendale PA*, p. 731.

HEN 1984 – Heninger S.G. (1984) Hydrothermal experiments on the andalusite - sillimanite equilibrium, M.Sci. thesis, The Pennsylvania State Univ, p. 42.

HIJ 1968 – K. Hijikata (1968) Unit-cell dimensions of the clinopyroxenes along the join CaMg-Si$_2$O$_6$–CaFe^{3+}AlSiO$_6$, *J. Faculty Sci. Hokkaido Univ. Ser. 4, Geol. Mineral.* **14**, 149.

HIL 1956 – Hill P.M., Peiser H.S., Rait J.R. (1956) The crystal structure of calcium ferrite and β calcium chromite, *Acta Cryst.* **9**, 981.

HIL 1985 – Hillert M., Jansson B., Sundman B., Ågren J. (1985) A two-sublattice model for molten solutions with different tendency of ionization, *Metall. Trans. A.* **16A**, 261.

HIL 1989 – Hillert M., Sundman B., Wang X. (1989) A thermodynamic evaluation of the Al$_2$O$_3$-SiO$_2$ system, *TRITA-MAC-0402 Mater. Res. Center R. Inst. Technol. Stockholm Sweden*.

HIL 1990–1 – Hillert M., Sellesby M., Sundman B. (1990–1) An assessment of the Ca-Fe-O system, *Metall. Trans. A*, **21A**, 2759.

HIL 1990–2 – Hillers M., Sundman B., Wang X. (1990–2) An assessment of the CaO-SiO$_2$ system, *Metall. Trans. B.*, **21B**, 303.

HIL 1991 – Hillers M., Sundman B., Wang X., Barry T. (1991) A reevaluation of the rankinite phase in the CaO-SiO$_2$ system, *Calphad* **15**, 55.

HOF 1956 – Hoffmann A., Fischer W.A. (1956) Bildung des Spinells FeO·Al$_2$O$_3$ und seiner mischkristalle mit Fe$_3$O$_4$ und Al$_2$O$_3$, *Z. Physik. Chem.* **7**, 80.

HOL 1960 – Holmquist S.B. (1960) Two new complex calcium ferrite phases, *Nature.* **185**, 604.

HOL 1971 – Holdaway M.J. (1971) Stability of andalusite and the aluminium silicate phase diagram, *Am. J. Sci.* **271**, 97.

HOL 1985 – Holland T.J.B., Powell R. (1985) An internally consistent thermodynamic dataset with uncertainties and correlations: 2. Data and results, *J. Metamorph. Geol.* **3**, 343.

HOR 1967 – Horibe T., Kuwabara S. (1967) Thermo-analytical investigation of phase equilibria in the Al$_2$O$_3$-SiO$_2$ system, *Bull. Chem. Soc. Jpn.* **40**, 972.

References Part I

HOR 1978 – Horowitz H.S., Longo J.M. (1978) Phase relations in the Ca-Mn-O system, *Mater. Res. Bull.* **13**, 1359.

HOW 1978 – Howald R.A., Eliezer I. (1978) The thermodynamic properties of mullite, *J. Phys. Chem.* **82**, 2199.

HUA 1995 – Huang W., Hillert M., Wang X. (1995) Thermodynamic assessment of the CaO-MgO-SiO$_2$ system, *Metall. Mater. Trans. A.* **26A**, 2293.

HUG 1967 – Hughes H., Roos P., Goldring D.C. (1967) X-ray data on some calcium-iron-oxygen compounds, *Mineral. Mag.* **36**, 280.

HUC 1971 – Huckenholz H.G., Yoder H.S. (1971) Andradite stability relations in CaSiO$_3$-Fe$_2$O$_3$ join up to 30 kb, *Neues. Jb. Miner. Abh.* **114**, 246.

HUC 1974 – Huckenholz H.G., Lindhuber W., Springer J. (1974) The join CaSiO$_3$-Al$_2$O$_3$-Fe$_2$O$_3$ of the CaO-Al$_2$O$_3$-Fe$_2$O$_3$-SiO$_2$ quaternary system and its bearing on the formation of granditic garnets and fassaitic pyroxenes, *Neues. Jb. Miner. Abh.* **121**, 160.

HUC 1978 – Huckenholz H.G., Ott W.D. (1978) Synthesis, stability and aluminium-iron substitution in gehlenite-ferrigehlenite solid solutions, *Neues. Jb. Miner. Abh.* **12**, 521.

ILE 1979 – Iler R.K. (1979) *The Chemistry of silica: solubility, polymerization, colloid and surface properties and biochemistry of silica.* Wiley, p. 896.

ILL 1985 – Illinets A.M., Malinovskii Y., Nevskii N. (1985) Crystal structure of the rhombohedral modification of tricalcium silicate, Ca$_3$SiO$_5$, *Sov, Phys. Dokl.* **30**, 191.

IML 1971-1 – Imlach J.A., Dent Glasser L.S., Glasser F.P. (1971-1) Excess oxygen and the stability of "12CaO.7Al$_2$O$_3$", *Cem. Conc. Res.* **1**, 57.

IML 1971-2 – Imlach J.A., Glasser F.P. (1971-2) Sub-solidus phase relations in system CaO-Al$_2$O$_3$-Fe-Fe$_2$O$_3$, *Trans. j. Br. Ceram. Soc.* **70**, 227.

INO 1982 – Inoue K., Ikeda T. (1982) The solid solution state and crystal structure of calcium ferrite formed in lime-fluxed iron ores, *Testsu Hagane* **68**, 2190 (in japanese).

INS 1935 – Insley H., Ewell R.H. (1935) Thermal behavior of the kaolin minerals, *J. NBS.* **14**, 615.

IWA 1936 – Iwase K., Nisioka U. (1936) The equilibrium diagram of the ternary system CaO-Al$_2$O$_3$-SiO$_2$, *Sci. Rep. Tōhoku Imperial Univ. Ser. 1* **26**, 441.

JAS 1975 – Jasieńska St., Tomkowicz D., Dargel L. (1975) X-ray diffraction study of CaF$_3$O$_5$, *Physica Status Solidi (a)* **29**, 665.

JER 1999 – Jerebtsov D.A., Mikhailov G.G. (1999) Phase diagram of CaO-Al$_2$O$_3$ system, *Ceram. Int.* **27**, 25.

JOH 1965 – Johnson R.E., Muan A. (1965) Phase equilibria in the system CaO-MgO-Iron oxide at 1500 °C, *J. Am. Ceram. Soc.* **48**, 359.

JOH 1967 – Johnson R.E., Muan A. (1967) Activity composition relations in solid solutions of the system CaO-feO-SiO$_2$ in contact with metallic iron at 1080 °C, *Trans. TMS-AIME.* **239**, 1931.

JOS 1977 – Jost K.H., Ziemer B., Seydel R. (1977) Redetermination of the structure of β-dicalcium silicate, *Acta Cryst.* **B33**, 1696.

JOS 1999 – Joswig W., Stachel T., Harris J.W., Baur W.H., Brey G.P. (1999) New Ca-silicate inclusions in diamonds - tracers from the lower mantle, *Earth Planet. Sci. Lett.* **173**, 1.

JOS 2003 – Joswig W., Paulus E.F., Winkler B., Milman V. (2003) The crystal strucure of CaSiO$_3$-walstromite, a special isomorph of wollastonite II, *Z Krist-Cryst. Mater.* **218**, 811.

KAE 1976 – Kaestle G. (1976) Zur berechnung von phasen diagrammen oxidischer mehrstoffsystem mit hilfe von mathematische thermodynamish modellen, PhD. Thesis, Tech. University, Clausthal.

KAP 1980 – Kaprálik I., Hanic F. (1980) *Trans. Brit. Ceram. Soc.* **79**, 128.

KAR 1985 – Karpinskii O.G., Arakcheeva A.V. (1985) Crystal structure of ternary hexagonal ferrite phase Ca$_{3.565}$Fe$_{0.06}$F$_{14.25}$O$_{25}$, *Sov. Phys. Dokl.* **30**, 439.

KAR 1987 – Karpinskii O.G., Arakcheeva A.V. (1987) Crystal structure of triple hexagonal Ca-ferrite Ca$_3$Fe$_{14.82}$O$_{25}$ α CFF phase, *Sov. Phys. Crystallogr.* **32**, 28.

KAU 1979 – Kaufman L. (1979) Calculation of quasibinary and quasiternary oxide system - II, *Calphad* **3** 27.

KEN 2001 – Kendal T. (2001) Wollastonite a review, *Ind. Min.* **411**, 63.

KIN 1951 – King E.G. (1951) Heats of formation of crystalline calcium orthosilicate, tricalcium silicate and zinc orthosilicate, *J. Am. Chem. Soc.* **73**, 656.

KIR 1973 – Kirkpatrick R.J., Steele I.M. (1973) Hexagonal $CaAl_2SiO_6$: a new synthetic phase, *Amer. Mineral.* **58**, 945.

KIR 1999 – Kirschen M., DeCapitani C., Millot F., Rifflet R.C., Coutures J.-P. (1999) Immiscible silicate liquids in the system SiO_2-TiO_2-Al_2O_3, *Eur. J. Mineral.* **11**, 427.

KLU 1987 – Klug F.J., Prochazka S., Doremus R.H. (1987) Alumina-silica phase diagram in the mullite region, *J. Am. Ceram. Soc.* **70**, 750.

KOC 1988 – Hoch M. (1988) Application of the Hoch-Arpshofen model to the SiO_2-CaO-MgO-Al_2O_3 system, *Calphad* **12**, 45.

KON 1956 – Konopicky K. (1956) Remarques relatives au diagramme d'équilibre SiO_2-Al_2O_3, *Bulletin de la Société française de Minéralogie et de Cristallographie* 1.

KUB 1967 – Kubaschewski O., Evans E.L., Alcook C.B. (1967) *Metallurgical thermochemistry (Materials Science & Technology Monographs)*, 4th edn. Oxford, Pergamon Press.

KUL 1994 – Kulkarni D., Prakash C. (1994) Structural and magnetic properties of $Ca_4Fe_8O_{19}$, *Bull. Mater. Sci.* **17**, 35.

KUS 1975 – Kusachi I., Henmi C., Kawaharza A., Henmi K. (1975) The structure of rankinite, *Mineral. J.* **8**, 38.

LAN 1956 – Langenberg F.C., Chipman J. (1956) Determination at 1600 °C and 1700 °C. Liquidus lines in $CaO.2Al_2O_3$ and Al_2O_3 Stability fields of the system CaO-Al_2O_3-SiO_2, *J. Am. Ceram. Soc.* **39**, 432.

LAN 1977 – Langer K., Lattard D., Schreyer W. (1977) Synthesis and stability of deerite, $Fe^{2+}{}_{12}Fe^{3+}{}_6[Si_{12}O_{40}](OH)_{10}$, and Fe^{3+}-Al^{3+} substitutions at 15–28 kb, *Contrib. Mineral. Petrol.* **60**, 271.

LAP 1970 – Laptev D.M. (1970) Calcul du diagramme de fusibilité du système CaO-SiO_2, *Izv. Vyssh. Uchebn. Zaved. Chern. Metall.* **12**, 13.

LAR 1953 – Larson H.R., Chipman J. (1953) Oxygen activity in iron oxide slags, *Trans. Soc. Min. Metall. Explor. Inc.* **197**, 1089.

LAR 1954 – Larson H.R., Chipman J. (1954) Activities of Fe, FeO, Fe_2O_3 and CaO in simple slags, *Trans. AIME, J. Met.* **6**, 759.

LEA 1956 – Lea F.M., Desch C.H. (1956) *The chemistry of cement and concrete*, Revised edn. Edward Arnold (Publishers) Ltd, London, p. 635.

LEE 1982 – Lee Y.E. (1982) Thermodynamics of the CaO-SiO_2 system, *Calphad* 6, 283.

LI 1997 – Li L., Tang Z., Sun W., Wang P. (1997) Calculation of phase diagrams of Al_2O_3-SiO_2-R_2O_3 systems, *Phys. Chem. Glasses B.* **38**, 323.

LI 1999–1 – Li L., Tang Z., Sun W., Wang P. (1999–1) Phase diagram estimation of the Al_2O_3-SiO_2-Gd_2O_3 system, *Phys. Chem. Glasses B.* **40**, 126.

LI 1999–2 – Li L., Tang Z., Sun W., Wang P. (1999–2) Phase diagram prediction of the Al_2O_3-SiO_2-La_2O_3 system, *J. Mater. Sci. Technol.* **15**, 439.

LIN 1964 – Lindsley D.H., Davis B.T.C., Macgregor I.D. (1964) Ferrosilite ($FeSiO_3$): synthesis at high pressures and temperatures, *Science.* **144**, 73.

LIS 1967 – Lister D.H., Glasser F.P. (1967) Phase relations in the system CaO-Al_2O_3-Iron Oxide, *Trans. J. Brit. Ceram. Soc.* **66**, 294.

LIS 1981 – Lisenokov A.A. (1981) Etude de l'interaction entre CaO et Al_2O_3 en présence de SiO_2, *Fiz Khiù Stekla.* 233.

LIU 1975 – Liu L.-G. (1975) Chemistry and structure of high pressure phase of garnets rich in almandine, *Nature* **255**, 213.

LIU 1979 – Liu L.-G. (1979) High pressure phase transformations in the system $CaSiO_3$-Al_2O_3, *Earth Planet Sci. Lett.* **43**, 331.

LIU 1991 – Liu K.C., Thomas G., Caballero A., Moya J.S., de Aza S. (1991) The microstructure and microanalysis of mullite processed by reaction sintering of kaolinite alumina mixture, *Proc. Inter. Meeting Mod. Ceram. Technol.* **1**, 177.

LYK 1980 – Lykasov A.A., Kozheurova N.V. (1980) The phase diagram of the Fe-Fe_3O_4-CaO system, *Inorg. Mater.* **16**, 754.

MAC 1937 – McMurdie H.F. (1937) Studies on a portion of the system: CaO-Al_2O_3-Fe_2O_3, *J. Res. Natl. Bur. Stand. (U. S.)* **18**, 475.

MAC 1969 – MacDowell J.F., Beall G.H. (1969) Immiscibility and crystallization in Al_2O_3-SiO_2 glasses, *J. Am. Ceram. Soc.* **52**, 17.

MAH 2010 – (2010) Mohapatra M., Anand S. (2010) Synthesis and applications of nano-structured iron oxides/hydroxides - a review, *J. Eng. Sci. Technol. Rev.* **2**, 127.
MAJ 1965 – Majumdar A.J. (1965) The ferrite phase in cements, *Trans. Brit. Ceram. Soc.* **64**, 105.
MAL 1981 – Malaman B., Alebouyeh H., Jeannot F., Courtois A., Gérardin R., Evrard O. (1981) Préparation et caractérisation des ferrites de calcium $CaFe_{2+n}O_{4+n}$ à valeur fractionnaires de n (32,52) et leur incidence sur le diagramme Fe-Ca-O à 1120 °C, *Mater. Res. Bull.* **16**, 1139.
MAO 2004 – Mao H., Selleby M., Sundman B. (2004) A re-evaluation of the liquid phases in the $CaO-Al_2O_3$ and $MgO-Al_2O_3$ systems, *Calphad* **28**, 307.
MAO 2006 – Mao H., Hillert M., Selleby M., Sundman B. (2006) Thermodynamic assessment of the $CaO-Al_2O_3-SiO_2$ system, *J. Am. Ceram. Soc.* **89**, 298.
MAR 1935 – Martin F., Vogel R. (1935) The system iron, iron oxide calcium orthoferric, *Arch. Eisenhüttenwes.* **8**, 249.
MAS 1995 – Massiot D., Trumeau D., Touzo B., Farnan I., Rifflet J.-C., Douy A., Coutures J.-P. (1995) Structure and dynamics of $CaAl_2O_4$ from liquid to glass: a high temperature ^{27}Al-NMR time resolved-study, *J. Phys. Chem. A.* **99**, 16,455.
MAT 2009 – Matsuura H., Kurashige M., Naka M., Tsukihashi F. (2009) Melting and solidifying behaviors of the $CaO-SiO_2-FeO_x$ slags at various oxygen partial pressures, *ISIJ Int.* **49**, 1283.
MCH 1974 – Mchedlov-Petrossyan O.P., Babushkin V.I. (1974) Thermodynamics and thermochemistry of cement, *6th Int. Congr. Cem. Chem. Moscow*, 5.
MCI 1937 – McIntosh A.B., Rait J.R. Hay R. (1937) The binary system $FeO-Al_2O_3$, *Royal Tech. College Jour. Glasgow.*, **4**, 72.
MEY 1980 – Meyers C.E., Mason T.O., Petuskey W.T., Halloran J.W., Bowen H.K. (1980) Phase equilibria in the system Fe-Al-O, *J. Am. Ceram. Soc.* 63, 659.
MID 1952 – Midgley C.M. (1952) The crystal structure of beta dicalcium silicate, *Acta. Cryst.* **5**, 307.
MID 1971 – H.G. Midgley (1971) Indexing of powder X-ray diffraction data for dicalcium silicates, *Trans. Brit. Ceram. Soc.* **70**, 61.
MID 1974 – Midgley H.G., (1974) The polymorphism of calcium orthosilicate, *6th Int. Congr. Cem. Chem. Mosc.* **I**, 63.
MIL 1986 – Millon E., Malaman B., Bonazebi B., Brice J.F., Gerardin E., Evrard O. (1986) Structure cristalline du ferrite hemicalcique $CaFe_4O_7$/Hemicalcic ferrite $CaFe_4O_7$ crystal structure, *Mater. Res. Bull.* **21**, 985.
MON 1975 – Mondal P., Jeffery J.W. (1975) The crystal structure of tricalcium aluminate, $Ca_3Al_2O_6$, *Acta Cryst.* **B31**, 689.
MOO 1952 – Moody K.M. (1952) The space group and cell dimensions of Rankinite, *Miner. Mag.* **30**, 79.
MOR 1935 – Morris L.D., Schole S.R. (1935) Revision in the system $Al_2O_3-SiO_2$, *J. Am. Ceram. Soc.* **18**, 359.
MUA 1955 – Muan A. (1955) Phase equilibria in the system $FeO-Fe_2O_3-SiO_2$, *Trans. Am. Inst. Min. Metall. Eng.* **203**, 965.
MUA 1956 – Muan A., Osborn E.F. (1956) Phase equilibria at liquidus temperature in the system $MgO-FeO-Fe_2O_3-SiO_2$, *J. Am. Ceram. Soc.* **39**, 121.
MUA 1957-1 – Muan A. (1957-1) Phase equilibrium relationships at liquidus temperatures in the system $FeO-Fe_2O_3-Al_2O_3-SiO_2$, *J. Am. Ceram. Soc.* **40**, 420.
MUA 1958 – Muan A. (1958) On the stability of the phase $Fe_2O_3.Al_2O_3$, *Am. J. Sci.* **256**, 413.
MUL 1986 – Müller D., Gessner W., Samoson A., Lippmaa E., Scheler G. (1986) Solid-state ^{27}Al NMR studies on polycrystalline aluminates of the system $CaO-Al_2O_3$, *Polyhedron* **5**, 779.
MUL 2003 – Müller D., Neubauer J., Götz-Neunhöffer F. (2003) Effect of phosphate rich raw materials on the crystallography and hydraulic properties of C_2S, *Proc. 11th ICCC Durban S. Afr.*1045.
MUM 1988 – Mumme W.G. (1988) A note on the relationship of $Ca_{2.3}Mg_{0.8}Al_{1.5}Si_{1.1}Fe_{8.3}O_{20}$ (SFCA) with aenigmatite group minerals and sapphirine, *Neues. Jb. Miner. Abh.* **8**, 359.

MUM 1995 – Mumme W.G., Hill R.J., Bushnell-Wye G., Segnit E.R. (1995) Rietveld crystal structure refinements, crystal chemistry and calculated powder diffraction data for the polymorphs of dicalcium silicate and related phases, *Neues. Jb. Miner. Abh.* **169**, 35.
MUM 2003 – Mumme W.G. (2003) The crystal structure of SFCA-II $Ca_{5.1}Al_{9.3}Fe^{3+}_{18.7}Fe^{2+}_{0.9}O_{48}$ a new homologue of the aenigmatite structure-type, and structure refinement of SFCA-type, $Ca_2Al_5Fe_7O_{20}$. Implications for the nature of the "ternary-phase solid-solution" previously reported in the $CaO-Al_2O_3$-iron oxide system, *Neues. Jb. Miner. Abh.* **178**, 307.
NAV 1982 – Navrotsky A., Peraudeau G., McMillan P., Coutures J.-P. (1982) A thermochemical study of glasses and crystals along the joins silica-calcium aluminate and silica-sodium aluminate, *Geochim. Cosmochim.* **46**, 2039.
NEW 1958 – Newkirk T.F., Thwaite R.D. (1958) Pseudoternary system calcium oxide monocalcium aluminate $(CaO-Al_2O_3)$ - dicalcium ferrite, *J. Res. Natl. Bur. Stand.* **61**, 233.
NEW 1966–1 – Newton R.C. (1966–1) Kyanite-andalusite equilibrium from 700 °C to 800 °C, *Science* **153**, 170.
NEW 1966–2 – R.C. Newton (1966–2) Kyanite-sillimanite equilibrium at 750 °C, *Science* **151**, 1222.
NIE 1972 – Niesel K. (1972) The importance of the alpha'L alpha'-H transition in the polymorphism of dicalcium silicate, *Silic. Ind.* **37**, 136.
NIK 2008 – Nikolic S., Hayes P., Jak E. (2008) Phase equilibria in ferrous calcium silicate slags: Part III. Copper-saturated slag at 1250 °C and 1300 °C at an oxygen partial pressure of 10–6 atm, *Metall. Materi. Trans. B.* **39**, 200.
NIS 1985 – Nishi F., Takéuchi Y., Maki I. (1985) Tricalcium silicate, $Ca_3O(SiO_4)$: the monoclinic superstructure, *Z. Kristallogr.* **172**, 297.
NIT 1983 – Nityanand N., Fine H.A. (1983) The effect of TiO_2 additions and oxygen potential on liquidus temperatures of some $CaO-Al_2O_3$ melts, *Metall. Mater. Trans. B.* **14B**, 685.
NOI 2003 – de Noirfontaine M.-N., Dunstetter F., Courtial M., Gasecki G., Signes-Frehel M. (2003) Tricalcium Silicate Ca_3SiO_5, the major component of anhydrous Portland Cement: on the conservation of distances and directions and their relationship to the structural elements, *Z. Kristallogr.* 218, 8.
NOV 1965 – Novokhatskii I.A., Belov B.F., Gorokh A.V., Savinskaya A.A. (1965) The phase diagram for the system ferrous oxide-alumina, *Russ. J. Phys. Chem.* **39**, 1498
NOV 1971 – Novak G.A., Gibbs G.V. (1971) The crystal chemistry of the silicate garnets, *Am. Min.* **56**, 791.
NOW 1951 – Nowotny H., Funk R. (1951) Ein Beitrag zum System: $Al_2O_3-Fe_2O_3-SiO_2$, *Radex Rundschau* **8**, 334.
NUR 1965 – Nurse R.W., Welch J.H., Majumdar A.J. (1965) The $12CaO.7Al_2O_3$ in the $CaO-Al_2O_3$ System, *Trans. Brit. Ceram. Soc.* **64**, 323.
OBS 1968 – Obst K.-H., Horn H.C., Stradtmann J. (1968) A Study on the solid-state phase equilibria in the system $CaO-FeO_n$ as an example of the application of the electron beam microanalyser, *Mikrochim. Acta (Wien), Suppl. Ill.* 147 (in German).
OBS 1969 – Obst K.-H., Stradtmann J. (1969) Das system kalziumoxyd—eisen (II)-oxyd als grundlage für untersuchungen zur kalkauflösung bei der stahlerzeugung, *Arch. Eisenhütenwesen* **40**, 615.
OBS 1970 – Obst K.-H., Fix W. (1970) *Arch. Eisenhütenwesen.* **41**, 703.
OEL 1955 – Oelsen W., Heynert G. (1955) Die reaktionen zwischen eisen-mangan -schmelzen und den schmelzen ihrer aluminate, *Arch. Eisenhütenwesen.* **26**, 567.
OKA 1991 – K. Okada, N. Otsuka and S. Somiya (1991) Review of mullite synthesis routes in Japan, *Am. Ceram. Soc. Bull.* **70**, 1633.
OLS 1951 – Ol'shanskii Y.I. (1951) Equilibrium of two immiscible liquids in the silicate systems of the alkaline earth metals, *Dokl. Akad. Nauk SSSR.* **76**, 93 (in Russian).
OLS 1948 – Oelsen H., Maetez H. (1948) Contribution to the metallurgy of the thomas process, *Arch. Eisenhütenwesen.* **19**, 111.
ONO 1970 – Onoda G.Y., Brown S.D. (1970) Low-silica glasses based on calcium aluminates, *J. Am. Ceram. Soc.* **53**, 311.

References Part I

OSB 1941 – Osborn E.F., Schairer J.F. (1941) The ternary system pseudowollastonite- åkermanite-gehlenite, *Am. J. Sci.* **239**, 715.
PAL 1988 – Palomo A., Blanco-Valera M.T. Vázquez T. (1988) Microstructures found in the solidification of compositions of the system $CaO-Al_2O_3-2Fe_2O_3$ and its variations by the presence of CaF_2, *Silic. Ind. (Bélgica)* V.IH, 21.
PAR 1952 – Parker T.W. (1952) The constitution of aluminous cement, *Proc. 3rd Int. Congr. Chem. Cem. London* 485.
PAT 2002 – Patrick T.R., Pownceby M.I. (2002) Stability of silico-ferrite of calcium and aluminum (SFCA) in air–solid solution limits between 1240 °C and 1390 °C and phase relationships within the $Fe_2O_3-CaO-Al_2O_3-SiO_2$ (FCAS) system, *Metall. Mater. Trans. B* **33**, 79.
PEL 1986 – Pelton A.D., Blander M. (1986) Thermodynamic analysis of ordered liquid solutions by a modified quasichemical approach—Application to silicate slags, *Metall. Mater. Trans. B* **17**, 805.
PHI 1958 – Phillips B., Muan A. (1958) Phase equilibria in the system CaO-iron oxide in air at 1 atm. Pressure, *J. Am. Ceram. Soc.* **41**, 445.
PHI 1960 – Phillips B., Muan A. (1960) Stability relations of calcium ferrites: phase equilibria in the system $2CaO \cdot Fe_2O_3\text{-}FeO \cdot Fe_2O_3\text{-}Fe_2O_3$ above 1135 °C, *Trans. Metall. Soc. AIME.* **218**, 1112.
POU 1977 – Pouchard M., Grenier J.-C. (1977) Un nouveau type de non stoechiometrie, ordonnancement des défauts d'oxygène au sein de perovskite lacunaire AMO_3-x, *C.R. Acad. Sci, Paris, série C*, **284**, 311.
POW 1998 – Pownceby M.I., Clout J.M., Fisher-White M. (1998) Phase equilibria for the Fe_2O_3 rich part of the system $Fe_2O_3\text{-}CaO\text{-}SiO_2$ in air at 1240 °C-1300 °C, *Instn. Min. Metall. (Sect. C: Min. Process. Extr. Metall.)*, **107**, C1.
PYT 1999 – Pytel K.A., Ciembronowicz E.M. (1999) Determination of the liquidus curves for high-melting oxide binaries of Al_2O_3, CaO and SiO_2 by means of the thermochemical data using the regular and subregular solution model, *Metall. Foundry Eng.* **25**, 93.
QIT 1988 – Qiti G., Wang S. (1988) The stability of laihunite-a thermodynamic re-analysis, *Sci. China. Ser. B Chem.* **31**, 1515.
RAA 1930 – Raaz F. (1930) Über den feinbau des gehlenit ein beitrag zur kenntnis der melilithe, *Sitzungsber Kaiserl Akad Wiss Math.-Naturwiss.* **139**, 645.
RAN 1915 – Rankin G.A., Wright F.E. (1915) The ternary system $CaO-Al_2O_3-SiO_2$, *Am. J. Sci. 4th Ser.* 1.
RAS 1982 – W.A. Rasim (1982) Entwicklung eines model zur beschreibung der in $FeO-Fe_2O_3$-$CaO-SiO_2$ Schlacken experimentall bestimmten Einfluss der komponenten auf das Fe2O3/FeO Verhältnis bei vorgegebenen Sauerstoffdrücken, PhD thesis, Clausthal.
RED 1948 – Redlich O., Kister A.T. (1948) Algebraic representation of thermodynamic properties and the classification of solutions, *Ind. Eng. Chem.* **40**, 345.
REE 1966 – Reeve D.A. (1966) Studies in the system $CaO-FeO-Fe_2O_3$, PhD thesis Birmingham (England).
REE 1967-1 – Reeve D.A., Gregory A.G. (1967-1) The system $CaO-FeO-Fe_2O_3$ at liquidus temperature, *Trans. Ins. Min. Metall. Sect. C: Min. Process. Extr. Metall.* **76C**, 268.
REE 1967-2 – Reeve D.A., Gregory A.G. (1967-2) Modifications of the oxygen potential diagram for the system Fe-Ca-O, *Trans. Ins. Min. Metall. Sect. C: Min. Process. Extr. Metall.* **76C**, 273.
REG 1979 – Regourd M. (1979) Polymorphisme du silicate tricalcique. Nouvelles données de la diffraction des rayons X, *C.R. Acad. Sc. Paris t.* **289**, 17.
RIB 1980 – Riboud P.V., Steiler J.-M. (1980) Données thermochimiques nécessaires aux bilans thermiques, *Techniques de l'ingénieur, Form M*, 1726.
RIC 1949 – Richardson H.M., Rigby G.R. (1949) The occurrence of iron-cordierite in blast-furnace linings, *Min. Mag.* **28**, 547.
RIC 1954 – Richards R.G., White J. (1954) Phase relationships of iron-oxide-containing spinels. Part I. relationships in the system Fe-Al-O, *Trans. J. Br. Ceram. Soc.* **53**, 233.
RIC 1967 – Richardson S.W., Bell P.M., Gilbert M.C. (1967) Kyanite sillimanite relations, *Carnegie Inst. Washington, Year book* **65**, 247.
RIC 1968 – Richardson S.W., Bell P.M., Gilbert M.C. (1968) Kyanite-sillimanite equilibrium between 700° and 1500 °C, *Am. J. Sci.* **266**, 513.

RIS 1977 – Risbud S.H., Pask J.A. (1977) Calculated thermodynamic data and metastable immiscibility in the system SiO_2-Al_2O_3, *J. Am. Ceram. Soc.*, **60**, 418.

RIS 1978 – Risbud S.H. Pask J.A. (1978) SiO_2-Al_2O_3 metastable phase equilibrium diagram without mullite, *J. Mater. Sci.* **13**, 2449.

ROB 1984 – Robie R.A., Hemingway B.S. (1984) Entropies of kyanite, andalusite and sillimanite: additional constraints on the pressure and temperature of the Al_2SiO_5 triple point, *Am. Mineral.* **69**, 298.

ROI 1964 – Roiter B.D. (1964) Phase equilibria in the spinel region of the system FeO-Fe_2O_3-Al_2O_3, *J. Am. Ceram. Soc.* **47**, 509.

ROL 1965 – Rollin M., Pham H.T. (1965) Phase diagrams of mixtures not reacting with molybdenum [$Ca(AlO_2)_2$–, $NaAlO_2$–and La_2O_3–Al_2O_3], *Revue Internationale des Hautes Températures et des. Refractaires* **2**, 175.

ROS 1974 – K. Rosenbach J. A. Schmitz (1974) Untersuchungen im dreistoffsystem eisen II -oxid - chrom III – oxid - tonerde, *Arch Eisenhüttenwesen.* **45**, 843.

ROY 1958 – Roy D.M. (1958) Studies in the system CaO-Al_2O_3-SiO_2-H_2O: III, new data on the polymorphism of C_2S and its stability in the system CaO-SiO_2-H2O, *J. Am. Ceram. Soc.* **41**, 293.

RUM 1981 – Rumyantsev P.F., Sakharov L.G. (1981) The crystallization of glass forming melts in system anorthite - gehlenite, *Phys. chem. glass (rus.)* **7**, 159.

SAA 1967 – Saalfeld H. (1967) Beitrag zur Kristallchemie des Dikalziumsilikates, Ca_2SiO_4, *Ber. Dtsch. Keram. Ges.* **44**, 421.

SAL 1982 – Salje E., Werneke C. (1982) The phase equilibrium between sillimanite and andalusite as determined from lattice vibrations, *Contrib. Mineral. Petrol.* **79**, 56.

SAL 1994 – Salem M.R., Elshereafy E., Abou Sekkina M.M. (1994) Thermophysical studies on the substitution of Al^{3+} in gallium iron garnets, *J. Mater. Sci.* **29**, 2939.

SCH 1937 – Schenk V.R., Layman A., Jenkel E. (1937) Equilibrium studies on oxidation reduction and carburization of iron XII, *Z. Anorg. Allg. Chem.* **235**, 65.

SCH 1947 – Schairer J.F., Bowen N.L. (1947) The system anorthite-leucite- Silica, *Bull. Geol. Soc. Finl.* **20**, 67.

SCH 1952 – Schairer J.F., Yagi K. (1952) The system FeO-Al_2O_3-SiO_2, *Am. J. Sci. Bowen.*, **42**, 471.

SCH 1953 – Schuhmann R., Powell R.G., Michael E.J. (1953) Constitution of the FeO-Fe_2O_3-SiO_2 system at slagmaking temperatures, *J. Metals (Trans. Amer. Inst. Min. Engrs)* **197**, 1097.

SCH 1973 – Schürmann E., Wurm P. (1973) Phase diagram and reduction equilibria of the ternary system Fe-Fe_2O_3-CaO between 550 °C and 1070 °C, *Arch Eisenhuttenwes* **44**, 637.

SCH 1974 – Scheel R. (1974) Investigation into the system lime iron oxide oxygen, *Arch Eisenhuttenwes.* **45**, 751.

SCH 1976–1 – Schürmann E. Kraume G. (1976–1) Melting equilibria in the system FeO-Fe_2O_3-CaO, *Arch. Eisenhüttenw.* **47**, 471.

SCH 1976–2 – Schürmann E., Kraume G. (1976–2) Phasengleichgewichte des schnittes FeOn-CaO im dreistoffsystem FeO-Fe_2O_3-CaO bei Eisensäzttigung, *Arch. Eisenhüttenw.* **47**, 327.

SCH 1976–3 – Schürmann E., Kraume G. (1976–3) Reduktiongleichgewichte im system FeO-Fe_2O_3-CaO oberhalb 1050 °C, *Arch. Eisenhüttenw.* **47**, 471.

SCH 1983 – Schmid H., de Jonghe L.C. (1983) Structure and non-stoichiometry of calcium aluminates, *Philos. Mag. A* **48**, 287.

SCH 1989 – Shulters J.C., Bohlen S.R. (1989) The stability of hercynite and hercynite-gahnite spinels in corundum- or quartz-bearing assemblages, *J. Petrol.* **30**, 1017.

SEL 1996 – Selleby M., Sundman B. (1996) A reassesment of the Ca-Fe-O system, *Calphad* **20**, 381.

SEL 1997 – Selleby M. (1997) An assessment of the Fe-O-Si system, *Metall. Mater. Trans. B.* **28**, 563.

SHA 1983 – Sharma S.K., Simon B., Yoder H.S. (1983) Raman study of anorthite, calcium Tschermak's pyroxene, and gehlenite in crysralline and glassy states, *Am. Mineral.* **68**, 1113.

SHE 1909 – Shepherd E.S. Rankin G.A. Wright F.E. (1909) The binary systems of alumina with silica, lime and magnesia, *Am. J. Sci.* **28**, 293.

SHE 2001 – Shen C.H., Lin R.S., Huang C.Y. (2001) Phase stability study of $La_{1.2}Ca_{1.8}Mn_2O_7$, *Mater. Res. Bull.* **36**, 1139.

SMI 1965 – Smith D.K., Majumdar A.J., Ordway F. (1965) The crystal structure of γ-dicalcium silicate, *Acta Crystallogr. B.* **18**, 787.

SMU 1969 – Smuts J., Steyn J.D., Boeyens J.C.A. (1969) The crystal structure of an iron silicate, iscorite, *Acta Crystallogr. B* **25**, 1251.

SNO 1942 – Snow R.B. McCaughey W.J. (1942) Equilibrium studies in the system $FeO\text{-}Al_2O_3\text{-}SiO_2$, *J. Am. Ceram. Soc.* **25**, 151.

SOR 1973 – Sorrentino F.P. (1973) Studies in the system $CaO\text{-} Al_2O_3\text{-}SiO_2\text{-}Fe\text{-}O_2$, PhD. thesis, University of Aberdeen.

SOR 2008 – Sorrentino F.P. (2008) Thermodynamic and modelling of the system Ca, Al, Si, Fe, O in the part relevant to high alumina cement, calcium aluminate cements: proceedings of the centenary conference, Avignon, IHS BRE Press, 2008, pp. 17-30.

SOS 1916 – Sosman R.B., Merwin H.E. (1916) Preliminary report on the system, lime: ferric oxide, *J. Wash. Acad. Sci.* **6**, 532.

SOU 1991 –Sourie A. (1991) *Mineralogy of aluminous cements and their reactivity*, PhD. thesis, University of Aberdeen, p. 159.

SOU 1991–2 – Sourie A., Glasser F.P. (1991–2) Studies on the mineralogy of high alumina cement clinkers, *Trans. J. Br. Ceram. Soc.* **90**, 71.

STA 1968 – Staronka A., Pham H., Rolin M. (1968) Etude du système silice alumine par la méthode des courbes de refroidissement, *Rev. Int. Hautes Temp. Réfract.* **5**, 111.

STE 1989 – Steele I.M., Pluth J.J. (1989) Crystal structure of $Ca_{5.3}Al_{10.7}Si_{5.3}O_{32}$: a stuffed tridymite structure. Geological Society of America, abstracts with program A45 (Vanman).

STE 1999 – Stebbins J.F., Poe B. (1999) Pentacoordinate silicon in high-pressure crystalline and glassy phases of calcium disilicate ($CaSi_2O_5$), *Geophys. Res. Lett.* **26**, 2521.

STO 1989 – Stößer R., Nofz M., Geßner W., Schröter Ch., Kranz G. (1989) Paramagnetic monitors (Mn^{2+}, Mn^{4+}, Fe^{3+} and O^{2-}) in the solid-state reaction yielding $12CaO.7Al_2O_3$ and other aluminates, *J. Solid State Chem.* **81**, 152.

STO 1995 – Stolyarova V.L., Shornikov S.I., Shultz M.M. (1995) High-temperature mass spectrometric study of the thermodynamic properties of the $CaO\text{-}Al_2O_3$ system, *Rapid Commun. Mass Spectrom.* **9**, 686.

SUZ 1968 – Suzuki K., Yamaguchi G. (1968) A structural study on α'-C_2S, *Proc. 5th Int. Congr. Chem. Cem. Tokyo* **1**, 67.

SVE 1982 – Svetic S., Jesenák K., Hrabě Z. (1982) The effect of atmosphere on the rate and mechanism of reactions in the system $CaCO_3\text{-}SiO_2$ at low temperature, *Silikaty* **26**, 289.

SWA 1946–1 – Swayze M.A. (1946–1) A report on studies of - 1. The ternary system $CaO\text{-}C_5A_3\text{-}C_2F$, *Am. J. Sci.* **244**, 1.

SWA 1946–2 – Swayze M.A. (1946–2) A report on studies of – 1. The ternary system $CaO\text{-}C_5A_3\text{-}C_2F_2$) the quaternary system $CaO\text{-}C_5A_3\text{-}C_2F\text{-}C_3S$, *Am. J. Sci.* **244**, 65.

TAK 1959 – Takeuchi Y., Donnay G. (1959) The crystal structure of hexagonal $CaAl_2Si_2O_8$, *Acta Cryst.* **12** 465.

TAK 1978 – Takeda Y., Naka S., Takano M., Shinjo T., Takada T., Shimada M. (1978) Preparation and characterization of stoichiometric $CaFeO_3$, *Mater. Res. Bull.* **13**, 61.

TAK 1980 – Takeda Y., Nakazawa S., Yazawa Y. (1980) Thermodynamics of calcium ferrite slags at 1200° and 1300 °C, *Can. Metall. Q.* **19**, 297.

TAO 2008 – Tao D. (2008) Prediction of component activities in the molten aluminosilicate slag $CaO\text{-}Al_2O_3\text{-}SiO_2$ by molecular interaction volume model, *J. Mater. Sci. Technol.* **24**, 787.

TAV 1936 – B. Tavasci (1936) Ricerche sul sistema $CaO\text{-}Fe_2O_3$, *Ann. Chim. Applicata.* **26**, 291.

TAY 1971 – Taylor H.F.W. (1971) The crystal structure of kilchoanite, $Ca_6(SiO_4)(Si_3O_{10})$, with some comments on related phases, *Mineral. Mag.* **38**, 26.

TAY 1993 – Taylor J.C., Aldridge L.P. (1993) Full-profile Rietveld quantitative XRD analysis of Portland cement: Standard XRD profiles for the major phase tricalcium silicate (C_3S: $3CaO.SiO_2$), *Powder Diffr.* **8**, 138.

TEO 1985 – Teoreanu I., Andronescu E., Cacoveanu L. (1985) Thermal phase equilibrium relationships in $2CaO.SiO_2\text{-}CaO.Al_2O_3$ system, *Rev. Roumain Chem.* **30**, 647.

TEW 1979 – Tewhey J.D., Hess P.C. (1979) The two phase region in the $CaO\text{-}SiO_2$ system: experimental data and thermodynamic analysis, *Phys. Chem. Glasses* **20**, 41.

TIM 1970 – Timucin M., Morris A.E (1970) Phase equilibria and thermodynamic properties of lime-iron oxides melts containing up to 30 percent SiO_2 at 1450 °C and, 1550 °C sections, Spec techn publ No 472, *Am. Soc. Test, Matl.* **25**.
TOR 1951 – Toropov N.A., Galakhov F.Y. (1951) New data on the system Al_2O_3-SiO_2, *Dokl. Akad. Nauk SSSR.* **78**, 299.
TOR 1953 – Toropov N.A., Galakhov F.Y. (1953) The mullite problem, *Voprosy Petrog Miner. Akad. Nauk SSSR.* **2**, 245.
TOR 1955 – Toropov N.A., Boikova A.I. (1955) Solid solutions of calcium aluminoferrites, *Russ. Chem. Bull.* **4**, 887.
TRO 1949 – Trömel G. (1949) Modifications of calcium orthosilicate Ca_2SiO_4, *Naturwissenschaften* **36**, 88.
TRO 1957 – Trömel G., Obst K.H., Konopiszky K., Bauer H., Patzak J. (1957) Untersuchen im system SiO_2:Al_2O_3, *Ber. Dtsch. Keram. Ges.* **34**, 397.
TRO 1966 – Trömel G., Obst K.H., Görl E., Stradman J. (1966) Das system Eisen (II) oxid calcium oxid- magnesium oxid als Grundlage für die Bestimmer der Verschlackung von Dolomit, *Tonind.-Ztg.* **90**, 209.
TUR 1961 – Turkdogan E.T. (1961) Oxygen potentials and phase equilibria in the Fe-Ca-O system, *Trans. Am. Inst. Min. Metall. Eng.* **221**, 546.
TUR 1962 – Turnock A.C., Eugster H.P. (1962) Fe-Al oxides: phase relationships below 1000 °C, *J. Petrol.* **3**, 533.
UBE 1990 – Uberoi M., Risbud S.H. (1990) Processing of amorphous calcium aluminate powders at < 900 °C, *J. Am. Ceram. Soc.*, **73**, 1768.
UDA 1969 – Oudalov Y.P., Medvedeva Z.S. (1969) Croissance de monocristaux d'aluminate de calcium dans le systeme CaO-Al_2O_3, *Mater. Res. Bull.*, **4**, 887.
VAN 1991 – van Hoek J.A.M., van Loo F.J.J., Metselaar R., de Haan J.W., van den Berg J.A. (1991) Formation of $CaO.10Al_2O_3$ with the β-alumina-type structure by means of ion exchange exchange, *Solid State Ion.* **45**, 93.
VID 1984 – Vidyasagar K., Gopalakrishnan J., Rao C.N.R. (1984) A convenient route for the synthesis of complex metal oxides employing solid-solution precursors, *Inorg. Chem.* **23**, 1206.
VOG 1933 – Vogel R., Martin E. (1933) The system ferrous oxide ferroferric oxide, *Arch. Eisenhüttenw.* **6**, 109.
WAN 1989 – Wang X., Hillert M., Sundman B. (1989) A thermodynamic evaluation of the Al_2O_3-CaO-SiO_2 system, Report, TRITA-MAC-0407, *Royal Inst. Techn., Stockholm*.
WEI 1966 – Weill D.F. (1966) Stability relations in the Al_2O_3-SiO_2 system calculated from solubilities in the Al_2O_3-SiO_2-Na_3AlF_6 system, *Geochim. Cosmochim. Acta.* **30**, 223.
WEI 1981 – Weisweiler W., Serry M.A. (1981) Microanalysis of reactions in the system SiO_2-Al_2O_3 at subsolidus temperature, *Ber. Dtsch. Keram. Ges.* **58**, 405.
WEL 1959 – Welch J.H., Gutt W. (1959) Tricalcium silicate and its stability within the system CaO-SiO_2, *J. Am. Ceram. Soc.* **42**, 15.
WEL 1960 – J.H. Welch (1960) A new interpretation of the mullite problem, *Nature.* **186**, 545.
WHE 1979 – Wheat T.A., Sallam E.M.H., Chaklader A.C.D. (1979) Synthesis of mullite by a freeze-dry process, *Ceramurgia Int.*, **5**, 44.
WIL 1967 – Willshee J.C., White J. (1967) An investigation of equilibrium relationships in the system MgO-FeO-Fe_2O_3 up to 1750 °C in air, *Trans. Br. Ceram. Soc.* **66**, 541.
WIL 1968 – Willshee J.C., White J. (1968) Equilibrium relationship in the systems FeO-Fe_2O_3-Al_2O_3 and FeO-Fe_2O_3-Cr_2O_3 up to 1750 °C in air, *Trans. Br. Ceram. Soc.* **67**, 271.
WIS 1955 – Wisnyi L.G. (1955) The high alumina phases in the system lime-alumina, Ph.D. Thesis, Rutgers University, New Brunswick, NJ, USA.
WOE 1991 – Woerman E. (1991) The application of mineralogy to high temperature metallurgy and materials processing, *Int. Congr. Appl. Mineral. (ICAM'91), Pretoria* **2**.
WU 1993 – Wu P., Eriksson G., Pelton A.D., Blander M. (1993) Prediction of the thermodynamic properties and phase diagrams of silicate systems - evaluation of the FeO-MgO-SiO_2 System, *ISIJ Inter.* **33**, 26.
YAM 1957 – Yamaguchi G., Miyabe H., Amano K., Komatsu S. (1957) Synthesis of each modification of C_2S and their certification, *J. Ceram. Soc. Jpn.* **65**, 99.
YAM 1963 – Yamaguchi G., Ono Y., Kawamura S., Soda Y. (1963) Differential thermal analysis and high temperature powder X-Ray diffraction of $2CaO.SiO_2$, *J. Ceram. Assoc. Jpn.* **71**, 9.

YAM 1966 – Yamaguchi G., Ono Y. (1966) Microscopic study of alite in Portland cement clinker, *Zement Kalk Gips.* **19**, 390.

YOD 1955 – Yoder H.S. (1955) Almandine garnet stability field, *Carnegie Inst. Washington Yearbook*, **54**, 97.

YOS 1970 – Yoshioka T. (1970) A new crystal with kalsilite-type structure on the $CaAl_2O_4$-SiO_2 join, *Bull. Chem. Soc. Jpn.* **43**, 2317.

ZAI 1990–1 – Zaitsev A.I., Korolyov N.V., Mogutnov B.M. (1990–1) Thermodynamic properties of $(\chi CaF_2 + \gamma Al_2O_3 + (1 - \chi - \gamma)CaO)(l)$ I. Experimental investigation, *J. Chem. Thermodyn.* **22**, 513.

ZAI 1990–2 – Zaitsev A.I., Korolyov N.V., Mogutnov B.M. (1990–2) Thermodynamic properties of $(\chi CaF_2 + \gamma Al_2O_3 + (1 - \chi - \gamma)CaO)(l)$ II. Model description, *J. Chem. Thermodyn.* **22**, 531.

ZAI 1991 – Zaitsev A.I., Korolyov N.V., Mogutnov B.M. (1991) Phase equilibria in the CaF_2-Al_2O_3-CaO system, *J. Mater. Sci.* **26**, 1588 (in Russian).

ZAI 1995 – Zaitsev A.I., Litvina A.D., Mogutnov B.M., Tsaplin A.A. (1995) Thermodynamic properties and phase equilibria in the system CaO-SiO_2-Al_2O_3, *High Temp. Mater. Sci.* **34**, 223.

ZAI 1997 – Zaitsev A.I., Mogutnov B.M. (1997) Thermodynamics of CaO-SiO_2 and MnO-SiO_2 melts, *Inorg. Mater.* **33**, 823.

Part II

Applications of the CaO–SiO$_2$–Al$_2$O$_3$–Fe Oxides Chemical System

Chapter 6

Applications to Hydraulic Binders

6.1 General Introduction

- A total of CaO, SiO_2, Al_2O_3, and iron oxides accounts for more than 95% of the typical bulk composition of the main hydraulic binders.
- CaO, SiO_2, Al_2O_3, and iron oxide form combinations that have hydraulic properties, that is; when they are finely ground, they react with water to produce insoluble compounds, giving rise to engineering and durability properties.
- The compounds within this system with such hydraulic properties are C_3S, C_2S, C_3A, $C_{12}A_7$, CA, and glass with specific chemical compositions, plus to a lesser extent, C_4AF, C_2AS, and CA_2.
- The use of these compounds and their mixtures (also commonly called cementitious materials) has led to the development of industrial binder for the construction, civil engineering, building chemistry, and refractory applications. They are principally, Portland cements (PC) and calcium aluminate cements (CAC).
- Calcium sulfoaluminate cements (CSA) are not considered in this book as Ye'elenite ($C_4A_3\$$) requires a source of sulphur to be synthesized at high temperatures leading to the study of 'the system' to which sulphur would be added. On the other hand, the book reports cases when a source of sulphur is directly used to produce a hydraulic binder. For example, PC that is produced by cogrinding PC clinker with gypsum.
- These hydraulic materials are commonly used as:
  - Mixtures with sand and water to form mortars used to bind together bricks and stones employed for the construction of walls.
  - Mixture with aggregates, sands and water to form concretes used for structural civil engineering applications (PC) or more complex formulations with admixtures to serve many specific applications, such as in building chemistry. Examples of these include self-levelling screeds and ceramic tile adhesives and these are often based on CAC.

- The properties of the binders can be modified, by use of admixtures or in blends to satisfy specific requirement, such as; fast setting and hardening, low heat release (blended cements), dimensional control (expansive cement), hot and pressure service life (oil well cement). These are classified as 'special cements'.
- Glass cements, based on CaO, SiO_2, Al_2O_3, and iron oxide are still at the research level.
- The manufacturing objective of the cement industry is to maximise the performance of the mineral phases they produce whilst minimising the energy needed for production. Performance can be defined by a set of properties, such as; mechanical strength, durability, and resistance to aggressive environment. For this purpose, it has been recognized that understanding of the formation of the mineral phases was essential.

6.2 Portland Cements (PC)

BEN 2002, HEW 1998, LEA 1956, TAY 1990, YOU 2005

6.2.1 *Characteristics*

6.2.1.1 Introduction

- Industrial clinker is generated by burning clay and limestone at around 1450 °C. The clinker is hard lumps or pellets that are co-ground with calcium sulphate to produce the more milled Portland cements (other, reactive ingredients can also be incorporated for some categories of Portland cement types as defined by CEN – EN 197-1; see sections 1.1.3.2, 11.3.1.2, 11.4.1.2, and 11.4.2.2). Thus, PC type CEM I (CEN – EN 197-1) is considered thereafter.
- The total of CaO, SiO_2, Al_2O_3, and Fe_2O_3 accounts for >95% of the typical bulk composition of Portland cement clinker.
- Depending on the proportion of iron and alumina, different types of clinker are manufactured: high C_3S Portland cement, Portland cement with elevated C_2S content, high or low C_3A Portland cement, white Portland cement, and high iron Portland cement. A chemical composition of a typical Portland cement clinker is given in table 6.1.
- The use of indices such as silica ratio, $SR = SiO_2/(Al_2O_3 + Fe_2O_3)$, alumina modulus, $AM = Al_2O_3/Fe_2O_3$ and lime saturation factor, $LSF = CaO/2.8 \times SiO_2 + 1.18 \times Al_2O_3 + 0.35 \times Fe_2O_3$ are a common industrial practice to design the raw material of OPC clinker with the following average values; $SR = 2.6$, $AM = 1.5$, and LSF 96%.
- The relationship of mineralogy, chemical composition, and indices is given in table 6.2.
- After burning, the clinker is cooled down rapidly and ground with calcium sulphate at the fineness required for the future application.

TAB. 6.1 – Chemical analysis of typical PC clinker (weight %).

CaO	SiO$_2$	Al$_2$O$_3$	Fe$_2$O$_3$	MgO	K$_2$O	Na$_2$O	TiO$_2$	SO$_3$
62–68	20–24	3–6	1–5	0.5–4.5	0.2–1.5	0–0.5	0.1–0.6	0–1.5

- The optimum quantity of SO$_3$ depends on several parameters such as the fineness (specific surface (SS) cm^2/g), the percentage of alkalis expressed as soluble Na$_2$O and the percentage of C$_3$A. The following formula gives good results for the optimum quantity of SO$_3$: SO$_3$ = 6.8 × 10^{-5} × SS × %C$_3$A or SO$_3$ = 1.2 (soluble Na$_2$O) + 0.2% (Al$_2$O$_3$) + 6.2 × 10^{-3} – 0.7 (weight %).

6.2.1.2 Identification of Clinker Phases

- Four phases are produced during burning:
    - A solid solution of C$_3$S called alite.
    - A solid solution of C$_2$S called belite.
    - A solid solution of C$_3$A.
    - A calcium aluminoferrite Ca$_4$(Fe$_{4-x}$Al$_x$)O$_{10}$

    (with $x = 2$, the composition is close to C$_4$AF).

- Figure 6.1 shows a polished section of a clinker observed by optical microscopy. Alite appears as small angular blue crystal. Belite occurs as a round, brown multistriated crystal. The interstitial matrix is formed by C$_3$A well differentiated from C$_4$AF (bright colour).
- In industrial Portland cement clinker, alite occurs as monoclinic M1 or M3 and contains about 4% minor elements. The presence of MgO favours the stabilization of M3 while the presence of SO$_3$ promotes the growth process of alite in favour of M3 at ambient temperature.
- Typical compositions of the four phases found in industrial clinker of PC are given by table 6.3.
- These solid solutions contain other elements than Ca, Si, Al, Fe, and O, not only due to impurities in the normal natural raw material, but also the growing replacement of these by secondary by-products and alternative combustible.
- In industrial clinker the α-polymorph of C$_2$S (belite) develops different types of crystal morphologies and microstructure, providing information on the processing conditions of clinkerization in the Portland cement kiln.
- The type of microstructure related to the stability range of the polymorph allows the determination of the burning conditions.
- For example, in figure 6.2, the cross-striation texture of belite occurs in association with the transition at temperatures higher than 1450 °C from α to α'H with six sets of twinned α'H lamellae in the host α phase. α'H polymorph inverts to β polymorph passing through α'L polymorph.

TAB. 6.2 – Relations between the mineralogy, chemical composition, and the indices.

Equation 1. Chemistry *versus* mineralogy (mass balance)
$C = 0.737 \times C_3S + 0.651 \times C_2S + 0.622 \times C_3A + 0.462 \times C_4AF$
$S = 0.263 \times C_3S + 0.342 \times C_2S$
$A = 0.378 \times C_3A + 0.2 \times C_4AF$
$F = 0.328 \times C_4AF$

Equation 2. Mineralogy *versus* chemistry (Bogue's formula)
$C_3S = 4.05814 \times C - 7.56977 \times S - 6.67768 \times A - 1.44069 \times F$
$C_2S = -3.05814 \times C + 8.569767 \times S + 5.03218 \times A + 1.085681 \times F$
$C_3A = 2.6455 \times A - 1.69377 \times F$
$C_4AF = 3.04878 \times F$

Equation 3. Indices *versus* chemistry
$LSF = 100 \times C/(2.802281 \times S + 1.6455 \times A + 0.355013 \times F)$
$AM = A/F$
$SR = S/(A + F)$

Equation 4. Chemistry *versus* indices
$C = 100[LSF(2.8022 \times SR \times (AM + 1) + 1.6455 \times AM + 0.35)]/[(AM + 1)$
  $(SR + 1) + LSF(2.8022 \times SR \times (AM + 1) + 1.6455 \times AM + 0.35)]$
$S = 100 \times (AM + 1) \times SR/[(AM + 1)(SR + 1) + LSF$
  $(2.8022 \times SR \times (AM + 1) + 1.6455 \times AM + 0.35)]$
$F = 100/[(AM + 1)(SR + 1) + LSF(2.8022 \times SR \times (AM + 1) + 1.6455 \times AM + 0.35)]$
$A = 100 \times AM/[(AM + 1)(SR + 1) + LSF(2.8022 \times SR \times (AM + 1) + 1.6455 \times AM + 0.35)]$

Equation 5. Indices *versus* mineralogy
$LSF = 100 \times (A + B \times C_3S)/(C + F \times C_3S)$
$SR = (0.263 \times C_3S + 0.349 \times C_2S)/(0.378 \times C_3A + 0.538 \times C_4AF)$
$AM = (0.378 \times C_3A + 0.21 \times C_4AF)/(0.328 \times C_4AF)$
$A = 7.602 \times SR(AM + 1) + 6.7187 \times AM + 1.4297$
$B = (AM + 1) \times (SR + 1) \times 0.01$
$C = 4.071 \times [2.8 \times (AM + 1) \times SR + 1.65 \times AM + 0.35]$
$D = -0.01 \times [2.8 \times (AM + 1) \times SR + 1.65 \times AM + 0.35]$

Equation 6. Mineralogy *versus* indices
$C_3S = [SR \times (AM + 1) \times (LSF \times 11.371719 - 7.56977) + (LSF - 1)$
  $\times (6.677669 \times AM + 1.440639)]/X$
$C_3S = 100 \times (4.05814 \times C - 7.56977 \times S - 6.67768 \times A - 1.44069 \times F)$
$C_2S = -3.05814 \times C + 8.569767 \times S + 5.03218 \times A + 1.085681 \times F$
$C_3A = 100 \times (2.6455 \times AM - 1.69377)/X$
$C_4AF = 100 \times 3.04878/X$
$C_4AF = 100 \times 3.04878/[(AM + 1)(SR + 1) + LSF$
  $(2.8022 \times SR \times (AM + 1) + 1.6455 \times AM + 0.355)]$
$X = (AM + 1)(SR + 1) + LSF(2.8022 \times SR \times (AM + 1) + 1.6455 \times AM + 0.355)$

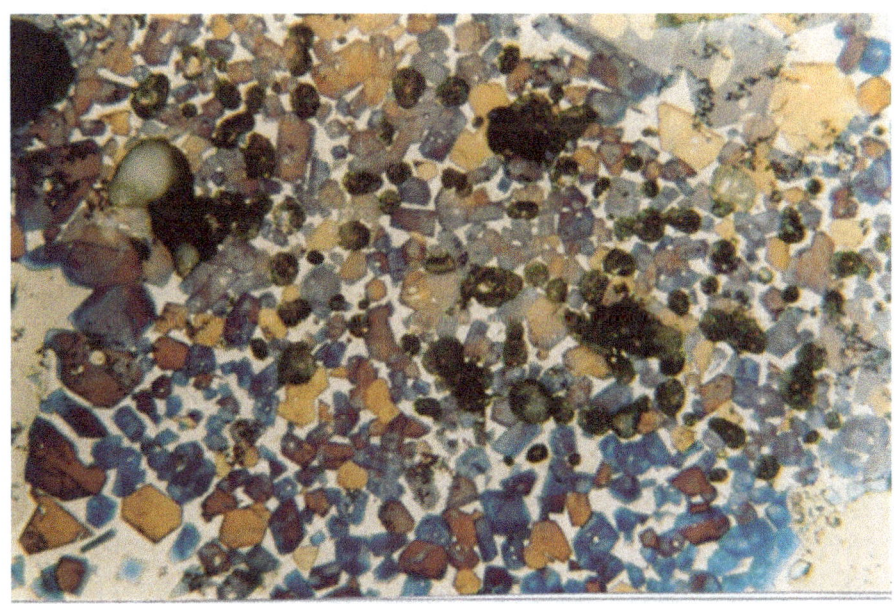

FIG. 6.1 – Polished section of a PC clinker (alite 57%, belite 90%, C_3A 9.8%, and C_4AF 8.6% formed at T° 1550 °C and subjected to fast cooling) observed by optical microscopy (magnification × 500) after etching by 2% NH_4Cl aqueous solution. Alite coexists with belite and a large amount of interstitial phase is observed.

TAB. 6.3 – Chemical analysis (average) of the four main phases of PC clinker.

Alite	$(Ca_{2.92}Na_{0.01}Mg_{0.05}Fe_{0.02})(Al_{0.04}Si_{0.96})O_5$
Belite	$(Ca_{1.97}Na_{0.01}K_{0.02}(Mg_{0.02}Fe_{0.02})(Al_{0.07}Si_{0.06}S_{0.01})O_4$
Calcium aluminate	$(Ca_{2.79}Na_{0.08}K_{0.03}Ti_{0.01}Mg_{0.09})(Al_{1.66}Si_{0.19}Fe_{0.18})O_4$
Calcium aluminoferrite	$(Ca_{1.97}K_{0.01}Mg_{0.02})(Al_{1.02}Si_{0.15}Ti_{0.05}Fe_{0.57}Mg_{0.21})O_4$

- When α-C_2S contains sufficient impurities, a remelting reaction occurs within crystals after the polymorphic change from α to α'H. During the remelting reaction belite crystals with Al/Fe ratio <1 produce a liquid which forms droplets on the lamellar boundaries. The rate of remelting reaction is very low. Belite crystals with Al/Fe >1 produce a liquid which readily spreads on the lamellae resulting in a high rate of reaction.
- Figure 6.3 shows belite crystals having a structure with plain or parallel striation. It occurred in a laboratory clinker burnt at 1350 °C during 12 h. When cooled below the stable temperature of α polymorph of C_2S, crystals of belite inverted to α'H polymorph without a change in composition and gave six sets of lamellae within the crystal.

FIG. 6.2 – Polished section of a PC clinker (alite 57%, belite 90%, C_3A 9.8%, and C_4AF 8.6%, formed at T° 1550 °C and subjected to fast cooling) observed by optical microscopy (magnification × 1000) after etching by 2% NH_4Cl aqueous solution. Cross striation of belite is observed.

- Figure 6.4 shows exsolution of impurities from belite during a slow cooling (FUK 1992, FUK 1993). It is distinguished from the type described in figure 6.2 by the presence of discrete particles along what are apparently traces of twinning planes and named dotted of belite. It also shows an orthorhombic form of calcium aluminate due to the presence of alkalis in the clinker.
- When α-belite contains sufficient impurities a remelting reaction occurs within crystals after the polymorphic change from α to α'H.
- The phase composition of a clinker can be calculated from the knowledge of its bulk composition and a model or measured by different techniques. The calculation (Bogue's calculation) assumes that equilibrium is attained and that uncombined lime measurement corrects the degree of disequilibrium (table 6.2 – equation 2).
- The principal techniques of measurement are: image analysis or point counting by optical or electron microscopy, X ray diffraction (Rietveld technique), and selective dissolution. A mathematical treatment based on the mass balance allows a good consistency of the results of the mineralogical composition of the clinker.

Applications to Hydraulic Binders

FIG. 6.3 – Polished section of a PC clinker formed at T° 1350 °C during 12 h and subjected to fast cooling, observed by optical microscopy (magnification × 1000) after etching by 2% NH_4Cl aqueous solution. Plain or parallel striation of belite is observed.

FIG. 6.4 – Polished section of a PC clinker formed at T° 1550 °C and subjected to slow cooling, observed by optical microscopy (magnification × 1000) after etching by 2% NH_4Cl aqueous solution. Dotted of belite and orthorhombic C_3A are observed.

6.2.1.3 Production of Clinker

6.2.1.3.1 Raw Mix Design
- The raw mix is constituted by a mixture of limestone and clay calculated to obtain SR = 2.6, AM = 1.5, and LSF 96%. LSF controls the C_3S/C_2S ratio, the silica ratio $SR = SiO_2/(Al_2O_3 + Fe_2O_3)$, controls the burnability and the alumina modulus $AM = Al_2O_3/Fe_2O_3$ fixes the C_3A/C_4AF ratio and impacts the burnability and the durability of the final cement.
- Table 6.2 gives the set of equations linking the main indices (SR, AM, LSF) and the chemical and mineralogical compositions of the clinker.

6.2.1.3.2 Clinkering Reactions
- During the production, the raw material (powder or pellets) is progressively heated to temperature up to 1450 °C.
- At temperatures above 500 °C, the thermal decomposition of clay such as kaolin ($Al_2Si_2O_5(OH)_4$) yields to a structurally disordered product called 'metakaolin'.
- At temperatures above 800 °C, limestone decomposes endothermically in CaO solid and CO_2 gas. The reaction is followed by the exothermic reaction of formation of C_2S.
- Between 800 and 1200 °C, the simultaneous presence of a gas phase containing reactive components such as CO_2, catalytic components e.g. H_2O and fine grained porous, chemically reactive products formed by decomposition leads to several reactions proceeding at unequal rates and giving metastable compounds in terms of the overall batch compositions such as calcium aluminates ($C_{12}A_7$), calcium aluminosilicates (C_2AS), and calcium ferrite (CF).
- The high local concentrations of volatiles e.g. chloride, sulphur oxide and fluoride, reached in modern fuel efficiency kilns may also play an important role in promoting reactions.
- Dicalcium silicate forms at much lower temperatures than tricalcium silicate and can be developed in the solid state whereas C_3S forms only in the presence of a liquid phases.
- Above 1300 °C, partial fusion occurs and results in the formation of 15/20% of an oxide melt. Eutectic occurred at 1338 °C with the composition, CaO 54.8, Al_2O_3 22.7, Fe_2O_3 16.5, and SiO_2 6.5 wgt %. At 1400 °C, the percentage of liquid phase is given by the formula (LEA 1956, HEW 1998):

  $L = 3.00 \times (Al_2O_3) + 2.25 \times (Fe_2O_3) + (MgO) + (K_2O) + (Na_2O)$ as wgt %.

- At the clinkering temperature which is typically 1400–1450 °C, the liquid phase facilitates chemical mass transport. It is possible to quantify the driving force leading to C_3S formation and to determine the role of minor components and mineralisers on the process kinetics.

The alite formation reaction consists of three elementary steps (MAK 1979, BES 1980):

(a) The dissolution of CaO and C_2S particle into the liquid phases.

FIG. 6.5 – Polished section of a PC clinker formed at T° 1550 °C and subjected to slow cooling, observed by optical microscopy (magnification × 1000) after etching by 2% NH_4Cl aqueous solution. Decomposition of alite is observed.

(b) The diffusion of Ca in the liquid supersaturated with C_3S.
(c) The precipitation of alite crystals of which the diffusion of CaO is the rate determining step.

- During cooling, the liquid phase crystallizes to yield fine grained phases which are interstitially distributed between the coarse-grained crystalline phases formed at clinkering temperature (C_3S and C_2S). Most of the ferrite and C_3A present in the clinker are formed by the crystallisation of this liquid during cooling.
- C_3S becomes unstable below 1250 °C with respect to CaO and C_2S. The factors favouring the thermal decomposition of C_3S and its solid solutions in the kiln include a slow cooling, the presence of certain chemical impurities, notably iron oxide in solid solution or molten sulphate, and the presence of water vapour and its pressure prevailing in the atmosphere (WOE 1960).
- Figure 6.5 shows the decomposition of alite in a laboratory burnt clinker slowly cooled from 1550 °C.

6.2.1.3.3 Thermal Balance
- The theoretical amount of energy required to produce 1 kg of clinker varies in the range 175–1850 J/kg. The endothermic decarbonation of limestone, the decomposition of clay (Kaolinite, Pyrophylite…), and the exothermal formation of C_2S are the significant heat exchanges.

TAB. 6.4 – Enthalpy of formation of calcium silicates and aluminates relevant to industrial binders measured by calorimetry of dissolution in $PbO-B_2O_3$ at 1173 K.

Mineral	C_3S	C_2S	C_3A	C_4AF	CA	C_2AS
Enthalpy of formation (kJ/mole)	−125	−119	+20	−47	−8	−130

TAB. 6.5 – Theoretical calculation of the thermal balance of clinker produced from calcium carbonate and kaolinite.

	kJ/kg clinker
Decarbonation (1782 kJ/kg) 1.185 kg of $CaCO_3$	2111
Decomposition of kaolinite (538 kJ/kg) 0.336 kg of kaolinite	180
Formation of 70% C_3S	−382
Formation of 12% C_2S	−83
Formation of 11% C_3A	8
Formation of 7% C_4AF	−8
Clinker (kJ/kg)	1826

- Table 6.4 shows the enthalpy of formation of the calcium silicate and aluminate relevant to the industrial binders.
- An example of theoretical calculation of the enthalpy of formation of 1 kg of clinker from a mixture of calcium carbonate and kaolinite is shown in the table 6.5. The range of variations is due to the difference in the enthalpy of decomposition of the different silica bearing compounds.
- In the case of industrial clinker, the total energy requirement includes energy lost in clinker, in exit gas and dust, air from the cooler, heat loss by convection and radiation and evaporation of water.
- Over the years, the evolution of the process has been more energy efficient, from wet to dry process and the introduction of preheaters and precalciners all with the aim of reducing energy consumption.
- Table 6.6 shows the range in energy requirements at different stages indicating that the energy requirement depends strongly on the process as the ranges between 3110 and 6070 kJ/kg of clinker.

6.2.2 PC Applications

6.2.2.1 Introduction

- The cement, ground at a defined fineness, is mixed with water, sand and aggregates to form a uniform wet rheology (consistency). The degree of fluidity of the wet mix is designed to facilitate its form filling function, ranging from highly fluid to a low flow, non-slumping consistency.

Applications to Hydraulic Binders

TAB. 6.6 – Heat requirement for clinker production (kJ/kg clinker).

Heat requirement for the different parts of the clinker production	kJ/kg clinker
Theoretical heat for chemical reaction	1750/1850
Dust in exit gases	10/20
Evaporation of water	20/2400
Heat lost in clinker discharge	30/100
Heat lost in exit gases	600/700
Heat lost in air from the cooler	400/500
Heat lost by radiation and convection	300/500
Total (kJ/kg)	3110/6070

- From the first minutes (hours), heat is released by exothermic cement hydration reactions as hydrates form and the wet composition becomes rigid, and this is termed setting. The concrete hardens and develops compressive and flexural strengths, whose values depend on the type of cement. During that time and later, dimensional changes occurs depending on the conditions of curing conditions (humidity, temperature etc....).
- Engineering and thermal properties, dimensional stability, and durability are the properties required when using concrete made with PC.
- The properties of the mortar and the concrete are characterised experimentally by standardized methods.
- The properties depend on the microstructures developed during the chemical reactions with water, the nature and the quantity of the final products of these reactions.

6.2.2.2 Hydration

6.2.2.2.1 Hydration of Portland Cement

- The reaction of water and Portland cement leads to the formation of calcium silicate hydrate 'C–S–H' and Portlandite, $Ca(OH)_2$, also written CH, and calcium sulfoaluminate hydrates (mono- and tri-calcium aluminates hydrates also named respectively AFm and AFt).
- A set of results obtained by calorimetry and analysis of the chemistry of the aqueous phase allows a division of the hydration in five distinct steps:

  (1) A first contact of cement with water with a very rapid exothermic dissolution of cement.
  (2) During mixing a first decrease in the reaction rate occurs and is ascribed to the formation of initial hydrate coating on surfaces of the C_3S grains and effectively greatly slowing the continued hydration leading to the induction period.
  (3) This induction period coincides with the transport and placing, a period during which the hydration rate seems to be very slow.

(4) Eventually the surface coating on the C_3S grains breaks down allowing an increase in reaction rate bringing about setting which is corresponding to the formation of sufficient hydrates to make the mix impenetrable to a Vicat test needle (a method of determining setting time).
(5) Hardening corresponding to a densification of the microstructure as more hydrates are precipitated finally leading to restrict the hydration rate.

- During hydration, the morphology of the C–S–H is modified (seen by TEM and SEM), from reticulated (lace-like) and needles radiating from the grains to crumple foil and dense gel or spherical agglomerates. Thus C–S–H evolves to a denser more crystalline material.
- The most reactive and most plentiful compound of Portland cement is alite (C_3S) and it is therefore not surprising that the hydration of PC shows some similarities with pure C_3S.
- The common procedure of research begins with the study of the hydration of individual pure compounds, the corresponding solid solution, and their interactions (simultaneous hydration of several compounds). The difference in the impact on the properties of the compounds studied individually or together is more important than the difference between the pure compound and its solid solutions studied separately.
- The main differences between Portland cement and pure phases of C_3S, C_2S, tricalcium aluminate, and calcium aluminoferrite are:

  (1) The presence of a minor element modifying the crystallographic structure.
  (2) The reactions with the other constituents present in Portland cement and mainly between C_3A and sulphate, calcium and alkalis.
  (3) The interactions with the hydration of alite.

- The main differences between the sequence of the hydration of Portland cement and that of pure C_3S is the interaction with mixture C_3A and calcium sulphate and the impact of this reaction on the rheology of the mix during the first contact of the cement with water (steps 1 and 2).
- This first period corresponds to the formation of Aft as reaction product of this interaction at first contact between cement and water when a very rapid exothermic dissolution of the cement occurs. The build-up of these reaction products on the surfaces slows everything soon after this initial stage and introduces the induction period.
- With the mixture still wet, the induction period allows the time needed for transport and placing, and this is followed by the acceleration period corresponding to the setting and hardening and the period of the development of the microstructure represented schematically by heat evolution (figure 6.6).

6.2.2.2.2 Hydration of C_3S

- The reaction of water and alite or C_3S leads to the formation of a calcium silicate hydrate 'C–S–H' and Portlandite according to the reaction:

$$C_3S + 5.3H \rightarrow C_{1.7}SH_4 + 1.3CH.$$

Applications to Hydraulic Binders

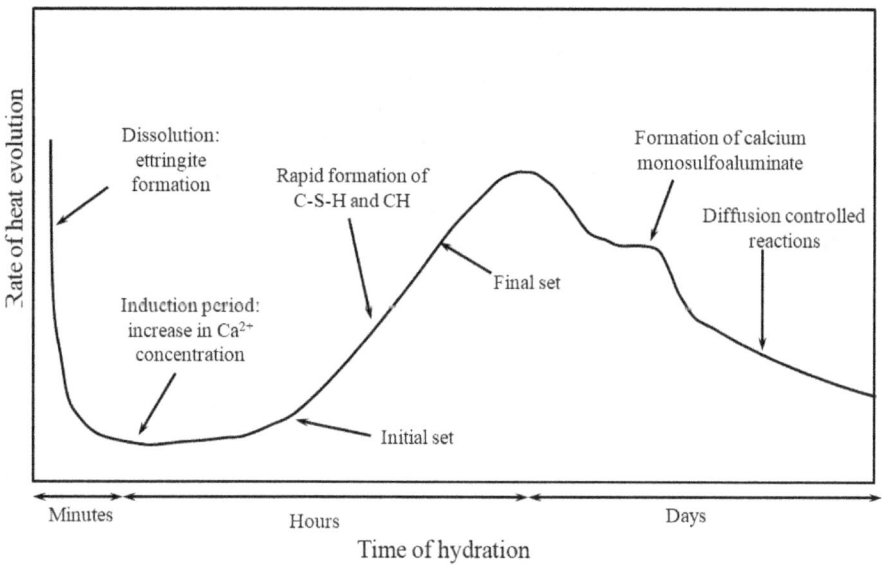

FIG. 6.6 – Schematic representation of heat evolution recorded during hydration of a cement (from Gartner et al. (GAR 2001, p. 91)).

6.2.2.2.3 Hydration of C_2S

- β-C_2S is the most plentiful polymorph of the 5 of belite (C_2S) in Portland cement clinker.
- The reaction of water and C_2S leads to the formation of C–S–H and Portlandite according to the reaction:

$$2C_2S + 4.3H \rightarrow C_{1.7}SH_4 + 0.3CH.$$

- All forms of C_2S produce a similar type of calcium silicate hydrate as well as Portlandite at different rates of hydration.
- The different structures of C_2S give different properties. β-C_2S is the most reactive variety although α'-C_2S appears to have a comparable hydraulicity. α-C_2S is weakly hydraulic and γ-C_2S hydrates very slowly. The hydraulicity of C_3S is the highest.
- The relation between the structure of the calcium silicates and the development of the hydration is not clear. The influences of the polymorphisms, the nature and amounts of foreign atoms have different effects.
- The difference in reactivity between C_3S and C_2S can be partially explained by the number of connections of CaOx polyhedra by common faces, the number of Ca neighbours of Ca atoms and the defects in the crystal lattice.
- The structure of the C_3S lattice is one of the main factors which determine its properties as a binding material. C_3S in the form of interstitial solid solution has a lower hydraulicity, while substitution type solid solution on the other hand has different reactivities depending on the size of the substituting ion.

- Substitution of Ca^{2+} by a smaller ion increases the hydraulicity of the solid solution and a larger substituting ion causes a decrease in hydraulic properties. The substitution of Si^{4+} by Al^{3+} increases hydraulicity (YU 1968).
- The much greater hydraulicity of β-C_2S in relation to γ-C_2S can be understood on the basis of β-C_2S containing irregular polyhedral coordination of Ca^{2+} ions (eight and nine fold) with longer Ca–O bonds that appears to more readily undergo structural rearrangement than in γ-C_2S where there is a more symmetrical coordination (six-fold) of Ca^{2+} ions and apparently stronger Ca–O bonds arising from the lower coordination number of Ca^{2+} ions present here.
- In order to reduce the cost of energy to manufacture PC clinker, an increase in the percentage of belite *versus* alite has been widely suggested (20% alite and 60% belite). The advantages are the reduction in the heat required for decarbonation and a lower temperature firing resulting in decreasing CO_2 emissions.
- To obtain the same performance as a standard PC, the shortcoming of the hydraulicity of belite (poor compared to C_3S) should be improved. Several solutions have been proposed: modification of the process (sol gel type, hydrothermal...) or addition of dopants.

6.2.2.2.4 Hydration of C_3A

- The reaction of water and C_3A leads to the formation of calcium aluminate hydrates according to the reactions:

$$2C_3A + 27H \rightarrow C_4AH_{19} \text{ (or } C_4AH_{13}) + C_2AH_8 \text{ (I) (hexagonal hydrates).}$$

$$C_4AH_{19} + C_2AH_8 \rightarrow 2C_3AH_6 \text{ (II) (cubic hydrates).}$$

- The high reactivity of C_3A can be explained by the mechanism and strength of various Al–O and Ca–O bonds. The structure of C_3A is cubic. There are 2 classes of Al and six classes of Ca atoms. It is built from rings of six AlO_4 tetrahedra of formula (Al_6O_{18}) surrounding holes of radius 1.47 Å and interconnected by calcium atoms. These rings are formed by a corner sharing of two oxygens per tetrahedron to give a structure with two non-bridged oxygens per AlO_4 tetrahedron where aluminium ion is present in four-fold coordination.
- Of the six types of Ca atoms in the structure, Ca_1, Ca_2 and Ca_3, occupying the corners of the pseudo-cells are each surrounded by 6 oxygen atoms in a distorted tetrahedral arrangement. The three others, Ca_4, Ca_5 and Ca_6 at the body-centre positions of the pseudo cells are irregularly coordinated with Ca–O bond.
- Without water, the Al_6O_{18} rings are the basis of the stability of the structure of C_3A but they are reactive with water.
- If Ca^{2+} or another cation goes into the central space in the Al_6O_{18} ring of the structure of $Ca_3Al_2O_6$, its hydration activity will be decreased.
- The reason for the high hydration activity of C_3A may lie in the reactivity of the CaOx polyhedra and the presence of the holes in the structure; the Ca_4–O bonds are weaker. Thus, the Ca_4–O bond will be the first to be broken when C_3A reacts with water.

Applications to Hydraulic Binders 225

- In presence of calcium sulphates that are present in PC, the reaction of water and C_3A leads to the formation of a calcium sulfoaluminate hydrates according to the reactions:

$$C_3A + 3C\$H2 \rightarrow C_6A\$_3H_{32} \text{ (ettringite) (III)}.$$

$$C_6A\$_3H_{32} + C_3A + 4H \rightarrow 3C_4A\$H12 \text{ (calcium monosulphoaluminate) (IV)}.$$

Note: $ corresponds to SO_3.

6.2.2.2.5 Hydration of Calcium Aluminoferrite

- C_4AF is not a well-defined stoichiometric compound, rather it is part of a solid-solution series called the ferrite phase.
- The course of hydration of the ferrite phase is very similar to that of C_3A.
- In the absence of gypsum, the reaction of water and calcium aluminoferrite leads to the formation of a calcium aluminoferrite hydrates according to the reactions:

$$3C_4AF + 60H \rightarrow 2C_4(A,F)H_{19} + 2C_2(A,F)H_8 + 4(F,A)H_3$$
$$\rightarrow 4C_3(A,F)H_6 + 2(F,A)H_3 + 24H.$$

- In the presence of gypsum, the hydration steps are the same as for C_3A. Iron substituted AFt is formed first, which later converts to AFm when gypsum is exhausted. An amorphous phase is also formed, supposedly aluminium substituted $Fe(OH)_3$.

6.2.2.3 Properties

- The mechanical properties of the concrete (strengths) are defined as the ability to resist failure. They depend on the mechanical properties of the hardened cement paste, of the aggregates and by the nature of the bond between them.
- It can be reasonably argued that:
  (1) Alite (C_3Sss) is responsible for the majority of the early strength of Portland cement.
  (2) Belite (C_2Sss) is responsible for the long-term strength.
  (3) C_3A is responsible for the rheological behaviour.
- The strength of the hardened cement paste is due to:
  (1) Interatomic forces within the individual particles of hydration products.
  (2) Atomic forces which bind the individual particles to each other and to the grains of still unhydrated cement.
  (3) The size and morphology of the hydration products.
  (4) The microstructure that develops (geometrical arrangement and density distribution of the hydration product).
- The strength of Portland cement is linked to the porosity of the cement paste according to the relation:

$$P_t = P_0 \exp(-K\,\sigma c)$$

where P_t is the total porosity, P_0 is the porosity at zero strength (60%), σc is compressive strength and K a constant equal to 5.37×10^{-3}.

- Deformation of hardened cement pastes occurs by:
  (1) Elastic deformation when external stress is applied.
  (2) Drying shrinkage and swelling deformation due to moisture change in the cement paste.
  (3) Creep deformation when prolonged stress is applied.
- A major property is the capability for a product or a structure such as a construction, to perform a function for which it was designed over a specified time. It quantifies the term of durability.
- Durability of PC is its ability to resist to weathering action chemical attack abrasion and other conditions of service.

6.2.2.4 Understanding of the Measured Properties

- The properties of the hardened cement paste depend on the development of the microstructure, achieved during the hydration of cement.
- A knowledge of the thermodynamic and kinetics of the chemical reactions involved during cement hydration is required for understanding the mechanism of the development of the microstructure and the quality of the concrete, the mortar and the cement paste.

6.2.3 Conclusions – Prediction of the Properties

- Some characteristics and properties of mortars and concretes are/can be determined as soon as the cement is manufactured (*e.g.* chemical composition, mineralogical composition, setting time, early strengths, etc....).
- Others require longer time before measurement can be made and may rely on test ages defined by national and international standards (*e.g.* compressive and flexural strengths at 3, 7, and 28 days, and also others such as durability testing...).
- It means that cement is generally used by the customer before all the test results of the batch of cement are known. This leads to a necessity to predict these results. For example, durability can be evaluated by accelerated experiments; strengths can be calculated using a model.
- Many models have been proposed for the mechanical strength prediction. It is known that the 28 days' compressive strength depends on the percentage C_3S, C_2S, C_3A and the fineness. As an example, the following linear equation can be used to link experimental data and the predicted results:

Rc_{28d} (MPa) = (A × %C_3S) + (B × %C_2S) + (C × %C_3A) + (D × %F) + E
with A, B, C, D, E determined statistically. F corresponds to fineness in cm^2/g.

- Among the data, the optical characteristics of the clinker have been considered in the following algorithm:

Applications to Hydraulic Binders

Rc_{28d} (MPa) = 253 + (6.4 × AS) + (21.9 × AB) + (4.0 × BS) + (21.5 × BC) with AS and AB being the size and birefringence of alite, BS and BC the size and colour of belite.

6.3 Calcium Aluminate Cements (CAC)

LEA 1970, MAN 1990, MAN 2008, TAY 1990

6.3.1 CAC Characteristics

6.3.1.1 Introduction

- The total of CaO, SiO_2, Al_2O_3 and Fe oxides account for more than 95% of the typical bulk composition of calcium aluminate clinker. TiO_2 and MgO are present in small quantities.
- The chemical composition of CAC varies over a wide range, providing cements with different mineralogical compositions and consequently suitable for a large variety of applications.
- Calcium aluminate cement was originally invented in 1908 as there was a need for a chemically resistant cement and it was shown that a cement based on calcium aluminate could fulfil that role.
- There are three major types of CAC developed industrially. They differ mainly by their content in iron oxide. They are classified here as iron-rich grades (Fe_2O_3 14–18%), low-iron grades (Fe_2O_3 1–3%) and iron free grades (Fe_2O_3 < 0.3%).
- Iron-free CAC is mainly used for refractory applications (see chapter 8).
- The chemical and mineralogical compositions are shown in table 6.7.
- The main properties developed by CAC are rapid hardening, resistance to chemicals and abrasion, refractoriness and can be used for cold weather concreting. Specific applications involve one or more properties together.

6.3.1.2 Identification Clinker Phases of Iron-Rich Calcium Aluminous Cement

- Given that the clinker contains 5 principle oxides, CaO, SiO_2, Al_2O_3, Fe_2O_3, and FeO, the maximum number of condensed phases in a stable assemblage over a range of pressures and temperatures is 5. Nevertheless, it is 6 if TiO_2 is taken into account.
- The number of mineral phases found in iron-rich CAC is higher than the maximum number of condensed phases in a stable assemblage. They are CA, $C_{12}A_7$, C_2S, C_2AS, calcium ferrite solid-solution (Brownmillerite), Wüstite, Perovskite (CT or $CaTiO_2$), spinel, and a quinary compound of composition ($Ca_{20}Al_{22.6}Fe^{3+}_{2.4}Fe^{2+}_{3.5}Si_{3.5}O_{68}$) having a characteristic of Pleochroite and a glassy phase.
- The presence of glass depends on the rate of cooling during manufacture of the clinker and is attributed to the metastability of the phase assemblage.

TAB. 6.7 – Chemical and mineralogical compositions and also properties of CAC *versus* OPC.

	PC	Iron-rich CAC	Low-iron CAC grade 50	Iron-free CAC grade 70
Chemistry				
CaO	68.8	38.5	36.5	29
SiO_2	22.6	4.5	7	
Al_2O_3	5.11	38	50.5	70.5
Fe_2O_3	3.5	13.5	2	0.1
FeO		3.5	1	
Σ	100	98	97	99.5
Mineralogy	C_3S/C_2S	$CA/C_{12}A_7/C_2S/$	CA/C_2AS	CA/CA_2
	C_4AF/C_3A	$C_2AS/C_4AF/P$	$\alpha A/C_2S$	
Fineness (cm^2/g)		300	400	370
Specific gravity (g/cm^3)	2.95/3.20	3.25	3.0	
Setting time (h)		3 h 20–5 h	3 h–3 h 30	2 h 05–3 h 30

Note: P for solid solution of Perovskite and Pleochroite.

- The clinker crystals obtained by rapid cooling are smaller and have more distorted morphologies than those obtained by slow cooling.
- A typical mineralogical composition of iron-rich CAC is; CA 61%, Brownmillerite 11%, C_2S 8%, Perovskite 8%, Spinel 2%, $C_{12}A_7$ 5%, C_2AS 3% and Pleochroite 2%. Figure 6.7 shows a polished section of a typical clinker by scanning electron microscopy.
- The chemical composition of the mineral phases present in this type of CAC shows that these mineral phases contain several impurities (table 6.8).
- Monocalcium aluminate takes iron (5–6%) and Si (0.5%) into solid solution and fast cooling tends to increase the extent of Fe and Si substitutions.
- The presence of a large fraction of Perovskite intergrowth into the ferrite phase has been observed by TEM. It is composed of two phases with very close chemistry and structures, Brownmillerite and a Perovskite type phase.
- Gehlenite (C_2AS) contains some Fe in the form of $C_2A_xF_{(1-x)}$ as it presents intergrowths with Pleochroite (SEI 1998, SCR 1990). It would be natural to regard the intergrowth as naturally metastable with respect to an idealised situation where the crystals are well formed and separated. That said, the ease with which mixed intergrowth formed during slow cooling suggests that the increase in free energy relative to their formation is low. The entropy of mixing offsets, at least in part, the rise in free energy resulting from interface formation, from intergrowth and from departure of ideal stoichiometry.
- Belite has a maximum Al_2O_3 content of 0.4% and 0.7% Fe_2O_3. It crystallises as the beta polymorph and does not present striation.

Applications to Hydraulic Binders

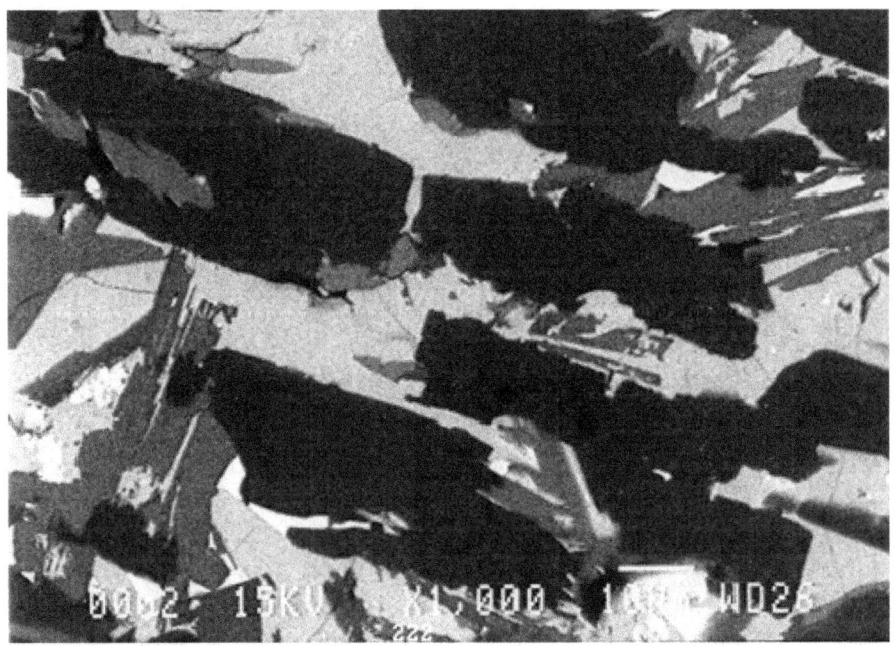

FIG. 6.7 – Polished section of a CAC observed by SEM (magnification × 1000).

- The general composition of spinel shows several possibilities of incorporations, and can be given by the formula; $(Mg^{2+},Fe^{2+},Fe^{3+})(Fe^{2+},Fe^{3+},Al^{3+})2O_4$.
- Pleochroite in the CaO–SiO$_2$–Al$_2$O$_3$–Fe oxides systems has the following general formula: $Ca^{2+}20(Al_{3+},F^{3+}{}_{32-2x})(Fe^{2+})SiO_{68}$.
- The chemical composition $Ca_{20}Al_{22.6}Fe^{3+}2.4Fe_{2+3}\cdot 5Si_{3.5}O_{68}$ was found in industrial CAC. It crystallizes as fibres suggesting a gradient of temperature for its formation (figure 6.8). It forms under reducing conditions and high silica contents.

6.3.1.3 Percentage of Clinker Phases

- The clinker compositions are very sensitive to the manufacturing process, cooling history and small variations in bulk composition.
- The phase composition of a clinker can be calculated from the knowledge of its bulk composition or measured by different techniques. The principal techniques of measurement are: image analysis or point counting by optical or electron microscopy, XRD (Rietveld method), and selective dissolution. A mathematical treatment based on the mass balance allows a good consistency of the results for the mineralogical composition of the clinker.

TAB. 6.8 – Chemical composition of the mineral phases found in industrial iron rich CAC.

Phase	Composition
Monocalcium aluminate	$Ca_{0.98}Mg_{0.02}Si_{0.01}Al_{1.88}Fe_{0.09}O_4$
Brownmillerite	$Ca_{3.90}Mg_{0.03}Si_{0.26}Ti_{0.27}Al_{0.94}Fe_{2.40}O_{10}$
Perovskite	$Ca_{1.39}Si_{0.07}Ti_{0.07}Al_{0.54}Fe_{0.11}O_3$
Mayenite ($C_{12}A_7$)	$Ca_{11.47}Mg_{0.2}Si_{0.06}Ti_{0.13}Al_{13.59}Fe_{0.25}O_{33}$
Pleochroite	$Ca_{20}Al_{22.6}Fe^{3+}{}_{2.4}Fe^{2+}{}_{3.5}Si_{3.5}O_{68}$
Gehlenite	$Ca_{3.67}Mg_{0.09}K_{0.01}Ti_{0.01}Al_{3.77}Fe_{1.17}Si_{1.39}O_{14}$
Belite	$Ca_{1.85}Si_{0.92}Mg_{0.05}K_{0.01}Ti_{0.01}Al_{0.1}Fe_{0.08}O_4$
Spinel	$MgAl_2O_4$ and $(Ca \cdot Fe^{2+})Fe^{3+}{}_2O_4$
Magnesio Wüstite	$(Fe_{0.91} \cdot Mg_{0.05})O$
CA_2	$Ca_{1.86}Mg_{0.06}K_{0.001}Ti_{0.03}Al_{7.92}Fe_{0.04}O_7$

- The prediction of the percentage of phases is possible when the cooling rate of the different fractions is known:

  ○ A mixture of CaO (42.9 wgt %), SiO_2 (4.6 wgt %), Al_2O_3 (37.7 wgt %), and Fe_2O_3 (14.7 wgt %) is sintered at 1200 °C until the equilibrium is reached. The phases detected are a solid solution of CA (containing Fe_2O_3), a solid solution of C_2AS (containing Fe_2O_3), a calcium aluminoferrite and C_2S with a proportion respectively of 38, 10, 45, and 7 wgt %.
  ○ If the same mixture is melted at 1450 °C and poured into two types of crucibles having different heights (10 and 5 cm), the solidified product is sawed to obtain slices of different thicknesses. In each slice corresponding to a specific cooling rate, the phases have been analysed by SEM (tables 6.9 and 6.10). It can be seen that the quantity of glass is higher in the cases of the small crucible where the rate of cooling is higher as well as in the bottom and the top of both crucibles.

- This experiment indicates that it is possible to estimate the phase proportions in the industrial case if the thickness of the product in mould is controlled. Nevertheless, the calculation based on the equilibrium condition gives a reasonably good approximation.
- Industrially, in order to achieve a good regularity of the properties, the production has been concentrated to very specific phase assemblages having well defined properties: CA, C_2S and C_4AF or CA, $C_{12}A_7$, C_2S and C_4AF or CA, C_2AS and C_4AF. Other phases such as calcium titanate, Pleochroite or glass are minimized by controlling the raw materials and the process as well (kiln atmosphere or rate of cooling).

6.3.1.4 Raw Mix Design and Formation of CAC

- Iron-rich CAC is manufactured by fusing a mixture of limestone (calcium carbonate) and ferruginous bauxite followed by rapid cooling. According to the patent of Bied (BIE 1909), the quantity of lime is defined to obtain the mixture C_2S/CA. The higher the alumina content, the lower the silica content and better is the quality of the final product.

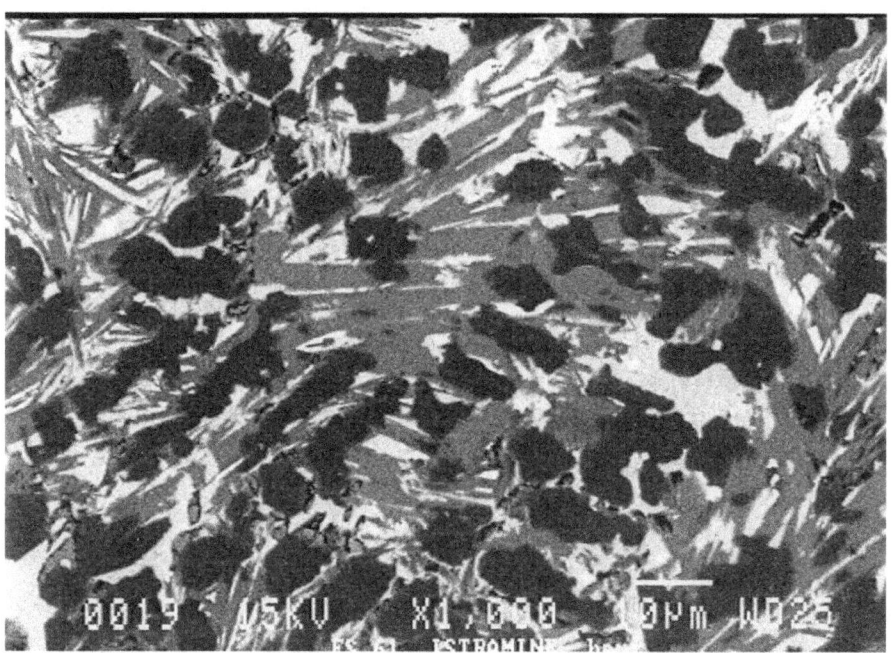

FIG. 6.8 – Polished section of a CAC observed by SEM (magnification × 1000); formation of Pleochroite.

- In the fusion process, the raw materials (blocks of bauxite and limestone) are progressively heated to temperatures up to 1450 °C such that total fusion occurs. After a residence time of about 20 min at this temperature, rapid cooling follows.
- Bauxite is the principal raw material source of alumina necessary for production of CAC. It contains alumina trihydrate (Gibbsite, Bayerite) or monohydrate (Boehmite and Diaspore).
- At atmospheric pressure, alumina trihydrate loses its water at about 250 °C and dehydration is complete at 500 °C. Transition alumina occurs at higher temperatures.
- Limestone decomposes in CaO solid and CO_2 gas at about 850 °C. The decomposition is endothermic so that heat must be supplied to those regions which are undergoing decomposition. At this temperature, reactions occur between solid phases; CaO, Fe_2O_3, and Al_2O_3 to give $C_{12}A_7$, C_2F, and C_4AF.
- Reactions among the main components, CaO, Al_2O_3, Fe_2O_3, and SiO_2 proceed at different kinetics.
- The product is liquid at 1450 °C.
- During cooling, CA crystallises as primary phase followed by calcium silicate or aluminosilicate and calcium ferrite.

TAB. 6.9 – Percentage of phases *versus* the cooling rate.

Cooling rate	Height of the crucible (cm)	Mineral composition of crystalline fraction (%)				Total Cryst.	Oxide analysis of glassy fraction (%)			
		CA	C_2AS	C_4AF	C_2S		CaO	SiO_2	Al_2O_3	Fe_2O_3
Moderate	8–10	35	7			42	47.7	5.3	21.6	25.4
Very slow	6–8	38	10	45	7	100				
Slow	4–6	35	7	30		72	49.3	11.1	22.2	17.5
Moderate	2–4	30	5			35	46.6	5.4	25.4	22.6
Fast	1–2	20				20	44.8	5.8	31.0	18.4
Very fast	0–1					0	42.9	4.6	37.7	14.7
Combined		29.6	5.8	15	1.4	51.8				

Note: Case of crucible 1 (10 cm height and 7 cm diameter).

TAB. 6.10 – Percentage of phases *versus* the cooling rate.

Cooling rate	Height of the crucible (cm)	Mineral composition of crystalline fraction (%)				Total Cryst.	Oxide analysis of glassy fraction (%)			
		CA	C_2AS	C_4AF	C_2S		CaO	SiO_2	Al_2O_3	Fe_2O_3
Moderate	4–5	35	7			42	45.2	5.0	25.7	24.1
Slow	2–4	30	7	30		67	47.1	9.3	28.5	15.2
Fast	1–2	20				20	44.8	5.8	31.0	18.4
Very fast	0–1					0	42.9	4.6	37.7	14.7
Combined		23.5	4.55	9	0	37.05				

Note: Case of crucible 1 (5 cm height and 7 cm diameter).

Applications to Hydraulic Binders

TAB. 6.11 – Chemical composition of exit gas of a reverberatory furnace producing iron-rich CAC.

CO	CO_2	H_2O	N_2	H2S	O_2
0.58	16.86	14.65	75.85	0.21	5.8

TAB. 6.12 – Theoretical heat requirement for CAC production by fusion (4.6% SiO_2, 38.7% Al_2O_3, 10% Fe_2O_3, 8% FeO, and 39.8% CaO).

	kJ/kg clinker
Decarbonation of limestone (1782 kJ/kg)	1163
Decomposition of bauxite AH_3 (1013 kJ/kg) 38.7 Al_2O_3	392
Decomposition of bauxite AF_3 (261 kJ/kg) 10% Fe_2O_3	26
Vaporization of water 249 J/g	221
Reduction in Fe_2O_3 (1660 kJ/kg) 8%FeO	142
Formation of 21% C_2AS (−774 J/g)	−100
Formation of 48% CA (−50 J/g)	−24
Formation of 30% C_4AFv(−97 J/g))	−29
Fusion	418
Total	2209

6.3.1.5 Process and Thermal Balance

- Different types of furnace have been used for the manufacture of iron-rich CAC. Originally an electric furnace and a water-cooled vertical furnace were used (water jacket process). In the case of the water jacket process, bauxite, and limestone mixed with coke were charged at the top of the furnace. A hot-air blast was introduced through a burner near the bottom of the furnace.
- A complete fusion occurred and molten cement poured continuously from a tap hole. In the present process, iron-rich CAC is manufactured in an L-shaped reverberatory furnace at about 1450 °C. The vertical part consisting of a rectangular stack, filled with blocks of raw material, is used as preheater. Moisture and CO_2 are driven off. Fusion occurs when the raw materials drop from the vertical stack into the horizontal part. The gas analysis at output is shown in table 6.11.
- During the process, a part of the ferric iron of the bauxite is reduced to ferrous oxide.
- The liquid is poured out of a tap-hole into a mould and permits to cool it down relatively slowly, normally the pegs are tipped in a heap where the clinker cools further.
- The clinker is then crushed in ball mills to the required surface area (>300 m^2/kg).

TAB. 6.13 – Total energy requirement for CAC industrial clinker production by fusion depending on the process steps.

Heat requirement for the different parts of the clinker production	kJ/kg clinker
Theoretical heat for chemical reaction	1650/2300
Heat lost in cooling water	400/600
Heat lost in clinker discharge	1400/1750
Heat lost in exist gas	800/1000
Heat loss by radiation and convection	400/600
Total	4500/5800

- The reduction in the clinker (% of Fe^{2+}) is due to the release of oxygen during the fusion and because a low oxygen partial pressure is maintained in the zone of melting (although the presence of excess oxygen in the exit gas).
- Theoretical heat requirement for CAC production can be estimated from thermodynamic database while the thermal balance of the reverberatory L shaped furnace is extrapolated from equivalent equipment used in glass and steel industry. The data shown in the table 6.12 correspond to a production from a raw mix composed of limestone and bauxite to get a CAC containing 4.6% SiO_2, 38.7% Al_2O_3, 10% Fe_2O_3, 8% FeO, and 39.8% CaO leading to the theoretical composition; 48% CA, 21% C_2AS, and 30% C_4AF.
- The total energy requirement for industrial clinker includes the energy lost in clinker, in exit gas and dust, the air from the cooler, the heat loss by convection and radiation and evaporation of water. The evolution of the process aims at reducing these energy consumptions by recovering energy from the clinker. Table 6.13 shows the range of heat requirement depending on the process; 4500–5800 kJ/kg.

6.3.2 CAC Applications

6.3.2.1 Introduction

- Rapid hardening and chemical resistances are the main properties of CAC.
- The properties of CAC depend on the way in which the phases present in the cement react, for instance hydrates are formed with water, or ceramic bonds on heating.
- The study of the evolution of the mineralogy is a key point to understand the properties of CAC.
- It can be reasonably said that CA is responsible for most of the mechanical strength of CAC.
- The methods used to follow the hydration of CAC are sequential (identifications of the hydrated phases, porosity measurement...) or continuous (evolution of the electrical conductivity, of the thermal behaviour *versus* time...).
- The hydration of the compounds present in the CAC leads to the following hydrates depending on temperature: CAH_{10}, C_2AH_8, C_3AH_6, AH_3 (amorphous $Al(OH)_3$ or Gibbsite) and C_2ASH_8 (Stratlingite).

Applications to Hydraulic Binders

TAB. 6.14 – Data relevant to the major hydrates found in the CaO–Al_2O_3–H_2O system.

	Lattice	Sp G	a (Å)	b (Å)	c (Å)	Angle (°)	d (g/cm³)
α1 C_4AH_{19}	Hexagonal	R2b C	0.58		6.4	120	1803
α2 C_4AH_{19}	Hexagonal	P63/m	0.58		2.1	120	1802
C_4AH_{13}	Hexagonal	R–	0.57		9.5	120	2046
CAH_{10}	Hexagonal		0.83		1.6		1730
C_2AH_8			0.57		1.1		1950
$C_2AH_{7.5}$	Monoclinic	C2/c	0.99	0.6	4.2	97	1943
C_2AH_5	Hexagonal	R3bc	0.57		5.2	120	2042
C_2ASH_8	Hexagonal	R3b	0.57		3.8	120	1936
C_3AH_6	Cubic		1.26				2530
AH_3							

- Table 6.14 summarises the data relevant to the major hydrates found in the CaO–Al_2O_3–H_2O system.

6.3.2.2 Hydration of Industrial Cements

- The reactivity calcium aluminates decreases with the ratio C/A of the mineral phase. Thus, in the case of CAC the most hydraulic phases are CA and $C_{12}A_7$.
- The hydration of industrial cement follows the hydration mechanism of its main compound (CA). The hydraulicity of calcium alumino-ferrite, C_2S and C_2AS is of a lesser importance.
- The initial reactivity of $C_{12}A_7$ is higher than that of CA and it governs the early hydration and rheological properties. C_2AH_8 rapidly forms at room temperature and could lead sometimes to a stiffening of the cement paste.
- The ion concentrations increase quickly in the aqueous solution leading to a pseudo equilibrium (called 'induction period') whose length is determined by temperature, water to cement content and cement composition. The composition of the aqueous phase is 24 mM/l of Ca^{2+} and 20 mM/l $Al(OH)_4$. It is supersaturated with respect to AH_3, CAH_{10}, C_2AH_8, and C_3AH_6, thus a nucleation process starts and then the critical crystal size is reached. Then, a rapid growth of the crystals occurs leading to a decrease in the ionic concentrations and allowing more cement phases to dissolve and then more hydrates to precipitate (BAR 1974, SOR 1995).
- The presence of $C_{12}A_7$ and all the related effects, explain the drastic shortening of the induction period of the CAC cement usually lasting for one to a maximum of three hours while the induction period exhibited by CA ranges between 10 and 24 h.
- At room temperatures of less than 10 °C, CAH_{10} forms by hydration of CA:

$$CA + 10H \rightarrow CAH_{10}.$$

- Between 10 and 50 °C, C_2AH_8, and AH_3 form according to:

$$2CA + 11H \rightarrow C_2AH_8 + AH_3.$$

- Above 50 °C, C_3AH_6 forms according to:

$$3CA + 12H \rightarrow C_3AH_6 + 2AH_3.$$

- C_3AH_6 is the only phase in the $CaO-Al_2O_3-H_2O$ system that is thermodynamically stable. Consequently, CAH_{10} and C_2AH_8 transform into C_3AH_6 in the so called 'conversion' reaction:

$$3CAH_{10} \rightarrow C_3AH_6 + 2AH_3 + 18H \quad (1)$$
$$3C_2AH_8 \rightarrow 2C_3AH_6 + AH_3 + 9H \quad (2)$$

- The conversion of CAH_{10} (1) is associated with a release of free water (60%), along with a decline in the volume of solids (*i.e.* a dramatic increase in porosity).
- The conversion of C_2AH_8 (2) is associated with a release of free water (37.5%), along with a decline in volume of solids (*i.e.* an increase in porosity).

6.3.2.3 Properties

- The rheology of calcium CAC pastes exhibits a Bingham type behaviour with low yield values and a plastic viscosity (BAN 1995).
- The setting starts at about 2–6 h and ends within 12 h.
- The mechanical strength of CAC increases rapidly and the maximum strengths are obtained within 1–2 days.
- The mechanical strengths depend on the percentages of CA and of SiO_2. For silica content less than 5%, the level of strengths remains constant at a high value. For SiO_2 >5%, the strength decreases considerably.
- For temperature below 50 °C, the strength starts to decline after reaching a maximum value. This phenomenon is attributed to the conversion (transformation of the CAH_{10} and C_2AH_8 primarily formed to C_3AH_6. The volume of the solids becomes smaller whereas the volume of the pores increases. This in turn results in a decline in strengths.
- The decline in strengths ends with the completion of the conversion reaction. Moreover, strengths can increase again at longer ages as the water released by conversion may complete the hydration of the remaining unhydrated cement grains.
- The rate of conversion depends on the temperature and the water/cement ratio.
- From engineering point of view, codes of practice and recommendations of placing and curing, prevent long term degradation of CAC concrete and many works witness its durability. Nevertheless, at present the use of CAC in structural application is banned in most countries.
- The total heat of hydration of CAC is in the range 450–500 kJ/kg; 70–90% are released during the first 24 h.

- The chemical shrinkage due to the difference in volume of calcium aluminate hydrates compared to the initial volume of CAC and water represents 16% of volume in the formation of CAH_{10} and 25% of volume in the case of C_3AH_6.
- The modulus of elasticity of CAC concrete is higher than PC of the same strength whereas creep is of the same order of magnitude.
- Porosity of hardened unconverted CAC pastes is lower than that of PC at the same W/C ratio. This has a favourable effect on permeability and along with it, on the corrosion and freeze–thaw resistance. However conversion leads to the porosity becoming higher than PC after conversion.
- The properties of the CAC mortars and concretes can be modified by the addition of admixtures or fillers. As an example, lithium salts are strong accelerators while gluconate acts as retarder.
- The resistance of CAC to chemical attack is due to the quality of the hydrates (especially the absence of CH compared with PC), the permeability which is the main factor determining the resistance of CAC concrete to chemical attack:

    ○ CAC concrete is resistant to diluted acid solutions (pH > 4).
    ○ CAC concrete exhibits very good resistance to ground water containing sulphates and in sea water.
    ○ Atmospheric CO_2 reacts with CAC concrete to form carbonated AFm phases (calcium hemicarboaluminate ($C_4A\underline{C}_{0.5}H_x$)) and calcium monocarboaluminate ($C_4A\underline{C}H_x$), calcium carbonate and $A\bar{H}_3$. These later constituents fill the existing pores and cause a reduction in the permeability leading to an improved durability.

- Many ways are explored to minimise water to cement ratio (thus the porosity, the pore size distribution and the permeability of concrete being one of the important factors in its chemical resistance), to define the stability range and structure of the calcium aluminate hydrates substituted by foreign elements, or to change the phase assemblages in order to move away from the CaO–Al_2O_3–Fe_2O_3 system by adding silica, carbonate and sulphate.
- Table 6.15 shows the relationship between the properties of CAC and its applications.

6.4 Special Cements

BAR 1983, LEA 1956, TAY 1990, HEW 1998

6.4.1 Fast-Setting Cements

6.4.1.1 Introduction

- Fast-set cements are binders having a setting time shorter than the one of Portland cement or calcium alumina cement (about 2 h).

TAB. 6.15 – Relation between the properties of CAC and its applications.

	Industrial flooring	Kiln devices	Tunnelling	Pipes	Sewers	Airport runways
Rapid hardening	X		X		X	X
Chemical resistance	X			X	X	
Abrasion resistance	X	X		X	X	X
Refractoriness		X				
Cold weather resistance			X			X

- It is convenient to distinguish flash-set cement with a setting time less than 5 min, very fast cement with a setting time between 5 and 10 min and fast-set cement with a setting time higher than 10 min.
- Reduced setting time can be obtained by different ways;
- A selection of phases among $C_{12}A_7$, $C_{11}A_7CaF_2$, C_3A, and $CaSO_4$ as bearing components such as $CaSO_4$ and CSA (calcium sulfoaluminate) are involved in fast-set cement.
- Mixtures of CAC and PC provide solution to control the setting time by:

    ○ A physical way (modification of the particle size distribution).
    ○ A chemical way (acceleration of the rate of reactions by the use of accelerators such as calcium chloride, sodium phosphate).
    ○ A structural way by selection of the hydrates and the modification of their structure.

- Table 6.16 shows the chemical and mineralogical analysis of fast setting cement. The total of CaO, SiO_2, Al_2O_3, and Fe oxide accounts for >80% of the typical bulk composition of cement with different setting times. The remaining is mainly SO_3 bearing compounds.

6.4.1.2 Flash-Setting Cements (Setting Time Less Than 2 min)

- Natural fast-setting cements such as Prompt cement belong to this family.
- Fast-setting natural cement (standard NF P15-314) is produced by burning at low T° (1000 °C) in a shaft kiln, a natural raw meal rich in alumina. The product requires a period kept in a silo to stabilise the uncombined lime. The setting time is about 2 min. It can be retarded using citric acid.
- The first detected hydrate is ettringite followed by C_4AH_{13} which is the main product of hydration.
- Table 6.17 shows the compressive strength on mortar as a function of time for both natural fast setting cement and a mixture of CAC/PC.

TAB. 6.16 – Chemical and mineralogical analysis of various fast-setting cements and some mixtures.

Composition								
SiO$_2$	18.58	13.33	5.91	18.93	20.46	19.01	9.70	14.23	10.31	6.60
Al$_2$O$_3$	8.99	18.77	20.73	9.52	13.40	11.83	28.86	21.95	5.08	31.17
Fe$_2$O$_3$	3.29	6.61	7.36	3.70	2.57	2.69	1.86	2.54	2.06	5.99
CaO	60.15	51.44	46.93	60.55	61.04	60.09	58.01	59.35	56.70	48.93
MgO	2.74	0.57	0.52	0.56	0.67	2.83	0.52	0.91	0.00	0.61
SO$_3$	2.66	1.01	17.77	3.19	0.91	2.58	0.41	1.02	25.77	6.70
K$_2$O	0.93	0.63	0.27	0.57	0.48	0.68	0.08	0.00	0.00	0.00
Free lime	2.34	0.00		2.03						
Total	99.83		99.48	100.00	99.53	99.71	99.44	100.00	99.92	100.00
C + S + A + F	91.01		80.92	92.70	97.47	93.63	98.43	98.07	74.15	92.69
Phase										
		C$_3$S	C$_3$S	C$_3$S	C$_3$S	C$_3$S	C$_3$S	C$_3$S	C$_4$A$_3$$\$$	C$_4$A$_3$$\$$
		CA	CA	CA	C$_2$S	C$_{12}$A$_7$	C$_{12}$A$_7$	C$_{12}$A$_7$	C$_2$S	C$_{12}$A$_7$
		C$_2$S	C$\$$		C$_{12}$A$_7$	C$\$$		C$_2$S	C$\$$	

TAB. 6.17 – Compressive strengths (MPa) for mortar (cement/sand = 1/3 made with natural fast setting cement and a mixture of CAC/OPC (standard NF EN 206/CN).

	15 min	30 min	1 h	2 h	1 day	7 days	28 days	90 days
Natural cement	2.5	3.2	3.6	4.5	7	10	12	15
CAC/PC	5	6		8	9	14	16	18

- Concrete made with natural fast setting cement at 600/800 kg/m^3 has a good resistance to acids.
- Cements with setting time less than 2 min find application in the following:
  - Control of water movements in galleries, caves and sewers.
  - Working in flowing water, water proofing, entrapment of spring water.
  - Caulking of boats, positioning of stops and temporary marks.
  - Instantaneous space filling, laying out of fence pegs, installation of joist, lintels.
  - Immediate temporary mechanical support.

6.4.1.3 Very Fast Setting Cements (Setting Time Between 5 and 10 min)

- Mixtures of Portland cement (PC) and calcium aluminate cement (CAC) belong to this family.
- When CAC is added in increasing quantity to PC, the setting time of the mixtures is rapidly reduced until an instantaneous set is obtained for a defined proportion. The same phenomenon is also observed when PC is added to CAC.
- The setting time of PC/CAC mixtures present a minimum for different PC/CAC ratios depending on PC composition and also the composition of the cementitious material. Figure 6.9 illustrates this behaviour.
- Table 6.18 shows the hydrates detected by XRD *versus* the PC/CAC ratio during the hydration of different PC/CAC mixtures at room temperature.
- It is shown that the reduction in setting time occurs when the precipitation of C_4AH_x (major hexagonal hydrate formed at room temperature in PC/CAC mixtures) is the most important:
  - The branch corresponding to the acceleration of PC can be explained by the simultaneous presence of $Al(OH)_4^-$ released from CA and Ca^{2+} released by C_3S to precipitate C_4AH_x.
  - The branch corresponding to the acceleration of CA can be explained by the diffusion of sufficient quantity of Ca^{2+} released by C_3S to precipitate rapidly C_4AH_x from CA hydration.
- The formation and quantity of hexagonal hydrates depends on the nature, the proportion, the fineness of PC and CAC, the percentage of SO_3 in PC and the temperature.
- Addition of sulphate ions (SO_4^{2-}) increases the setting time by forming ettringite instead of C_4AH_x.

Applications to Hydraulic Binders

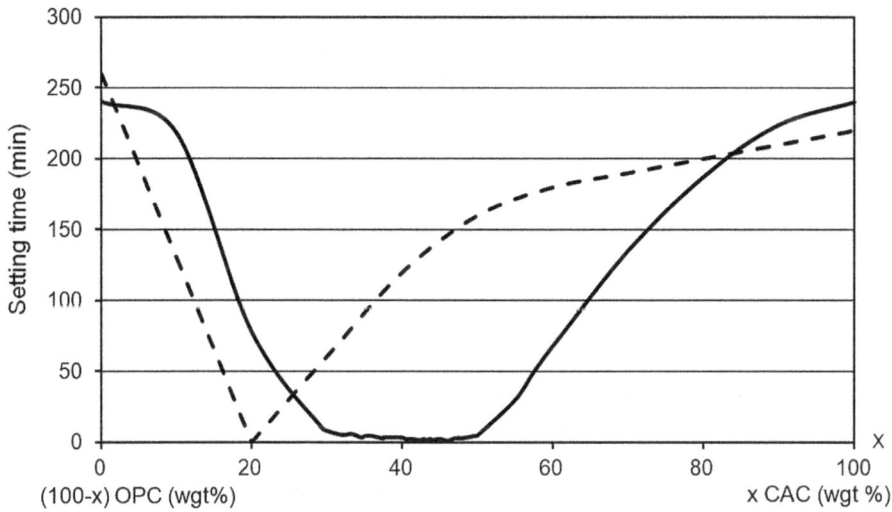

FIG. 6.9 – Setting time of mixtures of CAC/PC; pure paste (dashed) and concrete (plain).

TAB. 6.18 – Hydrates detected by XRD for hydration of different proportions of CAC and PC at room temperature.

% CAC	100	80	60	40	20	0
% PC	0	20	40	60	80	100
	CAH_{10}	CAH_{10}	C_3AH_6	C_4AH_{13}	CH	CH
	C_2AH_8	C_2AH_8	C_4AH_{13}	C_2ASH_8	C–S–H	C–S–H
		C_2ASH_8	C_2ASH_8	CH		
			CH			

- Initially, both CA and C_3S dissolve but C_3S reaction rate becomes very slow as soon as Al ions exceed some moles in the aqueous phase. Then, CA hydrates more and more rapidly with higher amounts of Ca^{2+} liberated by C_3S. The precipitation of C_4AH_x reduces progressively the concentration of Al ions.
- When Al concentration is very low, C_3S hydration rate starts to increase and the concentration of Ca^{2+} increases and C–S–H is formed in noticeable amounts.
- Finally, high concentrations of Ca^{2+} are reached due to the hydration of the remaining C_3S and CH is precipitated.
- Mixtures of pure CA and C_3S behave the same way as mixtures of pure CA and C_3A. The short setting time is unlikely due to the presence of hydrated phases, such as ettringite.
- Cements with setting time between 5 and 10 min find applications in:

- Drying up moist walls: vaults, tanks and reservoirs, cement rendering in water.
- Grouting and pipe, castings, installation of flashing of drain pipe.
- Flight of stairs, balustrades, reveals of window or doors, sink washstands, radiators, tanks of water, heater supports and all the minor aspect of sanitary works (*i.e.* electrical, plumbing, woodworking, etc....).

6.4.1.4 Cement with a Fast Setting (Setting Time Between 20 and 80 min)

- A wide family is produced by ternary mixtures of calcium sulphate, calcium aluminate cement (containing CA or/and $C_{12}A_7$) with Portland or natural cement.
- The addition of gypsum to a mix of 75% CAC and 25% PC gives fast setting cement and a very high initial mechanical strength due to the rapid formation of ettringite.
- The mix of 18% natural cement, 80% OPC and 2% gypsum gives a setting time of 30 min. The mechanical strength increases slowly with time as CA and C_3S are the main phases that hydrate.
- Calcium aluminate clinker containing $C_{12}A_7$ is produced by sintering bauxite and limestone at 1300 °C. The addition of $C_{12}A_7$ bearing compound to Portland cement gives a cement with a setting time of 15 min.
- Alite-fluoroaluminate cement or belite-fluoroaluminate cement that contains calcium fluoroaluminate ($11CaO_7 \cdot Al_2O_3 \cdot CaF_2$) in combination with calcium sulphate exhibits very fast setting and initial strength development owing to the formation of ettringite.
- The presence of alkali leads to the formation of $Al_2O_3 \cdot CaX_2$ (where X is F^- or Cl^-) that gives a better control of set. The addition of mineralisers (*i.e.* calcium chloride) gives a wider sintering temperature and facilitates the firing and the formation of $11CaO_7 \cdot Al_2O_3 \cdot CaF_2$ and C_3S.
- The presence of sulphate in the raw material during firing improves the early strength of the cement. With addition of calcium sulphate, the setting time is 30 min. These cements are the basis of the regulated-set or jet-set cements characterized by high early mechanical strengths.
- Cements that contain $C_4A_{33}\$$ in combination with calcium sulphate exhibit fast setting and fast early mechanical strengths development. These cements contain $C_4A_{33}\$$, C$, beta C_2S and free lime. They are manufactured by firing at 1380 °C, a mixture of kaolin and calcium sulphate or a mixture of kaolin, limestone and calcium sulphate.
- A cement produced by firing a mixture of clay, bauxite, gypsum and calcium fluoride at 1280 °C contains $C_{11}A_7CaF_2$ and $C_4A_3\$$.
- Cements with a fast-set find applications in:

  - Reinforcement of masonry wall carried out at low temperatures, grouting of rocks between tides.
  - Installation of fine dividing walls, sealing in bricks support at hollow ashlars.
  - Sealing of inspection covers and manhole.

Applications to Hydraulic Binders

TAB. 6.19 – Example of the chemical composition of a slag-based geopolymer (wgt %).

CaO	SiO$_2$	Al$_2$O$_3$	Fe$_2$O$_3$	K$_2$O	Na$_2$O	MgO
24	42	17	0.5	15	0.5	1

- Door and window frame.
- Moulding of window ledges setting walls, finishing and façade.
- Anchorage, gunning shotcreting.

6.4.2 Geopolymers

DAV 2015

6.4.2.1 Introduction

- Geopolymers are inorganic, typical ceramic materials that form long-range covalently bonded, amorphous networks.
- It essentially involves the geopolymerisation of alkaline soluble silicate with aluminosilicate such as metakaolinite, blast-furnace slag, fly ash, etc....
- The total of CaO, SiO$_2$, Al$_2$O$_3$, and Fe$_2$O$_3$ account for about >85% of the typical bulk composition of geopolymers. The remaining is mainly alkali salt.
- Table 6.19 shows an example of the chemical composition of a slag-based geopolymer.
- The category of geopolymer materials comprises:

  ○ Metakaolin based geopolymer binder of chemical formula (Na,K)–Si–O–Al–O–Si–O) with Si:Al ratio ranging between 1.5 and 2.5.
  ○ Silica based geopolymer binder (Na,K)–n(Si–O–)–(Si–O–Al–) with Si:Al ratio ranging between 15 and 40.
  ○ Sol gel based geopolymer binder (Na,K)–(Si–O–Al–O–Si–O–) with Si:Al ratio = 2.

6.4.2.2 Structure of Geopolymer

- Geopolymers are tetrahedral frameworks linked by shared oxygen atoms as poly (sialate) or poly(sialate-siloxo) depending on the SiO$_2$/Al$_2$O$_3$ ratio in the system. The tetrahedral aluminosilicate unit is condensed with alkali metal balancing the charge associated with tetrahedral Al.
- The polymeric framework of Geopolymer is similar to zeolites but the geopolymer is amorphous instead of crystalline like zeolite. It starts to crystallize at temperatures above 500 °C.

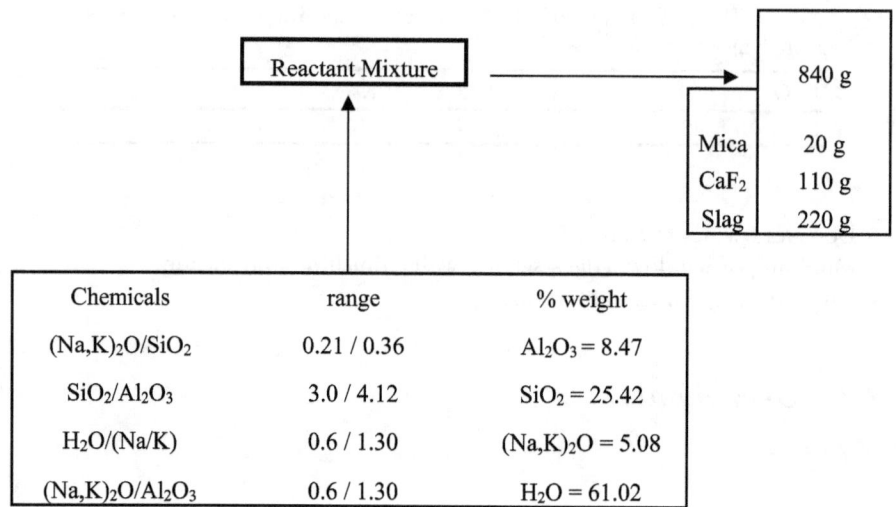

FIG. 6.10 – Reactant mixtures to form geopolymers. Note: The 840 g of reactant mixture contain 17.3 mol of water, 1.438 mol of potassium oxide, 4.45 mol of silicon dioxide and 1.08 mol of Al_2O_3 (DAV 2015).

6.4.2.3 Geopolymer Synthesis

- Geopolymers are synthesized from a two-part mix consisting of an alkaline silicate solution (often soluble) and solid alumino material (figure 6.10). In this process, many small molecules known as oligomers combine to obtain a covalent bonded network (geopolymerization).
- The geopolymerization with metakaolin in alkaline medium involved four main phases:

  (1) Alkaline depolymerization of the layered structure of the calcined kaolinite (metakaolin).
  (2) Formation of monomeric and oligomeric species including the ortho-sialate $((OH)_3-Si-O-Al-(OH)_3)$.
  (3) In the presence of soluble potassium silicate (waterglass) cyclic structure forms whereby the hydroxide is liberated by condensation reaction and can react again.
  (4) Polycondensation into higher oligomeric and polymeric 3D network.

- The geopolymerization kinetics for Na–poly(sialate-siloxo) and K–poly(sialate-siloxo) are slightly different. This is probably due to the different sizes of Na^+ and K^+ cations.
- Geopolymer cement is a binding system that hardens at room temperature.
- High early strength compositions are obtained by the blending of a mineral geopolymer referred to as a polysialate blast furnace slag.

TAB. 6.20 – Main applications of geopolymers depending on their Si/Al ratio.

2D network	Polymeric character	Polymeric character	3D network	
$20 < Si/Al < 35$	$Si/Al > 3/1$	$Si/Al < 3/1$	$Si/Al = 2/1$	$Si/Al = 1/1$
Fire resistant	Sealant for industry from 200 to 600 °C	Fire protection	Low CO_2 cements	Bricks ceramics
	Tooling for SPF aluminum	Fiberglass composite Foundry equipment	concrete	Fire protection
		Heat resistant composites from 200 to 1000 °C tooling for Ti proc	Radioactive waste encapsulation	

- The addition of ground blast furnace slag to the polysialate geopolymer accelerates the setting time and improves the compressive strength. The early high strength cement is obtained by adding to a reactant mixture consisting of alumino silicate oxides (Si_2O_5, Al_2O_3) with the aluminium cation in four-fold coordination, strong alkalis such as sodium or potassium hydroxide and a sodium potassium polysilicate solution to a given amount of ground BF slag.

6.4.2.4 Geopolymer Application

- Geopolymers may be used for:
  ○ Fire and heat resistant materials, thermal insulations, foams coating and adhesives.
  ○ Low energy ceramics tiles, refractory items, thermal shock refractories.
  ○ High tech resin systems, paints, binders and grouts.
  ○ Biotechnologies (materials for medical applications).
  ○ Foundry industry, tooling for the manufacture of organic fibre composites.
  ○ Composites for infrastructures repair and strengthening, fire resistant and heat resistant high-tech carbon fibre composites for aircraft interior and automobile.
  ○ Radioactive and toxic wastes containment.

- The Si/Al atomic ratio in the poly(sialate) structure determines the properties and application field. A low Si/Al ratio initiated a 3D network that is very rigid. A Si/Al ratio higher than 15 confers linear polymeric character on the geopolymer. Table 6.20 summarizes the main applications depending on the Si/Al ratio (DAV 2015).

6.4.3 Oil Well Cements

Ben 2002

6.4.3.1 Introduction

- Oil well cements are used for production of oil and gas in order to secure the metal casings and liners, out of which the oil and gas eventually flow.
- Basically oil well cement is a Portland cement whose quantity of phases is modified (less C_3A). Depending on the conditions (pressure and temperature) different types of cements are used. In all cases, the function of oilwell is to prevent the migration of gas and oil from the liners and metal casing and thus prevent the pollution of the working zone. Oil well cements protect the casing from external pressure that may be able to collapse and from possible damage due to corrosive gasses and water.
- The conditions of use required special properties. A slurry of cement is pumped to a depth below 6100 m and the temperature in use is about 200 °C.
- Oil well cements are also used to line water wells, waste disposal wells, thermal recovery wells, to line steel tubular goods like metal piping and for grouting the bases of production rigs and platforms.
- Nine classes of oil well cements are standardised by the American Petroleum Institute:

  (1) Class A is an ordinary Portland cement.
  (2) Class B is a sulphate resistant Portland cement.
  (3) Class C is intended for use on a depth of 6000 ft (1829 m) when conditions require a rapid hardening Portland cement.
  (4) Class D is intended to use from surface to a depth from 6000 to 10 000 ft (1829–3048 m) at moderate high temperatures.
  (5) Class E is intended to use from surface to a depth from 10 000 to 14 000 ft (3048–4267 m) at high temperatures and pressure.
  (6) Class F is intended to use from surface to a depth from 14 000 to 16 000 ft (4267–4877 m) at extremely high temperatures.
  (7) Class G intended for use as basic cement from the surface to a depth of 8000 ft (2438 m). As it is manufactured with accelerators and retarders, it can be used at a wide range of depth and temperatures.
  (8) Class H intended for use as basic cement from the surface to a depth of 8000 ft (2438 m) as manufactured with accelerator and retarder.
  (9) Class J cement is basically a combination of beta C_2S and quartz interground (60/40 by weight) suitable for application and high temperature geothermal wells, owing to its very low reactivity.

- Specifications for well cements divide oil well into three grades as following: ordinary (O), moderated sulphate resistant (MSR) and high sulphate resistant (HSR).

Applications to Hydraulic Binders

TAB. 6.21 – Typical oxides analysis of API class G and H of Portland cements (weight %).

CaO	SiO$_2$	Al$_2$O$_3$	Fe$_2$O$_3$	MgO	SO$_3$	K$_2$O
64.77	22.43	4.76	4.76	1.14	1.67	0.08

TAB. 6.22 – Typical composition (wt %) and specific surface area of API classes of Portland cements.

API classes	C$_3$S	C$_2$S	C$_3$A	C$_4$AF	SSA (cm^2/g)
A	53	24	8	8	1600/1800
B	47	32	5	12	1600/1800
C	58	16	8	8	1800/2200
D&E	26	54	2	12	1200/1500
G&H	50	30	5	12	1600/1800

- The chemical and mineralogical analyses are shown in tables 6.21 and 6.22. The total of CaO, SiO$_2$, Al$_2$O$_3$, Fe$_2$O$_3$ account for >95% (wt.) of the typical bulk composition of the oil well cement clinker.

6.4.3.2 Characterization of Oil Well Cement Utilisation

- Oil well cement must be performed at elevated temperatures and pressures both of which increase with depth.
- The maximum temperature encountered at the bottom of deep wells may reach 250 °C and may even exceed 300 °C in geothermal wells. Under these conditions the temperature of the slurry during pumping may reach 180 °C (bottom hole static temperature). The pressure to which the cement slurry is exposed is equal to the hydrostatic load plus the pumping pressure and may reach 150 MPa.
- Oil well cements are used mainly to restrict the movements of fluids between formations at different levels. In this way, productive oil and gas bearing formation may be sealed from water bearing layers. In the case of multilayer deposits, the production formations may be sealed off from each other. The hardened slurry is also intended to support and protect the casing.
- The cementing work is carried out by pumping the cement slurry down the steel casing of the well and up the annular space between it and the surrounding rock.
- More specialized techniques include squeeze cementing in which the slurry is forced through a hole in the casing into a void or into a porous rock and plugging in which the casing is blocked at a specified depth.
- The following features are the main desired characteristics of oil well cements:

   ○ The viscosity of the cement slurry must be low enough to allow easy pumping. The slurry must stay sufficiently mobile for the whole time needed to complete the pumping procedure, which usually does not exceed 3–4 h.

- In the oil industry, the pumpability of the cement slurry is usually assessed by the thickening time test. In this test, the consistency is measured with a high temperature high pressure specific consistometer in which the temperature and pressure expected to occur during pumping are simulated.
- After setting the slurry must attain a required strength at an acceptable speed. The rate at which the material develops its strength is relevant as it determines the time needed before drilling operations can continue. A retrogression of strength is likely to occur under conditions existing in the well.
- After being placed, the set cement slurry must exhibit sufficient resistance to the flow of liquids and gas. To achieve good performance in this respect, the volume changes that take place during setting and hardening have to be negligible.
- The hardened cement slurry must be sufficiently resistant to chemical attack especially by sulphates. In some oil fields, corrosion by overheated CO_2 rich steam may be critical, and must be taken into consideration.
- The available cement must be uniform to ensure predictable and reproducible behaviour of the slurry made from such binder.

6.4.3.3 Class G and H Oil Well Cements

- Class G and H are the most extensively used for cementing oil and gas well. They can be tailored to cope with a wide range of well cementing conditions by the addition of suitable additives to their slurries.
- The requirements are identical for both classes except the water/cement ratio to be used in the test performed in the determination of the performance characteristic of these cements.
- In general, Class H cements are more coarsely ground than Class G cements. The specific surface area of Class H is in the range 220–300 m^2/kg whereas that of Class G is in the range 270–350 m^2/kg.
- Class H is designated for lower water/cement ratio and hydrates more slowly.
- Both G and H oil well cements are moderated sulphate resistant (MSR) and high sulphate resistant (HSR) which differ mainly in their C_3A content. Class G and H oil well cements are low in C_3A, high in C_4AF, and sulphate resistant Portland cement. The free lime should be less than 0.5%.
- In producing Class G and H oil well cements, iron oxide must be added to the raw mix to produce more ferrite phase at the expense of calcium aluminate. The required amount of Fe_2O_3 is greater when HSR grade cement rather than MSR grade is produced.
- The properties of slurry made with Class G and H oil well cements may be modified by a variety of additives in order to cover a wide range of well depths, temperatures, pumping times and other requirements.
- Different density materials (low and high) may be added to the mix to optimize the bulk density of the mix. To lower the density, bentonite is an option and (2–12 wt% may be used, but it increases the water demand. Other materials may be added to the slurry to lower its density such as diatomaceous earth, expanded

Applications to Hydraulic Binders 249

perlite, etc. To increase the density of the slurry, finely ground hematite (Fe_2O_3), baryte ($BaSO_4$), Ilmenite ($FeTiO_3$), and Galena (PbS) may be used.
- The flow properties of the slurry may be improved by adding suitable dispersing agents such as condensated sulphonated naphthalene formaldehyde.
- Defoamers and deaerators are added to oil well cement slurries to prevent excessive foaming and air entrapment.
- Thickening time and progress of hydration may be controlled by accelerators ($CaCl_2$...) or retarders (sugar, citric acid...).
- Gypsum or other expansive agents such as calcium or magnesium oxide may be added to generate expansion and improve the sealing properties of the hardened material.

6.4.3.4 Understanding the Mechanism of Hydration

- Cement hydration tends to change when the temperature and pressure increase.
- As the temperature increases:

  (1) Belite (C_2S) and aluminoferrite (C_4AF) phases become more reactive with respect to the alite (C_3S) and tricalcium aluminate (C_3A) phases.
  (2) The setting time normally decreases.

Between 70 and 90 °C

- The thickening increases before decreasing when the temperature increases again. This can be explained by the fact that belite and aluminoferrite become more reactive relatively to C_3S and C_3A.
- An increase in the aluminoferrite hydration retards the hydration of alite and creates a thickening time longer than at room temperature. Then, the reactivity of all the cement phases tends to increase with temperature rise and the normally expected thickening increases restart after the end of the threshold temperature is overtaken.
- C–S–H phase formed at elevated temperatures has a higher C/S ratio and contains higher amounts of incorporated aluminium and sulphate ions.

Up to About 100 °C

- Belite hydration is accelerated by elevated temperature more effectively than C_3S.
- The strength of the hardened slurry at short hydration times is obtained from C_3S whereas after longer hydration times the contribution of C_2S to the strength development increases significantly. Above 70 °C, AFt phase becomes unstable and AFm phases are formed as the sole calcium aluminate sulphate hydrate.
- The hydration chemistry changes as crystalline hydration products tend to supersede the poorly crystalline/amorphous ones and the pattern of hydration behaviour is radically changed.
- To avoid strength retrogression occurring typically above about 100 °C, appropriated amounts of silica in the form of silica sand, silica flour or finely ground quartz have to be added to the slurry.

- Addition of about 35/40 wt % of silica can prevent strength retrogression by allowing the formation of 11 Å Tobermorite ($C_5H_6H_6$) below 200 °C. The formation of this phase is associated with good strengths and low permeability.

Between 100 and 150 °C

- Poorly crystalline C–S–H, commonly referred to as C–S–H II is produced.
- If there is no protection against strength retrogression, C–S–H II which has a high compressive strength and low permeability, transforms into the crystalline alpha C_2SH that has low compressive strengths and high permeability. It converts into tricalcium silicate hydrate, Jaffeite (C_6S_2H) which also has low compressive strengths and high permeability.
- Tobermorite is gradually transformed into Xonotlite (C_6S_6H) and Gyrolite ($C_8S_4H_8$). Both of these phases have lower compressive strengths and somewhat higher permeability than Tobermorite but these changes are much lower than when alpha C_2SH and $C_6S_2H_3$ are formed. Unlike the later, they do not cause strength retro regression and durability problems per se.
- Tobermorite can remain up to 250 °C since conditions do not arise under real hydrothermal situations.

Between 150 and 200 °C

- Under hydrothermal conditions, there is little aluminoferrite or few alumina phases present or any residual calcium hydroxide. The latter is effectively absorbed into the reacting media to form various crystalline calcium silicate hydrates. The former largely end up in solid solution within the hydrothermal calcium silicate hydrates. Some Hydrogarnet and Stratlingite are observed in the temperature range 100–200 °C.
- Increasing pressure also accelerates the rate of hydration. These have not been as extensively studied as temperature increase.
- Both pressure and temperature rise have a strong effect upon the hydration behaviour of oil well downhole particularly under high temperature, high pressure (HT-HP) conditions.
- Mix of chemical additives including heavy weight additives, a retarder, a strength retrogression inhibitor and a fluid controller are usually necessary to ensure that a good cementing work is carried out.
- Strength retrogression caused by α-C_2S and $C_6S_2H_2$ is normally not too severe to maintain the casings and/or liners in the annular space between the borehole wall and the casing liner.
- A compressive strength of 5 MPa should be adequate for such application. The compressive strength attained should normally be well above this value even with strength retrogression. However, the concomitant increase in permeability of the hardened cementitious products from around 0.1 mD (millidarcy) for C–S–H II about 10–100 mD for α-C_2SH and $C_6S_2H_3$, is sufficient to have a deleterious effect on the durability of the cement placed in the annular space. Thus, the presence of significant quantities of these two hardened products is best avoided.

- Addition of about 35–40 wt % of silica flour or silica sand can prevent strength retrogression by allowing the formation of Tobermorite ($C_5S_6H_5$) a phase which produces both high compressive strengths and low permeability.

At 200 °C and above 250 °C

- Tobermorite converts into $C_6H_2S_3$ (Jaffeite) but the strength retrogression is not severe and consequently the casings in the annular space between the borehole walls and the casing is maintained.
- Tobermorite can remain stable up to 250 °C since equilibrium conditions do not normally arise under real hydrothermal situations.
- Above 250 °C, Truscotite forms from Gyrolite and from any residual Tobermorite, although some Gyrolite may persist above this temperature.
- Truscotite also has lower compressive strength and higher permeability than Tobermorite but it is sufficient to avoid strength retrogression and durability issues in normal circumstances.
- Finely above 250 °C, $C_7S_{12}H_3$ may also be formed which is weaker and less permeable than Gyrolite.
- Any aluminoferrite derived phases have been independently established and it must be assumed that almost all Al and Fe ions become incorporated into the crystalline lattice of the existing calcium silicate hydrates.

About 400 °C

- Class G and Class H cements have an effective upper limit for usage under normal downhole conditions of 400 °C. This arises because above 400 °C, they become less stable as binder owing to the presence of high shrinkage; 5–10% or more under these conditions.
- Truscotite and xonotlite which are the most stable high temperature binders formed by Class G and Class H cements also decompose above 400 °C.
- As a consequence of this decomposition coupled with the high shrinkage experienced, the disintegration of the hardened takes place.

6.4.4 Expansive Cement

6.4.4.1 Introduction – Mechanism of Expansion

- Expansion of cement and concrete is a process in which chemical is converted into mechanical work that overcomes the cohesion of the system; expansive force results from crystallisation pressure produced by the precipitation of minerals during the chemical reactions concerned.
- Two conditions are necessary for the occurrence of crystallisation pressure:
  (1) Confined crystal by growth of solid products.
  (2) High supersaturation with respect to solid products that precipitate.
- Generally, the reduction in the activity of any reactants of the chemical reactions concerned will decrease pressure and expansion of cement and concrete.

- In contact with water, some chemical products react to yield new constituents containing water. This transformation is accompanied by modification of the volume. Even though the total volume of solids increases, the volume of the formed hydrates is less than the sum of the volume of the original cement and water consumed in the hydration reaction leading to a chemical shrinkage.
- As long as the paste is still plastic the chemical shrinkage results in a corresponding decrease in its external volume.
- After setting, the paste does deform; the external shrinkage represents a fraction of the total chemical shrinkage.
- Depending on the external curing condition, the hardened paste loses water and exhibits a drying shrinkage. The resulting tensile stress may cause cracks in the hardened concrete.
- Expansive cements are binders that generate expansive stresses in the hardened paste in the course of hydration counteracting the tensile stresses generated by chemical and drying shrinkages.
- These shrinkage-compensated cements are applied to the production of crack free concrete structures such as multi-storey car garages or water storage tanks.
- If the expansive stresses generated initially are higher than needed just to balance the tensile stress caused by drying, the hydrated material remains under stress even after it has completely dried.
- The 'self-stressing' cements are employed in the production of prestressed concrete products and structures.

6.4.4.2 Expansive Cements Based on Portland Cements

- The chemical reactions that can cause expansion of hardened concrete can be used to make cements of varying degrees of expansion *e.g.* shrinkage compensating cement and self-stressing cement.
- The total of CaO, SiO_2, Al_2O_3, Fe_2O_3 account for >85% of the typical bulk composition of expansive cements (table 6.23).
- Three types of cement designated by K, L, and M, are produced industrially. They are characterized by different alumina bearing compounds as the principal sources of alumina ions in the hydration reaction:

    ○ Type K is based on $C_4A_3\$$.
    ○ Type M contains CA + $C_{12}A_7$ (Calcium aluminate cements).
    ○ Type S contains a high C_3A – PC clinker. $C_4A_3\$$ is burnt separately at 1300 °C and interground with PC clinker with additional calcium sulphate.

- The formation of ettringite is the chemical reaction that produces a controlled expansion of restrained concrete. The reactions described table 6.24 are responsible for its expansion.

Applications to Hydraulic Binders

TAB. 6.23 – Chemical analysis of expansive cements (weight %).

Type	CaO	SiO_2	Al_2O_3	Fe_2O_3	MgO	SO_3	LOI
Shrinkage-compensating	62.85	19.24	5.83	1.94	3.48	6.65	1.54
Self-stressing	59.26	15.02	8.64	1.65	2.78	12.65	1.90

TAB. 6.24 – Chemical reactions involved during hydration of Type K cement.

Reaction 1	$C_4A_3\$$ + Type M cement	+ $8C\$H_2$	+ $6C$	+ $80H$	$\Longrightarrow$	$3C_6A\$_3H_{32}$
Reaction 2	CA + Type S cement	+ $3C\$H_2$	+ $2C$	+ $26H$	$\Longrightarrow$	$C_6A\$_3H_{32}$
Reaction 3	C_3A	+ $3C\$H_2$		+ $26H$	$\Longrightarrow$	$C_6A\$_3H_{32}$

6.4.4.3 Expansive Cements Based on Calcium Oxide Hydration

- Expansive cements based on calcium oxide are in essence Portland cement with a content of hard burn free lime in high enough amount to produce the required expansion based on the hydration of lime:

$$CaO + H_2O \rightarrow Ca(OH)_2.$$

- Cement compositions with high content of hard burn free lime are suitable as demolition agent owing to the relatively high stresses generated in the hydration expansion process.

6.4.5 Dental Cements

6.4.5.1 Introduction

- Dental cements are a group of materials acting to bond preformed restoration and orthodontic attachment in or on the tooth, including temporary restoration of teeth, use for cavity liners and bases to protect the pulp sedation or insulation and cementic fixed prosthodontic appliances.
- The ideal cement characteristics are:

  o The particles of powder should be very fine and mix easily with liquid.
  o Be tacky viscous when mixed to provide good adhesion between it and the canal wall when set.
  o Have adequate working and setting times.
  o Provide a good marginal seal to prevent marginal leakage. Thus, it should make a hermetic seal without shrinkage upon setting.

TAB. 6.25 – The main classes of dental cements.

Cement	Properties				
	Film thickness (μm)	Setting time (min)	Solubility (wt %)	Compressive strengths (MPa)	Modulus of elasticity (GPa)
Zinc phosphate	25–35	5–14	0.2 max	80–100	13
ZnO eugenol	25–35	2–10	1.5	2–25	
Polymer reinforced	35–45	7–9	1	35–55	2–3
EBA Alumine	40–60	7–13	1	55–70	3–6
Zinc poycarboxylate	20–25	6–9	0.06	55–85	5–6
Glass ionmer	25–35	6–9	1	90–140	7–8
Polymer based	20–60	3–7	0.05	70–200	4–6
MTA/C_3S		160		40	

- Have a good retention (bond between cement and restorative materials).
- Have low film thickness (ideal 25 μm) and high strengths in tension and compression.
- Have good thermal and chemical resistances.
- Be insoluble in saliva and tissues fluids.
- Be bacteriostatic or at least not encourage bacterial growth.
- Be non-irritant in acidic media.
- Be well tolerated by the periapical tissue.
- Have some opacity to X-rays for diagnostic purposes on radiographs.
- Have good aesthetics. It should not discolour tooth structure.
- Be soluble in common solvents if necessary.

6.4.5.2 The Main Classes of Dental Cements

- The main classes of dental cements are cements based on phosphate, phenolates, polycarboxylate, acrylic, glass ionomer, and mineral trioxide aggregates. Table 6.25 summarizes their main characteristics.

6.4.5.2.1 Phosphate Cements
- Zinc phosphate cement: the powder is mainly zinc oxide fired at high temperatures to reduce the reactivity. The liquid is an aqueous solution of phosphoric acid containing 45–64% H3PO4 and 30–55% water.
- The setting reaction is as follows:

$$\text{Zinc oxide} + \text{Phosphoric acid} \rightarrow \text{amorphous zinc phosphate.}$$

- The freshly mixed zinc phosphate is acidic with a pH between 1 and 2.

- The working time is between 5 and 14 min.
- The compressive strength ranges between 80 and 110 MPa with a powder/liquid ratio between 2.5 and 3.5 (the minimum strength for adequate restoration is about 60 MPa).

6.4.5.2.2 Silicophosphate Cements

- The powder is composed of blends of 10–20% by weight zinc oxide and silicate glass containing 12–25% fluoride. The blends are mixed or fused and then ground. The liquid is phosphoric acid containing 45% water.
- The setting reaction is represented as follows:

  ZnO + alumino silicate glass + Phosphoric acid → zinc aluminosilicate phosphate gel.

- The setting time is 5–7 min and the compressive strengths range between 140 and 170 MPa.
- Because of the acidity of the mixture and the prolonged low pH (4–5 after setting), pulpal protection is necessary on all vital teeth.

6.4.5.2.3 Phenolates Based Cements

- Zinc oxide – eugenol cements

  o The powder is pure zinc sometime containing silica as filler. The liquid is purified eugenol.
  o The setting reaction is as follows:

  Zinc oxide + eugenol + water → zinc eugenolate.

  o Setting time ranges from 2 to 10 min.
  o The compressive strength is low (7–40 MPa).
  o The solubility is moderate (1.5% after 24 h).

- Reinforced zinc oxide – eugenol cements

  o The solid is zinc oxide with 10–40% finely divided resin (PMMA, polystyrene, polycarbonate). The liquid is eugenol.
  o Setting time is 7–9 min and the compressive strengths range between 35 and 55 MPa. Because of the presence of the resin, the solubility is lower than that of zinc eugenol materials.

- The main advantages of these materials are the minimal biological effects, good initial sealing properties, and adequate strengths for final cementation of restorations. The principal disadvantages are the lower strengths and a moderate solubility.

6.4.5.2.4 EBA (Ethoxy Benzoic Acid) and Other Chelate Cements

- The powder is mainly ZnO containing 20–30% aluminium oxide and polymeric reinforced agents such PMMA. The liquid is eugenol and 50–66% EBA.

- These materials have been used for the cementation of inlays, crowns and fixed partial dentures, for provisional restoration and as base of lining materials.
- EBA cements are very easy to mix and have a long working time, good flow characteristics and low irritation to the pulp.

6.4.5.2.5 Calcium Hydroxide Chelate Cements
- The action of calcium hydroxide as a pulp-capping material that facilitates the formation of reparative dentine appears to be largely attributable to its alkaline pH, antibacterial effect and protein-lysis effect.
- These cements are used as liners in deep cavity preparations.
- Setting time is about 2 min and the compressive strengths at 7 min is 6 MPa.

6.4.5.2.6 Polycarboxylate Based Cement
- Zinc Polycarboxylate based cements are adhesive dental cements that combine the strength properties of the phosphate system with the biologic acceptability of the ZnO eugenol materials.
- The powder is zinc oxide, the liquid is a solution of polyacrylic acid or an acrylic acid copolymer with other organic acid such as itaconic acid leading to the following chemical reaction;

$$\text{Zinc oxide} + \text{polyacrylic acid} \rightarrow \text{zinc polyacrylate}.$$

- Setting time is 9 min at 37 °C and the compressive strength ranges from 55 to 85 MPa.
- They are used for the cementation of cast alloys, porcelain restorations and orthodontic band as cavity liners or base materials. They can also be used as provisional materials.

6.4.5.2.7 Acrylic Resin Cements
- In the case of acrylic resin cements, the powder is a finely divided methyl metacrylate polymer or copolymer containing benzoyl peroxide as the initiator. The liquid is a methyl metacrylate monomer containing an amine accelerator.
- The mix must be used immediately because working time is short. Moreover, these cements do not have an effective bonding to tooth structure in the presence of moisture.
- Acrylic resin cements are used for the cementation of restorations, facing and provisional crown.

6.4.5.2.8 Dimethacrylate Cements
- They are a combination of an aromatic dimethacrylate with other monomers containing various amounts of ceramic filler. They are basically similar to composite restorative materials.
- The powder is generally a finely divided borosilicate or silica glass with fine polymer powder and an organic peroxide initiator. The liquid is a mixture of bis-GMA and/or dimethacrylate monomer containing an amine promoter for polymerization.

Applications to Hydraulic Binders

- Monomer conversion is incomplete and thus manipulation is critical to get the optimum physical properties. Maximum properties are generally reached about 10 min after polymerization. Compressive strengths range between 100 and 200 MPa. These cements have high strengths, low oral solubility and micro-mechanical bonding to enamel, dentin, alloys, and ceramic surfaces. Disadvantages include the need for meticulous and critical techniques, more difficult sealing, possible leakage, and pulp sensitivity.
- They are used for bonding crown (usually porcelain) fixed partial denture, inlay, veneers and in direct resin restoration.

6.4.5.2.9 Glass Ionomer Cements
- The powder is finely ground calcium fluoro-silicate glass with a particles size around 40 μm.
- Typical composition by weight: SiO_2 43%, Al_2O_3 33.1%, CaO 11.4%, and Fe_2O_3 7.5%.
- The liquid is a 50% aqueous solution of polyacrylic-itaconic acid or other polycarboxylic acid copolymers that contain 5% tartaric acid.
- The setting time is in the range of 6–9 min. The compressive strength increases over 24 h to lie between 90 and 140 MPa.
- Glass ionomer are used for the cementation of cast alloys and porcelain restoration and orthodontic bands as cavity liners or base materials. They are also used as restorative materials especially for erosion lesions.
- In hybrid ionomers, the acid base setting reaction in glass ionomer cements has been modified by the introduction of water-soluble polymers and polymerizable monomer into the composition. Among the advantages are dual cure, fluoride release, higher flexural strengths than glass ionomer cements and the ease of handling.

6.4.5.2.10 MTA (Mineral Trioxide Aggregates)

CAM 2005, PAR 2010–1, PAR 2010–2, TOR 2010, ROB 2008

6.4.5.2.10.1 Introduction

- In 1995, a tooth filling material and its method of use was patented. It is based on a mixture of 80% of a product chemically and mineralogically similar to Portland cement Type 1 and 20% of Bi_2O_3 acting as radio pacifier for recognition on radiographs.
- The product was modified to comply with the regulation of the U.S. Food and Drug Administration and was developed industrially under the name MTA (mineral trioxide aggregate). It contains mainly tricalcium silicate, dicalcium silicate and Bi_2O_3.
- Initially recommended as a root-end filling material, MTA has been used for pulp capping, pulpotomy, apexogenesis, apical barrier formation in teeth with open apexes, repair of root perforation and root filling canal.

6.4.5.2.10.2 Characteristics of MTA

- Different grades exist, depending mainly on the iron content leading to white or grey colour; low iron content (WMTA) and medium iron content (GMTA).
- The particle size of MTA ranges from 1 to 10 μm.
- The mean setting time of is 165 ± 5 min.
- The compressive strength of MTA is 25 and 31 MPa at 1 and 28 days, respectively.
- The mean radiopacity for MTA has been reported at 7.17.
- The difference in composition leads to different expansion behaviour of WMTA and GMTA stored in HSSS (Hank's balanced salt solution).
- The pH value is 10.2 after mixing and rises up to 12.5 after 3 h. One of the most important features of MTA is the antimicrobial properties. Most bacteria present in the root canal system grow best at pH around 6.5–7.5. At pH more than 9, the loss of biological activity (Hsieh) leads to reversibly or irreversibly inactivate cell membrane enzymes of these microorganisms.
- MTA is recognised as a bioactive material able to create an ideal environment for healing. When placed in contact with human tissue, the material induces the following sequence:

    (a) $Ca(OH)_2$ that releases calcium ions for cell attachment and proliferation.
    (b) Creates an antibacterial environment by its alkaline pH.
    (c) Modulates cytokine production.
    (d) Encourages differentiation and migration of hard tissues producing cells.
    (e) Produces HAP on the MTA surface and provides a biological seal.

- MTA develops a sealing ability and marginal adaptation. It has been tested by leakage studies (dye, fluid filtration, bacteria and bacteria by products) and by electron microscopy, in all potential applications (root end filling, perforation repair materials, coronaral and apical plug material, canal filling material and root canal sealer).

6.4.5.2.10.3 From MTA to C_3S

- The main shortcomings of MTA are:

    ○ The presence of minor elements (table 6.26).
    ○ The lack of reproducibility of the cement.
    ○ The low amount of the most reactive phase, tricalcium silicate (C_3S), that ranges between 50 and 66 wt % (taking into account 15–20% of radiopacifier).

- Research has led to the development of a new family of bioactive and biocompatible dental materials for endodontics uses that are laboratory made products based on pure tricalcium calcium silicate.
- When accelerated with 15% of calcium chloride the setting time of pure tricalcium silicate is between 6 and 10 min.
- Its density is 2.25 g/cm^3 and the porosity is about 7% of the hardened product.
- Early strengths, measured according to ISO 9917:1991 is more than 100 MPa at 37 °C and the flexural strength is 34 MPa at 2 h.

TAB. 6.26 – Minor elements found in MTA.

Elements	Al	Mg	Fe	Ba	Zn	Cr	Mn	Ti	V	As	Ni	Pb	Cu
ppm	18 300	11 100	30 800	12	38	21	36	165	21	6	6		8

- Calcium silicate cements are able to tolerate moisture (hydraulic materials) and to harden in the presence of biological fluids (blood, plasma, saliva, dentinal fluids).
- Calcium silicate cements possess bioactive behaviour *i.e.* it stimulates the formation of new Apatite containing tissue since they are bioactive materials able to develop Apatite on their surface in a short induction period (GAN 2011) C_3S derived cements have been introduced in dentistry as materials for different endodontics clinical applications.
- The ability of calcium silicate cements to induce reparative dentinogenetic has been well demonstrated in animals in which direct pulp capping was performed in mechanically exposed pulps.

6.4.6 Glass Cements

- Calcium alumino silicate glasses are produced by grinding a glass in the system CaO–SiO_2–Al_2O_3 to a fineness from 300 to 500 m^2/kg.
- The oxide composition of the glass lies in the range 45/55% CaO, 22/40% Al_2O_3 and 12/26% SiO_2.
- A glass of a given composition is produced by an appropriate mixture at 1600 °C followed by quenching in excess water.
- By mixing calcium aluminosilicate glass with water, C_2ASH_8 (Stratlingite) and C_3AH_6 precipitate, leading to the setting and the subsequent compressive mechanical strengths.

Chapter 7

Application to Metal Refining

7.1 General Introduction

- Slag is an artificial mineral which develops simultaneously during the manufacture of iron and steel, ferro alloys and non-ferrous metals (ferroalloys, copper, and zinc).
- The total of CaO, SiO_2, Al_2O_3, and Fe oxides accounts for more than 95% of the typical bulk composition of slags. It also contains minor components (Mg, S, P, Ni, Cr...).
- Slags can be divided into several groups. At the low oxygen pressure prevailing in the furnace, only thermodynamically stable oxides are present and the most important constituents are CaO, SiO_2, Al_2O_3, Fe oxides. The phase relations can be understood from a consideration of this chemical system. Another group concerns slags from various processes which are carried out at higher oxygen pressures. Oxides of lesser stabilities are present as constituents of the slag composition in addition to the stable oxides.

7.2 Slags from Iron and Steel Industry

7.2.1 Blast Furnace Slags (BFS)

7.2.1.1 Introduction

- Iron is produced from ores or scraps under reducing conditions. In addition to large amounts of iron oxides they also contain silicates and aluminates and to separate these they must be combined with lime.
- The facility for lime to combine SiO_2, Al_2O_3, and Fe_2O_3 gives a product easy to melt (low T° liquidus, low viscosity, and high basicity). This liquid solution is called slag.

DOI: 10.1051/978-2-7598-2480-9.c007
© Science Press, EDP Sciences, 2020

- Slag is immiscible with molten iron, so the silicates can be removed from iron by draining off the slag (80 kg of lime/ton of iron). The slag is tapped in pits for further utilization while the hot metal follows other treatments.
- The abundance of lime in the earth crust is a guarantee of the durability of the manufacture.
- Steel contains iron, carbon (less than 2%), manganese (less than 1%), and other elements added for specific applications during the final step of the process.
- Four processes are currently used for the production of steel (figure 7.1):
    (1) Blast furnace/basic oxygen process (BOP).
    (2) Direct melting of scrap (electric arc furnace).
    (3) Smelting reduction.
    (4) Direct reduction (DRI).

- In some cases, the metal is over oxidised and must be 'killed' by the addition of reducing agents such as Si, Al, etc.... The chemical composition is adjusted according to the required quality of steel (secondary metallurgy). Inclusions are removed (clean steel). The metal is cast (continuous casting) and shaped to form items, such as; bars, slabs, rail, tube, plates, etc.
- The various production units are integrated in steelworks, including sinter plant, pellitisation plant; blast-furnaces, coke oven plant, basic oxygen steel plant, electrical steel plant, and ladle steel plant.
- In integrated steelworks, blast furnace is the unit where the primary reduction in iron ores takes place leading to liquid iron 'hot metal'.
- Blast furnace requires physical and metallurgical preparation of the burden and reducing agent: coke, pulverised coal forming carbon monoxide and hydrogen which reduce iron oxide. Coke is produced from coal by means of dry distillation in a coke oven and has better physical and chemical characteristics than coal.

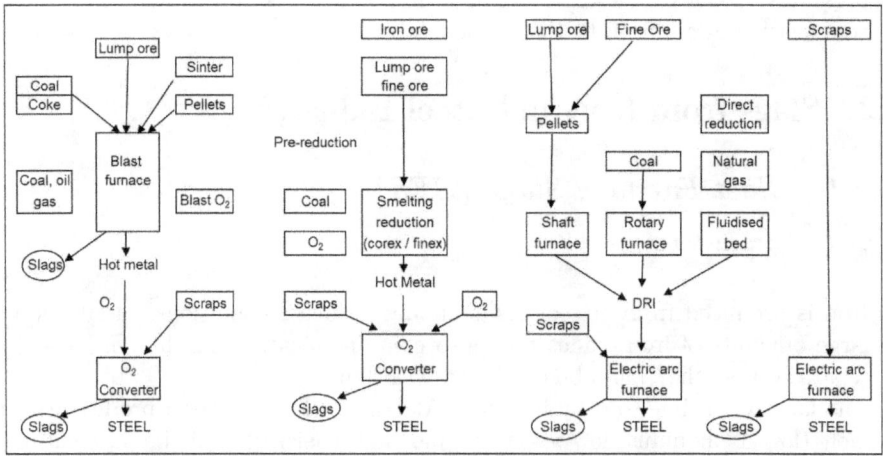

FIG. 7.1 – Processes currently used for the production of steel.

Application to Metal Refining

- Additional reducing agents/fuels are supplied by the injection of oil, natural gas and in few cases of plastics.
- A hot blast provides the necessary oxygen to form carbon monoxide (CO) that is the basic reducing agent for iron oxide.
- Iron ore fines are not suitable for direct use in blast furnace and must be agglomerated by sintering or pelletizing.

7.2.1.2 Agglomeration

- The total of CaO, SiO_2, Al_2O_3, and Fe oxides accounts for more than 95% of the typical bulk composition of agglomerates.
- The sinter consists of a mixture of fine ores, additives, iron bearing recycled materials from downstream operations such as coarse dust and sludge from BF gas cleaning, mill scale, casting scale, and coke breeze.
- Pellets are formed from the fines (<0.05 mm) of ore raw materials and additives into spheres (9–16 mm) using very high temperatures. This process is mainly carried out at the site of the mine or its shipping port.

7.2.1.2.1 Characteristics of the Agglomerates
- Sintering of pellets is relevant to the CaO–SiO_2–Al_2O_3–Fe_2O_3 system and phase relations help to establish the critical thermal and compositional parameters that control the bonding phase chemistry which in turn influences the physical characteristics of the pellet matrix.
- Fine ores (fraction <1 mm) are mixed with limestone fluxes (80/20 wgt %) and coke breeze and then heated to temperatures of around 1220–1300 °C. The composition of fine ores controls the sinter and thus pellet quality. The chemical analysis is 75.7% Fe_2O_3, 5.4% SiO_2, 0.35% Al_2O_3, and 18.45% $CaCO_3$.
- Sintering reactions convert the powdered raw material into a porous, strong cake composed of four main phases: iron oxide, glasses, dicalcium silicate, and SFCA.
- SFCA can be divided into two main types: the first one is a low-Fe form (named SFCA) that is characterized by a platy microstructure. The second is called SFCA 1 which appears needle like or acicular in cross section.
- Type and morphology of SFCA depend on the chemical composition of the sinter mixture and on the processing conditions of sintering.
- The chemical composition of many typical Hematite and Goethite iron ores and sinter pellet blends dictate that they will form a mixture of SFCA and SFCA 1 bonding phases.
- The typical temperatures used in sintering (1270–1300 °C) and pellet induration (1250–1350 °C) favour the formation of both SFCA and SFCA1 bonding phases up to 1300 °C. SFCA is favoured above 1300 °C.
- SFCA1 is the most desirable bonding phase in iron ore sinter and pellet since microstructures composed entirely of SFCA1 show higher physical strengths and higher reducibility than microstructures composed predominantly of SFCA (OLU 2007, DAW 1983).

- Alumina is an essential component in the production of SFCA and the amount of SFCA increases with increase in Al_2O_3. As temperature rises more alumina might be needed to obtain more SFCA.
- Iron ores containing Gibbsite ($Al(OH)_3$) as their source of Al_2O_3 are inferior to those containing Kaolinite. Alumina in the former is slow to react and hence less SFCA is formed particularly at low reaction temperatures. Furthermore, relicts of gibbsite are centres of weakness and result in weaker sinters. Since Al_2O_3 in Kaolinite is completely reacted in the sinters, no relict alumina particles could be detected. High silica content favours the optimum form which is fine and fibrous.
- Viscosity of the melt governs the crystallization of SFCA. High viscosity favours the formation of SFCA. Al_2O_3 increases the viscosity while MgO decreases it (EGU 1989).
- The free silica contained in the fraction below 28 mesh delays the formation of calcium ferrite and lowers the amount of liquid phase during sintering.

7.2.1.2.2 Mechanism of Agglomerates Formation

- Phase formation is dominated by solid state reaction, mainly in the $CaO-Fe_2O_3$ system. Studies of mixture of Hematite, Quartz, and Lime show that the reaction proceeds as follows when the temperature increases:

  (1) Hematite reacts with CaO at low temperatures (750/900 °C) to form calcium ferrite (C_2F).
  (2) C_2F reacts with Fe_2O_3 to give CF. CF increases proportionally with the alumina in the bulk sample.
  (3) The formation of SFCA and SFCA 1 begins at 1050 °C via a solid-state reaction mechanism.
  (4) At about 1160 °C, a melt is formed and Quartz dissolves into the melt. The presence of alumina increases the temperature at which both ferrite phases initially form. The formation of SFCA and SFCA 1 is more rapid in the high alumina systems.

- The presence of alumina increases the stability range of both SFCA phases lowering the temperature at which they begin to form.
- Though Al_2O_3 has been confirmed to be essential for the formation of SFCA, MgO inhibits its formation. However, there is no general agreement on how Al_2O_3 interacts with other components such as Fe_2O_3, CaO, and SiO_2 (EDS 1956, HUG 1967, JAS 1996, OLU 2007).

7.2.1.3 Properties and Formation of BF Slags

- The formation of types of blast furnace slags are summarized in figure 7.2.

7.2.1.3.1 Blast Furnace Slag Processing: High Temperature Formation

- The slags formed in the blast furnace are relevant to the $CaO-SiO_2-Al_2O_3$ system.

Application to Metal Refining

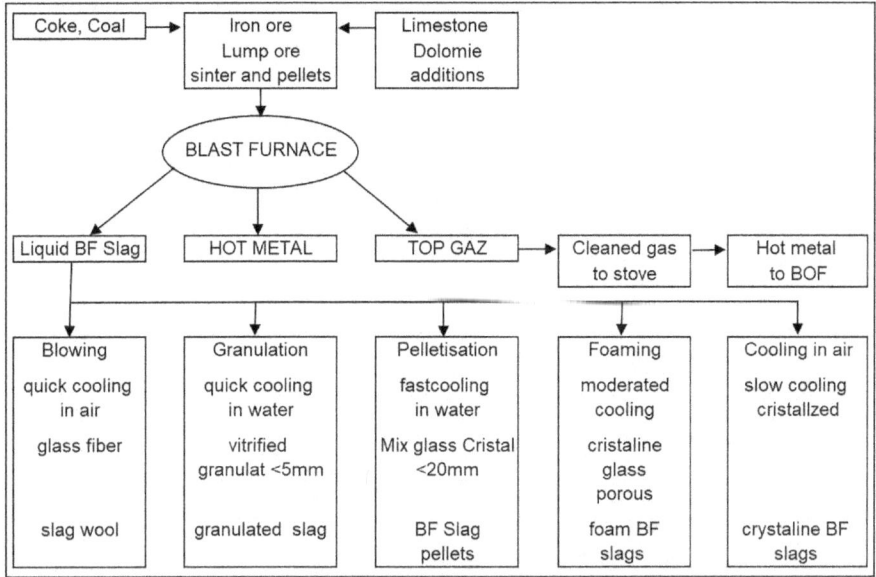

FIG. 7.2 – Formation and properties of BF slags.

- The optimum composition of slag can be estimated from the criteria of temperature, viscosity, and desulphurization potential.
- In the diagram of the CaO–SiO_2–Al_2O_3 system, the position of the isotherm at 1500 °C, shows a large potential area suitable for slags. The sulphur-removing capacity of the slag increases in the order $SiO_2 < Al_2O_3 < CaO$. The viscosity of the liquid decreases with increasing contents of CaO. For a given Al_2O_3, the viscosity of the liquid phase will be lower with lower silica content.
- The isoactivity curve indicates that at low silica concentrations, γ-SiO_2 is raised progressively as the alumina concentration increases. For values of α SiO_2 greater than 0.6, this trend is reversed and alumina additions lower γ-SiO_2 consistent with the amphoteric behaviour of alumina.
- The activities raised slightly as the temperature increase from 1500 to 1700 °C.

7.2.1.3.2 Blast Furnace Slag Processing: Cooling

- Depending on the way of cooling of the liquid slag, different types of products are obtained: granulation, pelletizing, and slag pit processes.
- Slag granulation involves pouring the molten slag through a high-pressure water spray in a granulation head located in close proximity to the blast furnace. For slag granulation fresh seawater in open or closed circuit can be used. The residual water is around 10%.
- In the pelletizing process, the molten slag is spread in a layer on a plate which acts as deflector. The slag is projected centrifugally into the air on a rotating

drum to complete the blowing up and cooling. The slag particles follow different trajectories according to their size which range from granulated sand to expended pellets.
- Slag pit process (slow cooling in air) involves pouring a thin layer of molten slag directly into slag pit. The molten slag is slowly cooled and crystallizes in open air. It contains Melilite and at lesser amounts Merwinite or Wollastonite.
- When cooled fast, the slag contains 90% of glassy phase.
- An average chemical composition of BF slags is given in table 7.1.

7.2.2 Converter Slags (Basic Oxygen Process or LD Processing)

7.2.2.1 BOP Processing

- The total of CaO, SiO_2, Al_2O_3, and Fe oxides accounts for more than 95% of the typical bulk composition of BOP slags.
- The objective in oxygen steelmaking is to burn (*i.e.* to oxidise) the undesirable impurities contained in the hot metal feedstock. Transformation of the liquid iron (hot metal) into steel is obtained by oxidation of carbon in a basic oxygen furnace (pear shaped converter). The main elements converted into oxides are carbon, silicon, manganese, and phosphorus. Sulphur content is reduced during pre-treatment of the hot metal.
- Carbon is reduced from 4–5% to 0.01–0.4%.
- The production of steel by the BOP process is discontinuous involving the following steps:

  (1) Transfer from the BF and discharged. The converter is charged with lime and after the molten metal (70–85%), the scraps are poured in. Oxygen is then blown on the surface of the liquid.
  (2) Pre-treatment of hot metal (desulphurisation, deslaging).
  (3) Transfer and oxidation in BOF (decarburisation and oxidation).
  (4) Secondary metallurgical treatment.
  (5) Casting.

- The slag is generated by the combination of lime (CaO) and the oxidised constituents of the iron during blowing oxygen into the melt (T° 1600 °C). Table 7.2 summarizes the main reactions.
- The liquid slag (BOS) is tapped in pits forming crystalline slag. An average chemical composition of LD slag is given in table 7.1.
- LD slag contains magnesio Wüstite solid solution ($Mg_xCa_yMn_{0.11}Fe_{0.6}O$) with $0.12 < x < 0.21$ and $0 < y < 0.14$, Lime solid solution ($Ca_{0.83}Mg_{0.08}Mn_{0.04}Fe_{0.05}O$), dicalcium silicate ($Ca_2Si_{0.91}P_{0.1}Fe_{0.01}O_4$), calcium aluminoferrite ($Ca_{2.2}Si_{0.03}Al_{0.04}Fe_{1.4}O_5$), Periclase ($Mg_{0.7}Ca_{0.03}Mn_{0.03}Fe_{0.2}O$) (CON 1981).
- The slags formed during blowing oxygen in BOP furnace are relevant to the phase diagrams of the SiO_2–FeO and CaO–SiO_2–FeO systems.

TAB. 7.1 – Average chemical composition of slags (weight %).

Oxide	BF	BF	LD	EAF	EAF	EAF low alloyed carbon steel
SiO_2	40.38	33–39	13.33	7.82	10/20	10–18
Al_2O_3	6.88	9–13	2.04	2.56	<10	3–8
CaO	35.64	39–42	45.89	56.83	30–40	25–45
MgO	13.56	6–9	8.05	5.3		4–13
FeO			15.96	22.41	15–35	
Fe_2O_3			5.95			
Fe						10–32
MnO			4.66	3.5		4–12
P_2O_5	1.04		1.96	0.55		<0.6
TiO_2			0.34	0.16		
Cr_2O_3			0.57			1–2
S	1.04	1.2–1.4	0.146	0.142		0.02
K_2O	0.40					
Na_2O	0.31					
LOI						
Free lime						

- The ternary FeO–Fe_2O_3–SiO_2 system is the simplest system which can represent adequately the chemical properties of the slag produced in steelmaking process and is important to understand the behaviour of siliceous refractories.
- The temperature of fusion of iron oxide in contact of silica is sensitive to gas composition; hence fusion may be accomplished in some cases by controlling the gas atmosphere without changing the temperature. The melting point of iron oxide in contact with silica has been determined as a function of temperature (DAR 1948).
- Liquidus surface and oxygen activity of liquid iron silicate slags in equilibrium with gamma iron ranging in composition from 0 to 40% (silica saturation) have been determined at temperatures from 1250 to 1400 °C.
- Equilibrium determinations were also made for the slag – gamma iron-Wüstite-gas and gamma iron-silica-gas systems.
- From the experimental and calculated data activities of FeO and SiO_2, it was found that the activity composition relationship deviates considerably from those to be expected for an ideal binary solution of FeO–SiO_2. However, the partial molal heat of solution of FeO from the activity temperature data was found to be equal to 0. Also, the partial molal heat of solution of liquid SiO_2 (metastable) in silica saturated slags was estimated to be 0 (SCH 1951).
- The experimental data combined with previously published information allow constructing a ternary diagram for the system showing the entire T°/composition range of stability of iron silicate slags (SCH 1953).
- The first slag to form is rich in FeO and SiO_2 and low in CaO. The composition lies in the Wüstite and Olivine primary phase region of the ternary system (T° 1200 °C).

TAB. 7.2 – Main reactions in BOF processing.

Oxidation BOP				
Carbon elimination				
	[C] +	[O]	→	[CO] off gas
	[CO] +	[O]	→	[CO_2] off gas
Oxidation of accompanying and tramp elements				
Desiliconation	[Si] +	2[O]	→	2(CaO–SiO_2)
Demanganese reaction	[Mn] +	[O]	→	(MnO)
Dephosphorisation	2[P] +	5[O]	→	3(CaO)–P_2O_5
Desoxidation				
Removal of residual oxygen through	[Si] +	2[O]	→	(SiO_2)
ferrosilicon aluminium	2[Al] +	3[O]	→	(Al_2O_3)

Note: [ ] dissolved in metal, ( ) contained in the slag.

- As CaO goes into solution, the slag moves into C_2S primary phase field. Liquidus T° increases to 1400 °C and above.
- The oxygen potential of the slags (10^{-5} atm) influences the liquidus isotherm. When PO_2 increases from 10^{-8} to 10^{-5} atm, the solubility of CaO at 1600 °C increases by 10%, starting from an acid oxidising slag in the early stages, to finish with a highly basic slag at 1600 °C (BOD 1972).
- The phase diagram shows that as the FeO content increases, the liquidus temperature of the slag increases. If this zone is approached before the temperature of the bath has risen high enough, then precipitation of C_2S can occur with the consequence of slag overflowing.
- Hence the FeO content of slag in the early stages must be maintained high enough to keep the liquidus temperature of the slag low until the slag temperature is high enough. This involves encouraging the formation of FeO by keeping the lance bath distance large enough.
- Two factors control the dissolution of lime dissolution in CaO–SiO_2–FeO system. The sintering of Lime and the formation of C_2S as intermediate compound near the dissolving Lime surface. The porous Lime shows higher reactivity of dissolution in CaO–SiO_2–FeO system (BAN 1982).
- When treating high phosphorus containing charge, two slags are necessary with an intermediated deslagging and the amount of phosphorus in the slag is high.
- For phosphorus removal, the activity of calcium oxide and the oxygen potential of the slag must be high while P_2O_5 activity coefficient must be low. The following formula gives the optimum content of lime:

% CaO = $1.41(P_2O_5 + \% Al_2O_3 + \% TiO_2) + 2.61\% SiO_2 - 0.005(\% FeO)$.

For desulfurisation, low oxygen potential is required, which is not the case in oxygen steel making processes, and another stage (secondary refining) is necessary.

7.2.2.2 Modification of the Composition of BOP Slag by Addition of Alumina (Bauxite)

- To avoid the precipitation of a layer of C_2S on the surface of lime, additions of fluorspar are usually made to the slag. From the isotherm at 1650 °C of the CaO–SiO_2–FeO system, it is shown that the unsaturated slag will be able to dissolving CaO without the intervention of calcium silicate only when the SiO_2/iron oxide ratio is less than 21/79 (WHI 1974).

7.2.3 EAF Slags: High Carbon Steel

- The direct smelting of iron bearing materials such as scrap is usually performed in electric arc furnace. The heat is generated by an electric arc. The main advantage is a close control of the temperature of the metal and the atmosphere over a wide range of oxygen potentials.
- Electric arc furnace slag is formed during melting steel scraps at 1600 °C by addition of limestone with some oxidised elements of the melt.
- The slag (EAF C) is air tapped and crystallises during cooling. An average chemical composition of EAF slag is given in table 7.1.
- EAF slag contains Gehlenite $(Ca_2(Al(Al,Si)_2O_7)$, Larnite (C_2S), Bredigite $(Ca_{14}Mg_2(SiO_4)_8)$, manganese oxides, Magnesioferrite $(MgFe_2O_4)$, and Magnetite (Fe_3O_4).
- For the production of carbon steel and low alloyed steels, the following operations are performed: EAF charging with scraps, melting, steel and slag tapping, ladle furnace treatments for quality adjustment, continuous casting and slag handling.
- The carbon steel is used for reinforcing bar, slabs, heavy rail, pipeline tube, heavy plates, stream generator tubes, ship building steel, boiler, and pressure vessel steel (LUX 2000).

7.2.4 Ladle Slag (Secondary Metallurgy) – Stainless – High Alloy Steel Production

- Electric arc furnace slag (EAFS) from high alloy production is formed during the manufacture of high alloy steel in different vessels, EAF, converter or ladle.
- Scraps or reduced iron sponge are melted with lime and alloys (ferro chromium, ferro nickel). Oxygen is blowing for mixing and chromium from the ferroalloy is oxidised and this oxidation is limited by the presence of silicon added as ferro silicon.
- Silicon oxidised as silica combines with lime to from a calcium silicate (the basicity index of the slag $CaO/SiO_2 = 1.2$). Liquid steel contains 1.5–5% of carbon.
- For high alloyed and special steels, the following operations are carried out: desulphurisation, degassing for the elimination of gases (N, H...), decarburisation (argon oxygen decarburization (AOD) or vacuum oxygen decarburisation (VOD)).

- The liquid steel is tapped in a vessel (AOD for example) and oxygen is blowing to oxidise carbon. Ferrosilicon is added to limit the oxidation of chromium and its dissolution in the slag. Lime is added to combine silica as calcium silicate and sulphur as calcium sulphide. The basicity index of the slag is 1.6.
- The slag is removed from the vessel and another amount of lime is added to increase desulphuration. The metal is removed and the final slag is recovered: 70 kg (per Ton of steel slag) with EAF (basicity 1.3), 90 kg with AOD (basicity 1.74), and 10 kg for ladle slag.
- The slag obtained is crystallised. A standard stainless steel contains 18% chromium and 8% nickel.
- Stainless steels are used for cold roller hoop and strip, steam generator tubing, nuclear fuel cladding tube, seamless tube, and pipe.
- Alloy steels are used for hot rolled coil, hot rolled hoop and strip, skelp, electrical sheet, bloom, billets, wire rod, round bars, spring steel.
- Tool steel tubes are used for production of component for armament, naval and aerospace industry.

7.2.5 Refining Under Reducing Slag

- For the production of high-quality steel, when very low sulphur content is required, refining is usually completed under a refining slag.
- After removing oxidising slag, reducing agents are added (Ferrosilicon, Al) and a reducing slag is formed by adding lime, coke and fluospar to the bath.

7.2.5.1 Desulfurization Agent: Case of LDSF

- Sulphide capacity of the melt $[C_s = (\% \, S) \sqrt{pO_2/pS_2}]$ of ternary system at 1500 and 1600 °C for the ternary can be estimated from the sulphide capacity of the melt of binary $CaO-Al_2O_3$ and $CaO-SiO_2$ systems and the chemical composition of the melt (TUR 1961).
- From the equilibrium data obtained with of $CaO-SiO_2-Al_2O_3$ melts, liquids or gasses containing sulphur, it is possible to assess the activity of lime in this melt (KAL 1960).
- Since 1970, there has been an increasing demand for steels with lower sulphur specification. Sulphur removal, as globularization of the remaining sulphide inclusion is necessary to obtain maximum toughness particularly in the transverse direction measured as Charpy-V notch measurement (CVN). Improvement in the transverse CVN is associated with improvement in ductility in the through thickness direction (Z direction).
- There are three main methods by which sulphur may be removed from blast furnace iron or molten steel:

    (1) Reaction with metallic addition (Mg rare earth elements).
    (2) Reaction with compound additives such as CaC_2, Na_2CO_3.
    (3) Reaction with fluid basic slags with low FeO contents.

Application to Metal Refining

- The following equation represents the removal of sulphur from the metal, by reaction with the slag. High sulphur capacity slags are primarily a way of producing low melting point options. The major component of which from the standpoint of sulphur removal is lime:

$$\{S\}_{metal} + (O^{2-})_{slag} \rightarrow (S^{2-})_{slag} + \{O\}_{metal}.$$

  $\{S\}_{metal}$ and $\{O\}_{metal}$ represent the elements dissolved in the metal and $(O^{2-})_{slag}$ $(S^{2-})_{slag}$ represent the elements dissolved in the slag

- If oxygen is linked to Ca as CaO, the reaction can be written:

$$\{S\}_{metal} + (CaO)_{slag} \rightarrow (CaS)_{slag} + \{O\}_{metal}.$$

- The equilibrium constant is written: $K = (a_{CaS} \cdot a_O)/(a_{CaO} \cdot a_S)$ where a_O and a_S represent, respectively, the activity of oxygen and sulphur in the metal and the slag.
- Alumina, silica or a combination of both are used to flux the lime. Desulfurization is better with slags that are heated at several hundred degrees above the melting point where their fluidity is reasonably high.
- In 1920, R. Perrin developed a process for the desulfurization of steel in the ladle. A synthetic lime alumina slag was melted in a submerged arc furnace and tapped into the ladle. Molten steel from a separated furnace was poured onto the slag. Desulfurization was found to be rapid and efficient.
- Based on this principle a method of producing a synthetic calcium aluminate slag was patented (LDSF) (SOR 1984).

7.3 Formation and Properties of Liquid Slags

- The slag should be completely liquid, at steelmaking temperature. The location of the liquidus surface must be determined in the corresponding diagram.
- Its composition must be adapted to the process using the thermodynamic properties of the chemical components of the system.
- Some uncontrollable variations of composition will inevitably take place, and must be within acceptable tolerances of the refining process (range of acceptable compositions).
- Its viscosity must be low enough to render favourable flow and diffusion of the species (determination of the structure of slag).
- Factors that determine the rate of chemical reactions, within the slag and between the slag and metal, including the erosion of the refractory lining, have to be examined.
- It should have high sulphur removing capacity.
- Molten slags are ionic in nature, consisting of positively charged ions (cation) and negatively charged complex silicates, aluminates, and phosphates (anion).
- In solid and molten silicates, each silicon atom is tetrahedrally surrounded by four oxygen atoms and each oxygen atom is bonded to two silicon atoms.

The silicate tetrahedron has 4 negative charges (the valency of Si is +4 and that of oxygen is −2).
- Addition of metal oxides such as FeO, CaO, to molten silicates brings about a breakdown of the silicate network.
- At low concentrations, Al_2O_3 behaves like a network modifying oxide and forms aluminium cation (Al^{3+}).
- At high concentrations, Al enters the tetrahedral structure, isomorphous with silicon.
- The activity of an oxide dissolved in molten slag forming M^{2+} and O^{2-} can be determined experimentally.
- Slag can be characterized by its basicity (B); a ratio of the concentration of the network breaker and the network former:

$$B = (\% \text{ CaO} + \% \text{ MgO})/(\% \text{ SiO}_2 + 0.84x\% \text{ P}_2\text{O}_5).$$

7.4 Slags from Non Ferrous Industry

- The $CaO-SiO_2-Fe$ oxides system represents the behaviour of slag found in the non ferrous metallurgy.

7.4.1 Copper Slag

- Copper slags are a by-product of the metallurgy of copper.
- Slags comprised of silica and iron oxides have particular practical importance in copper smelting, in the acid steelmaking process and accounting for the service behaviour of silica refractories.
- In copper smelting the converter slags approach the low silica limits while the reverberatory furnace slags have intermediate silica contents, perhaps approaching silica saturation in some circumstances. Liquid formed in the pores of silica refractories are close to silica saturation.
- Acidic steelmaking slags are low in ferric oxide and approach saturation with metallic iron.
- Oxidised slags such as cooper smelting slags, contain a large proportion of ferric oxide and approach the saturation of magnetite (LI 2012).
- The occurrence of both alumina and Hercynite in the same ingot may be due to an incomplete mixing of the deoxidiser throughout the melt.
- It is indispensable to obtain activities of three components in the slag for understanding and describing chemical reaction of those elevated temperature processes (TAO 2008).

7.4.2 Silico-Manganese Slag

- The silico-manganese slags come from the pyrometallurgical processing of ferro- and silico-manganese alloys.

- They contain significant proportions of MnO (10/20%) and are essentially vitreous with minor amounts of crystalline anhydrite, quartz and potassium sulphide. They are acidic.
- The use of silico-manganese slags in concrete is feasible. Up to 80 kg/m^3 does not affect the compressive strength. When ground at a fineness of 6000 cm^2/g, it can be incorporated in self-levelling concrete instead of limestone (TAN 1974, PER 1999).

Chapter 8

Application to Refractory Materials

8.1 Introduction

- The total CaO–SiO_2–Al_2O_3 system accounts for 95% of the typical bulk composition of the main industrial refractories sector. Some other products based on MgO, Cr_2O_3, ZrO_2, and SiC have also refractory properties but their application is outside the scope of this chapter.
- Refractory materials are generally described as non-metallic (ceramic) materials that are resistant to decomposition by heat, and retain strength and form at high temperatures in excess of around 550 °C, but in practice they are used for much higher temperature applications, typically between 1000 and 2000 °C.
- They are often designed to insulate not only hot reaction chambers from the surroundings in order to keep heat losses to a minimum, and thus to maintain the conditions of firing under control, but also many other applications.
- High melting point, chemical inertness at high temperatures and mechanical strengths are suitable for application as refractory. It is the case in the system CaO–SiO_2–Al_2O_3, for CaO (2572 °C), SiO_2 (1723 °C), Al_2O_3 (2020 °C) and their main combination; A_3S_2 (1850 °C), CA (1390 °C), CA_2 (1789 °C), CA_6 (1860 °C), C_3S (2150 °C), CS (1544 °C), C_3S_2 (1460 °C), C_2S (2130 °C), C_2AS (1590 °C) and CAS_2 (1553 °C).
- Large quantities of refractories are consumed by heavy industries (steel, cement, glass, and ceramic industries). Only oxides which are abundantly present in the earth crust and hence are available at a reasonable cost may be considered as potential raw materials for refractories.
- The oxides which singly or in combination are of greatest importance as refractory materials are in ranking of decreasing availability; SiO_2, CaO, Al_2O_3, MgO, Cr_2O_3, and ZrO_2.
- Refractory materials are used for lining kiln (rotary, vertical shaft, tunnel, reverberatory furnace, converter, incinerator…), transport vessel (ladle, tundish,

TAB. 8.1 – Type of refractory *versus* firing temperature.

Type of refractory	Firing temperature	Type of refractory	Firing temperature
Fireclay bricks	1250–1500 °C	Chrome/magnesia bricks	1550–1650 °C
Silica bricks	1450–1500 °C	High fire bricks	1700–1800 °C
High alumina bricks	1550–1800 °C	Dolomite bricks	1500–1550 °C

torpedo ladle...) or special application (burner protection, slide gate, cooling devices...).

- The choice of the refractory material is governed by the temperature of utilization and the chemical environment (table 8.1).
- Chemical reactions of the charge, furnace atmosphere, change in operating temperature, speed, and frequency of the temperature change, thermal properties requirements (heat conductivity, heat storage...) are important parameters for the selection of the most suitable refractories for a given use.
- The products based on silica, alumina, silico-alumina cover a large area of refractory applications as raw material, such as:

  o Naturally occurring minerals, *e.g.* siliceous, aluminous, silicoaluminous (Clay, Shale, Kaolin, Sillimanite, Kyanite, Andalusite, Bauxite).
  o Synthetic fused and sintered, *e.g.* Mullite, calcined alumina, Corundum.

- The refractory materials are prepared from several solid components and are possibly containing some air space. The distribution of pore sizes, density and permeability are structural properties leading to mechanical properties at room temperature (mechanical strength, elastic modulus) and at elevated temperatures (creep, hot modulus of rupture, torsion characteristics...).
- The properties of refractories are normally modified under the influence of temperature, and the following properties characterise the product: thermal expansion, permanent volume change, specific heat capacity, heat conductivity, thermal shock resistance, and corrosion resistance (effect of slag).

8.2 Raw Materials Based on Al_2O_3 and SiO_2

8.2.1 *Natural Raw Materials*

- Siliceous materials: quartzite (α-Quartz) is the material primarily used for the production of silica bricks. It undergoes a reversible change in crystal structure at 573 °C to form β-Quartz. This phenomenon (called the α to β Quartz inversion) is accompanied by a linear expansion of 0.45% which can lead to cracking of ceramic ware if cooling occurs too quickly through the inversion temperature.

Application to Refractory Materials

TAB. 8.2 – Raw materials based on Al_2O_3 and SiO_2 for refractory applications (weight %).

	SiO_2	TiO_2	Al_2O_3	Fe_2O_3	CaO	MgO	K_2O	Na_2O	P_2O_5	LOI
Silica (UK)	98	0.2	0.6	0.5			0.2			
Bauxite (Greece)	6.4	2.9	48.75	22.50	4.15	0.8	0.07	0.07	0.22	13.5
Ball clay	26.4	1.4	59.8	1	0.2	0.5	2.4	0.4		7.9
Boké (Guinea)	0.35	3.5	63.40	1.8			0.02	0.02	0.09	31.20
Kaolin	47.65	0.22	36.75	0.75	0.03	0.22	1.14	0.12	0.09	12.80

At higher temperatures still (temperature depending on what other oxides are present, but normally well above 870 °C) it converts to the other polymorphs of SiO_2, namely Cristobalite and Tridymite and these also exhibit temperature-induced inversions.

- Clay, shale, kaolin: they contain silica and 22–40% alumina (table 8.2).
- Sillimanite, Kyanite, and Andalusite: they contain 60% Al_2O_3. They are unstable and when they are heated; they undergo transformation to Mullite and silica with an increase in volume. Kyanite converts at 1325 °C with 16–18% increase in its volume. Sillimanite converts at 1530 °C with an increase in its volume of 7–8%. Andalusite converts at 1350 °C with an increase in volume of 3–4%.
- Bauxites mainly contain aluminium hydrates and iron oxides. Deposits with low iron oxide contents can be used as refractories. They contain Diaspore (α-$Al_2O_3 \cdot H_2O$), Boehmite (γ-$Al_2O_3 \cdot H_2O$), Gibbsite or Hydrargillite (γ-$Al_2O_3 \cdot 3H_2O$). When heated Diaspore converted to corundum at 450 °C, Boehmite and gibbsite to γ-Al_2O_3 at 280 °C.

8.2.2 Synthetic Raw Material

- Fused and sintered Mullite: in order to avoid the volume expansion which is to be expected when using Silimanite or Bauxite the Mullite is produced from Kaolin and calcined alumina either by sintering in a rotary kiln at 1800 °C or by fusing in arc furnace. Sintered Mullite ($3Al_2O_3 \cdot 2SiO_2$) contains crystallized Mullite with 72% Al_2O_3 and a product containing 77% Al_2O_3 ($2Al_2O_3 \cdot SiO_2$). A refractory is said to have been electrocast when it has been shaped by pouring the electrofused melted materials into a mould.
- Calcined alumina: it contains α-Al_2O_3 with a small amount of γ-Al_2O_3. Starting with bauxite it is refined by the Bayer process and calcined in rotary kiln at 1200–1300 °C.
- Corundum: synthetic corundum is produced by fusing bauxite or calcined alumina in an arc furnace. One brand (from bauxite) contains 95–97% Al_2O_3 and impurities (metallic iron, carbide, silicide…). Another brand containing 99% of Al_2O_3 is colourless and produced from calcined Al_2O_3.

8.3 Applying Refractory Materials

- Refractories can be applied either in shape form (bricks, blocs, monoliths...) or as cement to bond the individual bricks.

8.3.1 Brick and Monolithic Refractories

8.3.1.1 Silica Bricks

- A green (meaning unfired) silica brick is composed with quartz, lime, and an organic binder. It is fired by bringing to temperature very slowly to about 1450 °C (below the temperature of stability of Tridymite 1470 °C) for a few hours and then cooled very slowly.
- It consists mainly of quartzite but with several percent other oxides such as; alumina and alkalis. Other oxides are intentionally added during the manufacture of the bricks (such as lime 2–4%). Nevertheless, the refractory body becomes further contaminated during its service. Figure 8.1 shows the thermal expansion of the various phases of silica.
- It is necessary to convert most of the silica to Tridymite in order to obtain high mechanical strength because of the smaller volume change accompanying the Tridymite crystallographic inversion as compared to Cristobalite.
- Silica bricks have a tendency towards subsequent expansion, the degree of which depends on the remaining unconverted quartz.
- The quartz starting material first inverts to β-Quartz (commonly known as High Quartz) and then to Cristobalite with expansion of the grains. At this stage, the grain shape has hardly changed but on cooling the brick has low mechanical strength. On conversion of the Cristobalite to Tridymite, high strengths develop as Tridymite crystals grow across grain boundaries.
- The upper temperature limit of use of silica bodies is the liquid development that is the main cause of failure. It can be determined from phase diagrams. Alumina and alkalis are particularly harmful. Liquidus and solidus temperatures are drastically lowered by relatively small additions of these components.
- Silica brick is used right up to the point of failure through the melting of one of its constituents and by this causing a lowering of the temperature at which the brick erodes rapidly by melting. This will cause a corresponding lowering of the maximum temperature of usefulness of the brick. In the principal uses of these bricks, their resistance to melting and the resistance to the fluxing action of the constituents of slag are the main required properties.
- In the Cristobalite liquidus region of the $CaO-Al_2O_3-SiO_2$ phase diagram, the curve of the constant amount of liquid as function of composition at 1660 °C shows that a decrease in alumina content from 0.7 to 0.3% takes a significant difference in percent of liquid present when the silica refractories are being used at this temperature.

Application to Refractory Materials

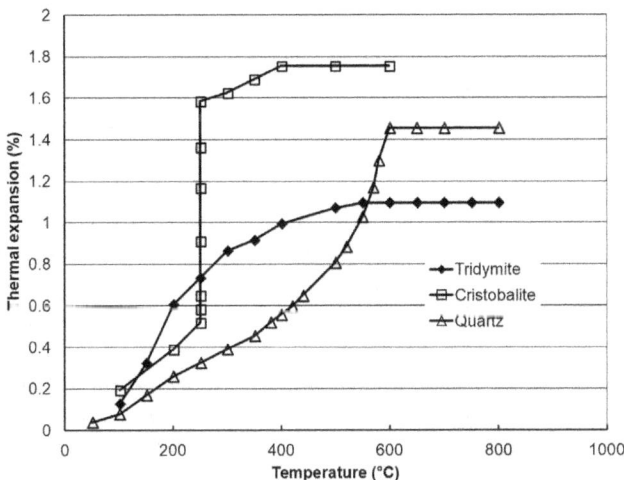

FIG. 8.1 – Thermal expansion (%) of the various phases of silica as a function of temperature.

- When alumina is present (0.5%) addition of 2–3% of lime increases slightly the refractoriness of the bricks.
- Iron is the principal fluxing material. As iron oxide penetrates the hot face of the brick, lime, and alumina tend to migrate back from the hot face leaving a surface composed of iron oxide and silica. The phases which form can be estimated from the $FeO\text{--}Fe_2O_3\text{--}SiO_2$ diagram.
- It is seen that the lowest temperature formation (solidus temperature) decreases from 1455 °C at the highest oxygen pressure (pO_2 in air = 0.21 atm) to a minimum at 1140 °C when oxygen pressure is 10^{-9} atm. In air, SiO_2 content of the liquid in equilibrium with silica at 1500 °C is equal 20% whereas at $pO_2 = 10^{-9}$ atm and the same temperature, it equals to 43%. The fluxing of silica brick by iron oxide is much more severe at low than at high oxygen pressure.
- The maximum temperature to which a silica brick in contact with iron oxide can be used without complete failure of the body is that of the liquidus within the region of the two liquids. This temperature is seen to increase when the oxygen pressure decreases from 1665 °C in air to 1700 °C at $pO_2 = 10^{-9}$ atm.

8.3.1.2 Alumina–Silica Bricks

- The composition of refractories lies as a first approximation in the binary $Al_2O_3\text{--}SiO_2$ system. If Al_2O_3 is <72%, the refractory is composed, after firing, of Mullite and silica and the liquid will develop at 1500 °C (Cristobalite–Mullite eutectic). If the percentage of Al_2O_3 is greater than 72%, the refractory is composed of Mullite and Corundum and the liquid will develop at 1840 °C (Mullite–Corundum eutectic).

8.3.1.3 Fireclay Bricks

- They contain 22–45% alumina and are classified according to the method of preparation: plastic formed, semi-dry pressed, and dry pressed.

8.3.1.4 High Alumina Bricks

- They contain 50–55% Al_2O_3 (alumina enriched bricks), 55–65% Al_2O_3 for Kyanite bricks, 60–70% Al_2O_3 for Sillimanite bricks, 72–75% Al_2O_3 for Mullite bricks, 75–85% for bauxite bricks and 65–99.5% for corundum bricks.
- The refractoriness is largely influenced by various oxides such as potash (common constituent of clay) and iron oxide. In order to evaluate reactions in the body composed of Al_2O_3, SiO_2 and iron oxide, the oxygen partial pressure of the atmosphere surrounding the refractory must be taken into account.
- Under strongly reducing conditions, a liquid will develop at a temperature of 1210 °C when a fireclay brick consisting initially of Mullite and Cristobalite absorbs a small quantity of iron oxide. If the brick is originally composed of Mullite and corundum, the liquid will not form until a temperature of 1380 °C. In air, the lowest temperatures of liquid formation are, respectively, 1380 and 1460 °C.
- The difference in the two behaviours is related to the manner in which iron oxide can be accommodated within the structures of the crystalline phases that are present. Mullite and corundum accommodate iron oxide in the ferric state Fe^{3+} and any substitution occurs under reducing conditions.
- Small quantities of iron cause a liquid to develop at temperatures above 1210 or 1380 °C. In air most of the iron is present as Fe^{3+}, thus a moderated amount of iron oxide can be absorbed by the alumina silica refractory in air without the formation of liquid phase even at temperatures above 1380 or 1460 °C.
- Because of the relatively low temperature of the Cristobalite–Mullite eutectic in the Al_2O_3 system, silica and high alumina refractories are incompatible.
- The resistance of alumina silica refractories to slag attack is generally poor because low liquidus and solidus prevail close to the Al_2O_3–SiO_2 join of the Al_2O_3–SiO_2–iron oxide system.

8.3.1.5 Monolithic Refractories

- Monolithic refractory is a term for all unshaped refractory products, typically a mass that is cast or projected. The word 'monolithic' is derived from the word monolith meaning 'big stone'. These materials are incorporated into a suspension that eventually harden to form a solid mass.
- Every type of bricks corresponds to a chemical composition of a monolithic product but in brick form.
- Monolithics can be sub-divided into several types:
    - Plastic refractories.
    - Ramming mixtures.
    - Fritting compounds for lining induction furnace.

Application to Refractory Materials

○ Loam sand for lining steel ladle.
○ Castables which are granular aggregates with addition of a hydraulic binder (they are poured behind shuttering).
○ Gunning mixture (fine grained preparation that is sprayed on the desired surface with compressed air).
○ Insulating compounds made from light fireclay serve as an insulating layer behind the wear lining of any desired furnace or even as the inner lining if no corrosion from slag or flue dust is expected.
○ Finished refractory shapes made from refractory concrete that is placed by the user in a comparatively short time.

8.3.2 Refractory Cement and Mortar

8.3.2.1 Manufacture of Low-Iron Calcium Aluminate Cements (Low-Iron CAC)

- The total of CaO, SiO_2, and Al_2O_3 accounts for >95% of the typical bulk composition of calcium aluminate clinker.
- Low iron calcium aluminate cements (low-iron CAC) were industrially developed because of their high refractoriness that depends on their chemical composition and mineralogy; the refractoriness can reach 1800 °C.
- Low-iron calcium aluminate cements are mainly used to bind refractory blocks, to protect burner lance or to design refractory devices for steel, glass or cement industry.
- The chemical composition and mineralogy of the main brands of low-iron CAC are given in table 8.3. The phases found in low-iron CAC are: CA, CA_2, C_2AS, α-alumina.
- Low-iron CAC can be produced in different ways.

  ○ Melting mixture of low-iron bauxite and limestone in a reverberatory furnace.
  ○ Sintering a mixture of lime and alumina in a small rotary kiln.
  ○ By a reduction process from ferruginous bauxite (through an electric arc furnace, such as; the Higgins process). In this process, the tips of the electrode are at a short distance above the molten bath. Heating is done by radiation from the arc developed between the electrode tips and the bath. The process that requires about 2815 kW/h is the following:

  $1.35t$ bauxite + $1t$ limestone + $0.1t$ coke → $1t$ of cement + ferrosilicon (Fe/Si).

  ■ Ferrosilicon is obtained as a by-product.
  ■ When the bauxite contains a large proportion of SiO_2, it must be reduced to prevent the formation of Gehlenite in the cement.

- The clinker of CAC made by fusion is not porous (figure 8.2) and very hard to grind.
- In the case of a production by sintering, the product is porous (50–70%). During sintering alumina changes from gibbsite to γ-Al_2O_3 (500 °C), δ-Al_2O_3 (800 °C), θ-Al_2O_3 (1050 °C) and finally α- to β-Al_2O_3 ($T > 1200$ °C). The degree of

TAB. 8.3 – Chemical composition and mineralogy of low iron CAC (weight %).

Oxide components	Percentage of Al_2O_3 in low-iron CAC			
	40%	50%	70%	80%
Al_2O_3	37.5–41.5	50.8–54.2	68.7–70.5	79/82
CaO	36.5–39.5	35.9–38.9	28.5–29.0	17/20
SiO_2	4.2–5.0	4–5.5	0.2–0.6	<0.4
Fe oxides	14.0–18	1.0–2.2	<0.4	<0.4
TiO_2	<0.4	<0.4	<0.5	<0.1
CA	47–57	64–74	54–64	
CA_2			Traces	
Al_2O_3 alpha			Traces	

- calcination of the bauxite impacts its reactivity with lime. Sintering is diffusion of lime into the alumina grain. Thus, C_3A may be observed outside alumina grain and CA_2, CA_6 or unreacted alumina can be detected within alumina grains.
- Figure 8.3 shows a micrograph of a polished section of a low iron CAC produced by arc furnace. Hexagonal crystals of CA_2 (dark grey) can be observed in a glassy matrix of CA.
- The refractories based on low-iron CAC are prepared by firing hydrated cement. During firing a complex dehydration, interaction and ceramic bonding process occurs.
- The interaction zone between polycrystalline α-Al_2O_3 and CaO–Al_2O_3 eutectic melt at 1530 °C produces a reaction zone that differs in several aspects from the one observed for CaO–Al_2O_3 solid state interaction.
- Table 8.4 shows the chemical composition of the main phases of the low-iron CAC.
- CA_6 is the predominant bonding phase derived from hydrated CA after firing and formation of ceramic bonds.
- Only a strongly textured CA_6 layer is clearly defined. Additional layers are mixtures of complex phases. Silica, a common impurity in sintered alumina, is rejected by the advancing CA_6 phase and accumulates in channels that provide an easy transport path enhancing reaction kinetics.
- The structure of CA_6 varies according to curing temperature of the castable. The combination of calcium and alumina is slightly exothermic but the loss of heat from the gas and convection is shown in table 8.5.

8.3.2.2 Hydration of Low-Iron Calcium Aluminate Cement

- The hydration of CAC depends on the temperature of curing.
    - At 20 °C, the hydration of calcium aluminate cement gives CAH_{10} (hexagonal crystal) at 6 h, C_2AH_8 (hexagonal crystal) at 8 h and C_3AH_6 (cubic crystal) at 2 days.

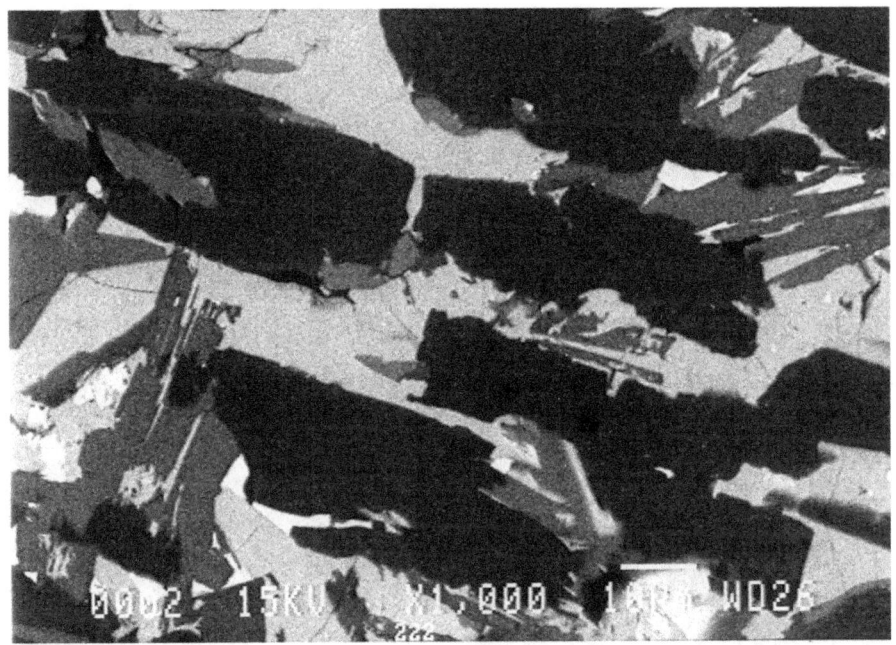

Fig. 8.2 – SEM micrograph of CAC clinker without any porosity.

○ At 40 °C, C_2AH_8 is formed first followed at 7 days by C_3AH_6 which is the dominating phase.
- The transformation of hexagonal into cubic is accompanied by an increase in the porosity and a decrease in mechanical strength (called conversion).
- The addition of silica bearing component (fume silica, slag...) modifies the kinetics of reaction and the hydrated constituents.

○ At 20 °C in the presence of silica, CAH_{10} is formed, but in lesser amounts than for neat cement. C_2AH_8 is also formed. The rate of the reaction decreases. C_2AH_8 reacts with silica to give C_2ASH_8. The hydration of the cement – silica blend is slowed down compared to the pure cement.
○ At 20 °C in the presence of slag, C_2AH_8 is formed first followed by C_3AH_6 and C_2ASH_8.

- When kept in hot water over a long period of time, calcium alumina cement loses progressively its mechanical strength whereas the 1/1 mixture of CAC and slag did not show any reduction in strength with time after a period extending up to 2 years. Such a mixture placed at 65 °C, indicates that the conversion is taking place faster but the same amount of C_3AH_6 is detected along with AH_3 and C_2ASH_8. This behaviour can be explained by the consumption of calcium ions for activating the slag which removes calcium ions from the pore solution and hence restricts the conversion of CAH_{10} to C_3AH_6. At 2 years, the main hydrates of a

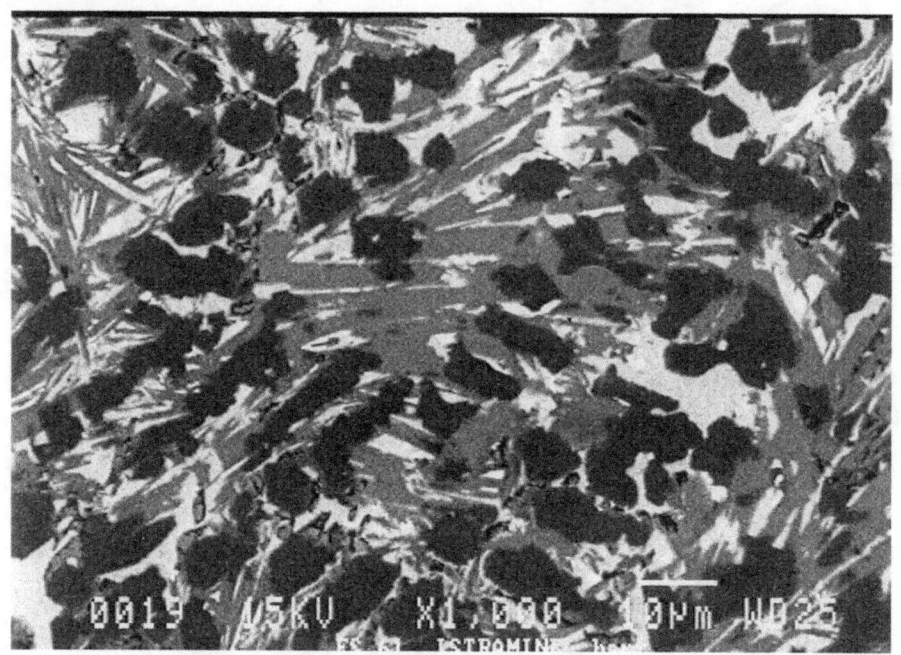

FIG. 8.3 – SEM micrograph of low iron CAC obtained in reducing conditions.

TAB. 8.4 – Chemical composition of the main phases of low-iron CAC (weight %).

	SiO_2	Al_2O_3	Fe_2O_3	CaO	K_2O	TiO_2	MgO	SO_3
CA	0.4	63.5	0.1	34.3	0.2	0.2	0.8	0.4
CA_2	0.3	76.6	0.1	21.2	0.1	0	1.4	0.1
C_2AS	15.8	43.1	0.3	38.6	0	0	1.6	0.5
Al_2O_3	0	98.9	0	0.8	0.1	0.1	0	0

slag – CAC mixture at a proportion 3/1 and W/C = 0.4 are C_2ASH_8, an amorphous 'outer' gel which is an intimate mix of alumina gel and C–S–H and an 'inner' gel which is rich in Mg^{2+}.
- Table 8.6 shows the influence of the fineness of the cement on mechanical strength and setting time obtained for a concrete (30% CAC, 9% water and tabular alumina 500 kg/m^3).
- When the T° increases, the mineralogical composition is modified from a hydraulic bond to a ceramic bond (table 8.7).

8.3.2.3 Relationship of Hydration Reaction to Castable Properties

- Low iron CAC is an important constituent of refractory castable.
- Refractory castables can be classified as cement castables (15/25% CAC), low-cement castables (5/10% CAC), and ultra-low cement castables (under 5% CAC).

TAB. 8.5 – Heat balance for the production of CAC (kJ/ton).

Theoretical heat for chemical reaction	1650/2300
Heat lost in clinker discharge	1400/1750
Heat lost in exit gases	800/1000
Heat lost by radiation and convection	400/600

TAB. 8.6 – Mechanical strengths in compression (Rc) and flexion (Rf) and setting time (min) of a concrete made with low iron CAC having different fineness.

Fineness (cm^2/g)	2500	3500	4500	5000	6000
Setting time (min)	165	220	240	225	330
Rc 6 h–20 °C (MPa)	22.1	27.2	30.7	32.2	27.3
Rc 24 h–20 °C (MPa)	27.2	39.4	43.6	46.9	53.1
Rf 800 °C (MPa)	3.4	3.9	4.1	4.7	4.4
Rc 800 °C (MPa)	42.5	59	64	68.5	72.5

TAB. 8.7 – Type of bond *versus* temperature.

T°	Mineralogy	Bond type
70 °C	$CA/CA_2/A$	Hydrated
200 °C	C_3AH_6/AH_3	Hydrated
800 °C	$C_{12}A_7/CA/A$	Ceramic
1100 °C	$CA/CA_2/A$	Ceramic
1200 °C	$CA/CA_2/CA_6/A$	Ceramic
1500 °C	$CA/CA_2/CA_6$	Ceramic

- Two castable properties govern the time scheduling a castable installation; the working time that a castable can be placed before unworkable or stiff and the time when the cast structure has sufficient strength for the mould to be stripped.
- The working time (the end of workability) occurs when there is sufficient flocculation and stiffening so that the castable will not move under vibration and thus cannot be properly consolidated any more. The hardening generally starts to occur as the hydration rate is fast enough (arising of the massive precipitation of hydrates) as seen by the increase in temperature.
- Figure 8.4 shows the link between these aspects of castable properties, workability and hardening time and the hydration process as followed by ionic conductimetry and exothermic profile (PAR 2004).

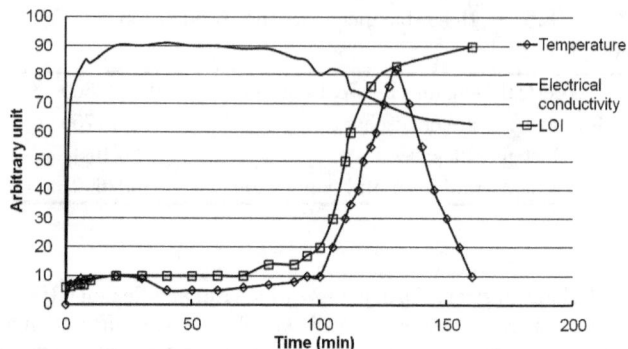

FIG. 8.4 – Hydration steps by various methods with regard to castable properties (PAR 2004).

8.4 Refractories Consuming Industry

- The end market for refractories (46 millions of tons worldwide) is first of all.
- The iron and steel industry is the main user of refractories (70%) followed by the cement industry (7%), glass, ceramics industry (6%), chemical industry (4%), non-ferrous metals industry (2/3%), and other industries like petrochemical, incinerator (6%).
- The refractory policy must be adapted to the process of manufacture. There are many sorts of kiln but rotary, vertical and shaft kilns are the most used.

8.4.1 Iron and Steel Industry

- Iron and steel industry utilizes several types of furnace (blast furnace, electric arc furnace, reheat furnace...) and equipment for transport or treatment (casting, degassing...) requiring protection against high temperatures.
- The traditional refractory lining used in blast furnace was made up of fireclay bricks with less than 45% alumina.
- The current trend is to use different materials in each zone of the furnace. These are chosen to suit the environment in each particular section due to the differences in temperature.
- At the top of the furnace (the stack) where the raw material is fed and the gas is removed, the temperature is below 300 °C. Below the stack, the temperature increases up to 1200–1250 °C in the shaft's lower section. At the bottom, the hearth acts as crucible to receive molten iron and slag and the temperature goes as high as 1750 °C.
- The main reasons for the wear undergone by the refractories in the lower part of the shaft are the chemical action of the slag and the vapour of alkalis and CO; the temperature fluctuations and the abrasive wear form the action of the descending charge and the liquid pig iron.

Application to Refractory Materials 287

- Alumina-based refractories have a good resistance to the action of the slag in blast furnace, erosion by liquid pig iron and failure at high temperatures. Moreover, they are inert with respect to CO. However, when used at high temperatures in alkaline medium, the corundum may be converted to β-alumina and the volume of refractories increases by up to 20%.
- Silicon carbide is used in the lower part of the shaft and bosh (*the lower part of a blast furnace, between the hearth and the stack*) but the main problem is to find a binder resistant to alkalis.
- Monolithics with similar alumina content to the bricks may also be applied in the higher section of the stack for a large part of lining.
- A complete change in the steel industry practice from the use of open hearth furnaces and the Bessemer converter, to basic oxygen furnaces (BOF) and electric arc furnaces also modified the refractory utilization. BOF and the electric arc furnaces use predominantly basic linings, although the roof of the arc furnace is usually lined with high alumina bricks (80–85% Al_2O_3).
- A fastest growing sector of the steel industry is continuous casting. The development of sliding gate valves is the most important method of tapping steel in casting process. Sliding gate valves are manufactured from high alumina refractories (tabular alumina 85–95% Al_2O_3 with incorporation of Mullite).
- Ladle, tundish and lining originally made from fireclay bricks are replaced by high alumina CAC. Monolithics castable refractories are also used.

8.4.2 Non-Ferrous Metal Industry

- The wide range of industry yields a wide range of refractory linings. As an example, aluminium industry uses fireclay, high alumina carbon and silicon carbide, while copper industry uses silica, magnesia, fireclay, and carbon.

8.4.3 Cement Industry

- The process used by the cement industry derives from the lime manufacturing industry. Rotary kilns are used for burning and additional equipment are added for cooling, preheating or precalcining. Most of these equipment also require refractory materials. Special cements such as calcium alumina cement may also be produced using reverberatory furnaces or electric arc furnaces.
- The lining of a modern cement kiln depends on the position in the kiln and on the process (wet, dry or half dry process...).
- A dry process kiln can be subdivided into around three zones characterized by different working conditions separated by transition zones. A specific refractory material is used for each zone.
- In calcining zone, characterized by abrasion between the kiln lining and the raw feedstock, the temperatures are in the range 800/1000 °C. Aluminous firebricks (42–83% Al_2O_3) are used in this zone from around 8 to 12 m in diameter.

- In the burning zone (between 2 and 5 m in diameter) characterized by corrosive chemical attack and a high temperature (1600 °C), dolomite-based refractories are used.
- The burner pipe is lined with abrasion resistant monolithic castable of >50% Al_2O_3.
- The discharge zone (extending up to 2 m in diameter) is subjected to high levels of abrasion. Chemically bonded or fired high alumina brick or monolithics are used.
- In a suspension preheater or calciners, raw material particles exchange heat with the outlet gas. Insulation bricks, bulk fibres, and monoliths consisting of low/medium aluminous refractories are used.
- The clinker is cooled out of the kiln in rotary satellite cooler or in grate cooler. In rotary coolers, the clinker is rapidly cooled in a rapid rotating cylinder. Abrasion resistance firebricks or monolithics are used (70–80% Al_2O_3). The lining of grate cooler is achieved with high alumina fireclay or monolithics.

8.4.4 Whiteware, Traditional Ceramic Industry

- Ceramics industry produces their own refractory and employs them in firing units that are most frequently employed than tunnel kiln.
- These kilns are used to produce whiteware, bricks, pipes tiles, electrical insulator, walls, crowns, car docks, setters, pusher, plates, and saggers (*the last three are ceramic applications*).

8.4.5 High-Tech Ceramic Industry

- Refractory castables are useful for the construction of flame detectors, inner walls of concrete pressure vessel surrounding serving as radiation shields and for making pads on which take off aircrafts.
- A mix of calcium aluminate and fused silica aggregates is useful in casting rocket nozzles and infilling the stainless steel honeycomb that provides protection for missiles during atmospheric entry.
- CA is generally explored as a catalyst support and could be used as host lattice which can be doped with suitable ions to develop solid-state lasers and high temperature ceramic sensors.
- The sintering characteristics are important for many of these potential applications and high temperature ceramic sensors.

8.4.6 Glass Industry

- The basic operation in glass manufacture is melting of component (silica, sodium carbonate, lime, borate, and other inorganic additives) into a homogeneous molten mass. The temperature is around 1500 °C.

- Glass industry uses silica, fireclay, magnesia, and zirconia to obtain tank furnace, wall, burner blocks, ports, feeders, and special glass tank.
- Sodacalcic glass represents 90% of the products and the refractories are mainly silica and Sillimanite.
- Furnaces producing flat glass product as window or plate glass and container glass operate continuously with the raw material entering at one end and glass leaving at the opposite end. Special glasses are made in batches in a hearth furnace. Refractories must be resistant to the highly corrosive effect of molten glass and to thermal shocks.

Chapter 9

Application of the Glassy Products

9.1 Introduction

BOY 1984, MOR 1954, SCH 1980, ZAR 1982

- CaO, Al_2O_3, SiO_2 and Fe oxides account for more than 80% of the chemical composition of the principal industrial glasses.
- Silica is the basis of most commercial glasses.
- Glasses are amorphous solids having a disordered non-crystalline structure and exhibiting a glass transition when heated towards the liquid state. X-ray and electron diffraction studies show that glasses lack long range order of the constituent atoms (absence of sharp X-ray peaks).
- Several methods can be used to obtain a disordered structure: fast cooling and mechano-activation are often mentioned.
- Glassy materials also occur naturally, such as Obsidian, often found in volcanic areas. This material which consists mainly of silicon dioxide, sodium and calcium oxide was used during antiquity to make arrowheads, spare heads and knives.
- Glasses do not have a sharp melting point and do not leave, like a crystal, a preferred direction.
- Glasses like liquids are isotropic. They are brittle and transparent.
- The glass industry is essentially a commodity industry inspite of the existence of some specialities such as optical glasses. Over 80% of the industry output is sold to other industries and the glass industry as a whole is very much dependent on the building, food and beverage industry.
- Glasses are used industrially for building materials (cladding materials), for domestics ware (bowls, vases, bottles, jar and glasses) because of their inertness against corrosive liquid.
- Glass products can be coloured for an extensive use in the manufacturing of art craft (stained glass window, glazed ceramics devices, etc.).

- Their properties of reflectivity and refractivity open applications as optical lenses, prisms and fine glassware.

9.2 Structure of Glass

- The difference between the properties of glass and the properties of crystallised material having the same chemical composition can be explained by the lack of order in the structure of glass.
- As example, vitreous silica glass that has a low thermal expansion and is resistant to chemical agent, can be used as furnace tubes and melting crucibles. Polymorphs of SiO_2 (Quartz, Cristobalite and Tridymite) do not have this property. Vitreous slags produced as a by-product of iron manufacture by quenching a liquid based on CaO, SiO_2 and Al_2O_3 are reactive when mixed with an activator (alkali, Portland cement...) while a product with the same composition cooled slowly to favour crystallisation is inert (see applications as hydraulic binder, slag and glass cements).
- Glass formation has been observed in a large number of systems but has been used in the case where the glassy state could be industrially processed easily (silicate, borate, phosphates).
- In crystalline form of silica and silicates, silicon is associated with four neighbours to form a tetrahedral structural unit. Silicon lies at the center of the tetrahedron with oxygen ions at the four corners. Adjacent tetrahedra shares corners.
- In the case of glassy structure, the tetrahedra are supposedly randomly arranged in the space (random network theory). The structural disorder is due to the variation in the Si–O–Si angle.
- The molecular arrangement in a glass is a frozen form of a random arrangement in a liquid.
- Addition of a monovalent ion in the form of oxide (soda silica glass) disrupts the continuity of the network and produces changes in the physical properties.
- In a glass, some oxides act as network forming oxides. Silicate, borate and phosphate act as network formers and easily form glass.
- Other oxides such as Na_2O act as network modifiers and disrupt the continuity of the network.
- Oxides such as Al_2O_3, BeO, TiO_2 and ZrO_2 are classified as intermediate because they can act as network formers or modifiers depending on the atomic environment they occupy.
- During heating, the glass becomes soft transforming progressively into a viscous liquid. The mechanical properties of glass transform progressively from an elastic solid to a viscous liquid.
- The properties of glass depend on the structure or only on the composition.
- The transport properties, such as electric conductivity, depend on the structure and minor elements.
- The properties not linked to the disordered structure depend on the nature of atoms and their interaction with the close neighbour (density, elastic constant, specific heat, electric permittivity). For this type of glass, their properties can be

TAB. 9.1 – Transport properties of CaO, Al$_2$O$_3$ and SiO$_2$.

	SiO$_2$	CaO	Al$_2$O$_3$
Specific volume (10^6 m^3/kg^{-1})	4.35	3.03	2.44
Thermal expansion coefficient (10^6 K^{-1})	2.67	16.67	16.67
Young modulus (10^{-8} Nm^2)	7	7	18
Heat capacity (J kg^{-1})	8	7.96	8.68
Thermal conductivity (Wm^{-1} K^{-1})	12.85	13.27	15.57

tailored by a linear composition adjustment. The property X is linked by a formula:

$$X = \sum C_i \cdot X_i.$$

Note: X_i being the value of the property of compound i and C_i its proportion in the glass.

- Table 9.1 shows the properties of CaO, Al$_2$O$_3$ and SiO$_2$ that are the main compounds of glasses.

9.3 Classification of Glass Products Containing CaO, Al$_2$O$_3$, SiO$_2$ and Fe Oxides

- The glassy products can be classified according to the type of utilization or/and on the chemical composition and the process. The cross-linking of the classification by use and by chemistry leads to some overlaps.
- The most widely used classification based on the chemical composition of glass gives the following groups; fused silica glass, soda lime glass, borosilicate glass, alumino silicate glass and high technical glass.
- Three main groups of chemical assemblages represent 95% of the industrial production of glass; soda lime silica glass, sodium silicate glass and alumino silicate glass. They correspond to the domestic use and the building and car sectors.
- These formulations and processes have to satisfy the following needs of the sector of the glass industry: container glass, flat glass, continuous filament glass fibre, domestic glass, special glass, mineral wool, high temperature insulation wool, frits and gaze.
- Table 9.2 shows the type of manufactured products relative to the properties required and the chemical compositions of the glass.
- Other types of glass are produced in small amounts for special use with different formulation and process; lead crystal glass, opal glass, glass ceramics, optical glass, phase separated and reconstructive glasses, photosensitive glasses, sealing glass and electronics.

TAB. 9.2 – Type of manufactured products relative to the properties required and the chemical compositions of the glass.

Commercial products	Properties	Chemistry
Container glass, float glass	Glass stability, low temperature	Soda lime glass
Lighting, fibre optics, tubing, labware	Inertness, expansion, refractoriness	High silica Glass, Vycor glass
Glass fibre, cook ware	Low expansion, fiberization	Boro silicate
Mineral wool	Fiberization	Alkali borosilicate, opal
Car, aircraft windshield, electron tube	Rheology	Alumina silicate glass
Cook ware	Low expansion	Glass ceramics

9.4 Products, Chemistry and Process

9.4.1 High Silica Glass – Vycor Glass

- High silica glasses are characterized by chemical inertness (acid durability), high thermal refractoriness (viscosity), low thermal coefficient of expansion, high thermal shock resistance and high optical transparency.
- Commercially available products are distinguished as 'clear' (*fused quartz*) or 'opaque' (*fused silica*).
- High silica glasses are the base of a variety of products; products for chemistry (labware, tubing...), optical fibres, products for optic (lenses, window, mirror...), lighting (high intensity discharge).
- The high silica system (95% SiO_2) comprises a family of glasses whose property range is determined by the trace of impurities or intentional dopants.
- A glass with this requirement is produced by selective leachability of a glass obtained by fusion of an alkali borosilicate at 1500 °C. After conversion into porous objects by acid leaching, the glass is sintered at 900/1200 °C into reconstructed, pore-free high silica glass (sometimes referred generically as Vycor glass, but this is a brand name of Corning glass – other similar brands exist).
- Pore-free high silica glass has many uses:

    ○ It is used in laboratory glassware as crucibles, dishes and tubing to construct instruments.
    ○ Pore-free high silica glass tubes protect throwaway thermocouples that are pushed through the flowing slag into interior of a molten metal.
    ○ It is used as a mould to melt glasses by making it the liner of refractory crucible space craft window, envelops for heat lamps and high intensity lights.
    ○ Because its good UV transmission, pore-free high silica glass is used for gas discharge lamps, germicidal lamps, dewatered Vycor transmit.

Application of the Glassy Products 295

9.4.2 Soda-Lime Glass

9.4.2.1 Container Glass

- Container glass is made using a soda lime formulation, melted in a fossil fired furnace and shaped by fully automated process.
- Because of its high viscosity at high temperatures, melting and working are difficult and require the addition of fluxes (*e.g.* $Na_2O/CaO...$).
- The main properties are the transmission of light (use as transparent articles) and its inertness (does not change the taste of the container contents).
- Glass containers are produced in a two-stage moulding process using pressing and blowing techniques.
- There are three main sectors:

  ○ The beverage sector including bottles for wine (still, sparkling, fortified), spirits, beer, ciders, flavoured alcoholic beverage, soft drink, fruit juice and mineral water (75% of the total glass packaging the container).
  ○ The food sector including milk, jams, and spreads, sauces and dressing.
  ○ Bottling and glass packaging for the pharmaceutical and cosmetics industries.

9.4.2.2 Flat Glass

- It is made with a soda lime formulation. The main properties are the high level of light transmission leading to its use as flat glass in building and automobile industry.
- There are two types of flat glass; rolled glass and float glass:

  ○ Rolled glass is formed by a continuous double roll process. Molten glass at about 1000 °C is extruded between water-cooled steel rollers to produce a ribbon with a controlled thickness and surface pattern.
  ○ In the manufacturing process of float glass, the molten glass is poured on to a bath of molten tin and it forms a ribbon with the upper and lower surfaces becoming parallel under the influence of gravity and surface tension.

- Most of the flat glass is produced by the float glass process. The most important markets for float glass are the building (insulated glazing...) and automotive industry (laminated windscreen, side and rear glazing and sunroofs).
- Container and flat glasses fall within a narrow range of composition (table 9.3) representing a compromise between stable glass formation, low-temperature liquidus and durability.

9.4.3 Sodium Borosilicate – Glass Fibres

- Modifications of the properties are achieved by the addition of different oxides. Low expansion is obtained by the addition of B_2O_3. B_2O_3 favours the fiberization of the glass melt yielding to the production of glass fibre.

TAB. 9.3 – Chemical composition of the principal types of glass products based on CaO, Al$_2$O$_3$, SiO$_2$ and Fe oxides (weight %).

Oxide	Container	Flat	Boro-silicate	Sodium borosilicate glass fibre			Light-bulbs	Lead crystal	Ovenware	Sodium vapour lamps	Solder-glass
				Type E	Type C	Type S					
SiO$_2$	72.2	72.8	80.1	54.5	66.0	65.0	71.5	54.0	81.5	5.5	5.0
B$_2$O$_3$			12.0	7.5	5.5	10.0		0.5	11.9	16.0	
Na$_2$O	13.8	12.7	4.5	0.4	8.5		15.5	0.2	4.7	9.5	
CaO	11.0	8.1	0.1	17.4	14.0		6.6				
MgO	0.1	3.8		4.4	3.0		2.8				
Al$_2$O$_3$	1.6	1.4	2.6	14.9	4.0	4.0	2.0	0.1	1.9	17.5	17.0
K$_2$O	0.5	0.8	0.3	0.2			1.0	12.2			
PbO								31.8			64.0
BaO			12.2							52.0	
ZnO											14.0

- Glass based on sodium borosilicate glass shows a high resistance to chemical corrosion and temperature change (low thermal expansion coefficient). Consequently they are much used for laboratory equipment, pharmaceutical containers, lightings, cookware (*the commercial brand Pyrex is an example*) and oven doors.
- Fibre glass applications are divided into three categories:
  (1) Those using discontinuous fibres (5/6 μm long) packed as batts or boards leading to thermal and acoustic insulations (buildings, transportation and appliances) but also good mechanical strengths and durability. A classic soda-lime silica glass is modified to improve water and acid durability and lower its viscosity.
  (2) Those using fabric or mat cover applications where fibre glass protected by an organic or metallic coating is used by itself not in a matrix glass mats. For example, this is used in batteries to prevent shorting of the electrode during discharge (resistance to acid). Drapes, fabrics, clothing, cable and wire insulation, sleeves, yarn and threads, screening.
  (3) Those intended for reinforcement using continuous or chopped fibres taking the advantage of high specific mechanical strength with chemical inertness, low-electrical conductivity, low-dielectric constant or transparency. The glass is coated with an organic substance to ensure good bonding between the glass and the material to be reinforced. They are used in plastics and polymer (boat, car, aircraft, consumer products, storage tanks, appliance housing, printed circuit boards, transparent panels), Teflon (fabrics structures), rubber (tire cord, tire), asphalt (shingles, road board) and mortar.
- Three types of glass have been developed for production of textile grade continuous glass fibre (E, C, S glasses). E glasses are lime alumina borosilicate compositions, developed to have high bulk electrical resistivity and high surface resistivity as well as good fibre forming characteristics. C glass was developed for acid resistance and S glass for high tensile strength.

9.4.4 Aluminosilicate Glass

- It is defined as a silicate glass containing alumina as a major constituent. Its main characteristic is a steep viscosity temperature curve.
- Alkali aluminosilicate glasses have high ion mobility lending themselves to unique applications.
- The calcium aluminosilicate system has been explored in more detail by characterizing eutectic glasses of the $CaO-Al_2O_3-SiO_2$ system (table 9.4).
- Alkaline earth aluminosilicate glasses have tight structure giving high electrical resistivity and low gas permeation (IMA 1968). Aluminosilicate glasses are used for car and aircraft windshields, frangible, electronic housing, missile dome covers, tape reel flange, aircraft mirror, pipette, centrifuge tubes. ($SiO_2 + Al_2O_3$

TAB. 9.4 – Characteristics of the glasses having the composition of the eutectics in the CaO–Al$_2$O$_3$–SiO$_2$ system.

	Eutectic	SiO$_2$	Al$_2$O$_3$	CaO	Expansion coef	Annealing point	Strain point	Viscosity at 1000 °C
	°C	wgt %	wgt %	wgt %	× 10^7/°C	°C	°C	P (poise)
SiO$_2$/A$_3$S$_2$/CAS$_2$	1345	70.4	19.8	9.8	33.1	883	831	1612
SiO$_2$/CAS$_2$/CS	1170	62.1	14.6	23.3	54.7	772	728	1323
CS/CAS$_2$/C$_2$AS	1265	42	20	38	76.5	781	746	1166
C$_2$S/CS/C$_2$AS	1310	41	11.8	47.2	88.7	770	738	
C$_2$AS/CAS$_2$/CA$_2$	1380	31.7	29.1	29.1	59.5	832	800	1230
C$_2$S/CA/C$_2$AS	1337	6.95	43.35	49.7	84	822	796	

80%). The most active application is the envelop for tungsten halogen lamp; electron tube envelop.

9.4.5 Special Applications

- $CaO + SiO_2 + Al_2O_3$ account to more than 80% of the chemical composition of the glass. The other oxides are B_2O_3 and alkalis.

9.4.5.1 Opal Glasses

- Opal glasses are essentially phase separated glasses with their opacity resulting from light refraction and internal scattering between the separated phases. Opal glass can be translucent or opaque. This phase separation may be either liquid/liquid or liquid/solid.
- Crystalline opal glasses contain about 3/10 vol % crystalline phase, whereas glass ceramics, by definition have 50 vol % or more crystalline phase (up to 90%).
- The overall opacity of an opal glass is controlled by three factors:
  (1) The refractive index difference between the two phases.
  (2) The degree of phase separation (number of crystals in accordance with the principle of crystal growth – Tamman theory) or the volume of separated phases.
  (3) The size and distribution of the separated phases.
- The most important use being in food service lighting and cosmetic packaging.

9.4.5.2 Glass Ceramics

- Glass ceramics are micro-crystalline solids produced by the controlled devitrification of glass. They are melted, shaped and then converted by heat treatment to a crystalline ceramics.
- The crystal precipitates are Wollastonite ($CaSiO_3$) or Diopside ($CaMgSi_2O_6$).
- The compositions leading to commercial applications are based on silica and other oxides: Li_2O, Al_2O_3, MgO, $F/Na_2O/K_2O$ and CaO.
- The manufacturing advantage of glass ceramics over conventional ceramics is the ability to use high speed plastics forming process developed in the glass industry (pressing, blowing and rolling).

9.4.5.3 Glazes Enamels, Frits

- Glass can be used as a coating for a ceramic substrates (for instance glazes on pottery items amongst others) or a metal substrate (porcelain enamel). It makes the substrate chemically more inert and impervious, more cleanable, more resistant to abrasion and scratching, mechanically stronger and aesthetically more pleasing to the touch and eye.

TAB. 9.5 – Composition based on SiO_2 of an ophthalmic crown glass.

	SiO_2	CaO	Al_2O_3	ZnO	K_2O	Na_2O	TiO_2
Weight %	68	9	2	3	8	9	Trace

- The coating material must be fused to a homogeneous viscous liquid at a temperature, low enough, that distortion of the substrate does not occur during firing of the coating does. It must bond with the substrate.
- The coefficient of expansion of the coating and the substrate must be related to each other so that excessive stresses and strains do not result from cooling the fired ware, causing defects such as spalling or crazing. For optimum performance the coating must be compressed.
- Many glazed formulations based on the $CaO-SiO_2-Al_2O_3-Fe_2O_3$ system exist. A representative chemical composition can be given as an example (wt %): 46.5% SiO_2, 4.4% Al_2O_3, 3.1% TiO_2, 11.8% Fe_2O_3, 9.5% CaO, 5.6% MgO, 1.5% K_2O, 2.6% Na_2O. Glazes are used on the green body of hard-paste porcelain, soft-paste porcelain, sanitaryware and ceramic tiles.
- To apply a glaze to a glass ceramic two conditions must be met. Firstly, the glaze must mature at temperatures below which the body will deform and second, the glaze must have a lower thermal expansion than the body in order to preserve or enhance the strength of the article.
- With a glass ceramics having a thermal expansion of 95×10^{-7} °C^{-1} and an upper crystallization temperature of 1100 °C, a lead, lime, alkali alumino-borate glaze (thermal expansion 65×10^{-7} °C^{-1}) can be applied and matured well below the point of deformation of the body. Moreover, due to the thermal expansion difference between body and glaze below the strain point of the glaze, a compression is developed at the surface of the body which increases the flexural strength by a factor of two.

9.4.5.4 Optical Glass

- Biological and medical applications need high-quality glasses for making optical instruments.
- The specific qualities required for high-quality optical glass are high transmittance in the visible spectrum and reduced chromatic aberration.
- The composition is largely variable which includes the presence of lead, boron, barium with a silica content rarely superior to 50% (table 9.5 as example).
- The coefficient of linear thermal expansion ranges from 35 to 150×10^{-7} °C^{-1}.
- Density values for range from 2.4 to 6.3 g/cm^3.
- Knowledge of the refractive index as a function of wavelength (dispersion of glass) is essential for the lens designer concerned with reducing chromatic aberration. Index and dispersion requirement for ophthalmic products are far less stringent than for optical instruments and have scientific well established requirements. They are mass reduced from glass blanks.

Application of the Glassy Products

- Glasses for active optical elements; the most widely used are the filter glasses. They are used in the form of flat plates positioned normal to the optical axis of an optical system. The glasses selected as base glasses are from alkali lime silicates, zinc or lead borate glasses in which colorants are introduced.

9.4.5.5 Microelectronics in the Information Processing Industry

- Glasses are used primarily as electric insulators, passivation layers protecting device against hostile environment and key process step components. They are used as thin films or layers.
- Major areas of information processing technology are covered; processing, transfer, storage, display and printing. Modification of the properties is achieved by addition of different oxides:

  o Lead borosilicate glasses are used for the construction of reagent bottles.
  o Combined with germanium, alumina (90%) is used as glass optical fibres that are required in the information transfer.
  o Barium borosilicate glasses are used in the storage units.

Chapter 10

Application of Ceramic Products

10.1 Introduction to Ceramics

- CaO–SiO_2–Al_2O_3–Fe oxides account for more than 80% of the chemical composition of the major ceramic products.
- Ceramics can be defined as the art and technique to manufacture useful and aesthetic products by forming and firing clay or similar materials.
- The term 'ceramics' is derived from the Greek 'Keramos' meaning 'burned earth' and is used to describe materials manufactured by the pottery industry.
- The term 'Ceramic product' is used for the inorganic material (with possibly some organic content) made up of non-metallic compounds made by a firing process. In addition to clay-based materials, today ceramics include a multitude of products with a small fraction of clay or even none at all. Ceramics can be porous or vitrified, glazed or unglazed.

10.2 Structure of Ceramics

10.2.1 Processing of Manufacture

- Ceramics production involves a set of transformations of a mixture of minerals into a product containing new crystallized minerals and glassy phase.
- As opposite as the manufacture of products obtained by fusion of the raw material until the liquid state is reached, shaped and cooled down, the ceramic products are obtained by first shaping to the raw mix and then firing and cooling down.
- In both cases, the structure of the final product can be totally glass or crystal or a mixture of glass and crystal.
- Crystallisation can be achieved by a slow cooling of the liquid or by devitrification of the glass.

DOI: 10.1051/978-2-7598-2480-9.c010
© Science Press, EDP Sciences, 2020

- The general layout of the manufacture of ceramic products can be summarized as follows: for single firing: mining of raw materials → designing the mix of constituent materials → mixing → shaping → drying → coating with glaze → firing → cooling.
- The general layout of the manufacture of ceramics products can be summarized as follows for a double firing: mining of raw materials → designing the mix of constituent materials → mixing → shaping → drying → firing → glazing → firing → cooling.
- The properties required for a ceramic product are: high mechanical strengths, wear resistance, long service life, chemical inertness, dimensional stability, non-toxicity, resistance to heat and fire, electrical resistance, etc....
- Firing aims at obtaining these properties. It can be done in tunnel kiln (continuous operation) or shuttle kiln (discontinuous firing).

10.2.2 Shaping

- The properties are strongly linked to the techniques of shaping and the thermal profile. The main techniques are moulding, casting, jiggering, jollying/turning, extrusion and pressing.
- Moulding process depends on the shape of the item to be manufactured. A process of forming can be done at once or in the assembling of different parts.
- Casting process is used for the production of complicated or non-rotary symmetrical articles. The low viscous casting slip is poured into a plaster mould which shapes the negative outside shape of the item. The plaster absorbs water from the slip so that the slip starts to stiffen at the slip plaster boundary. When the required thickness of the item has been achieved, the excessive slip can be poured out. The shrinkage of the item is sufficient to separate the item from the mould. After drying, the consolidated item can be handled. This process requires a high concentration of solid (65/75%) and can be realized by addition of defloculent (sodium silicate).
- Stages in the process such as; jiggering, jollying and turning require a plastic body of low water content (15–25%) provided by the vacuum press. Jiggering is used for moulding flatware, like saucers and plates whereas jollying is used for hollow wares such as cups, bowls and tureens. The one side is shaped by the mould and can be decorated by a relief. The opposite side is determined by the template and thus can only be of symmetrical shape. Turning is a process to manufacture big insulator by vacuum press. The water content is reduced to 12/18%.
- Extrusion applies to all items of extended shape having the same cross section. The outside contours of the item are determined by the shape of the orifice.
- Pressing technique depends on the item. The process of semi dry pressing is used for the manufacture of low voltage insulating components. Dry pressing can be used for tableware.
- Special processes such as 'Stick-up slip' are used to attach handles to cups of coffee pots.

Application of Ceramic Products

10.2.3 Physico-Chemical Changes During Firing

- Raw materials used in ceramic bodies are mainly mixtures of clay, quartz, feldspar and to a lesser extent some other minerals, such as; Nepheline, Syenite, calcium carbonate and Wollastonite.
- The residual moisture is driven off when heating the mixture from room temperature to 200 °C.
- When the temperature increases (from 300 to 500 °C), the structural water of the clay is released. Carbonates (calcite, dolomite) are decarbonated between 750 and 950 °C.
- The temperature, at which the liquid phase occurs, depends on the presence of fluxing constituents (alkalis, iron oxide) and is largely variable.
- Phase diagrams in the CaO–SiO_2–Al_2O_3–Fe_2O_3 system are used to predict the mineralogical evolution of the mixture during firing.

10.3 Classification of Ceramic Products Containing CaO, SiO_2, Al_2O_3 and Fe Oxides

10.3.1 Introduction

- The main objective of the ceramic industry is to manufacture useful and aesthetic products with a controlled porosity. A given porosity can be achieved by the right design of the thermal profile of firing and/or by combining different types of raw materials (kaolinite, clay, quartz, flux...). Porosity governs the mechanical strength, abrasion resistance, dimensional stability, resistance to water and chemicals and fire resistance.
- Ceramic products can be classified according to the type of utilization or the chemical composition and the process, sometimes both. The cross-linking of the classification according to use, and by the chemistry leads to some overlaps. Three broad families emerge (table 10.1):

  o Traditional ceramics include products with usefulness and aesthetic requirements in our environment. They are pottery, roof tiles, household ware, earthenware, stoneware, porcelain, vitreous china and sanitaryware.
  o Ceramic products find utilizations in the construction area (bricks, vitrified clay, pipes, refractory products, expanded clay aggregates, wall and floor tiles).
  o High technical ceramics require high specific or strategic attributes, such as; abrasive, radome, space shuttle protection, etc....

10.3.2 Traditional Ceramics

10.3.2.1 Pottery, Bricks and Roof Tiles

- Ceramics have been around since the dawn of civilization, and is indeed used to identify and age civilizations. At the origin of humanity, the most common raw

TAB. 10.1 – Broad families of ceramics.

Sector of application	Products	Type of ceramics	Firing T°C	Porosity
	Pottery	Earthenware		
		Faienza/Majolica	1150 °C	10%
	Household ceramics	Hard porcelain		>2%
	Cooking ware			
	Table ware	Soft porcelain	1200 °C	
		Bone china	1240 °C	
		White china		
		Fine white stoneware		
		Dense earthenware	1250 °C	
	Sanitaryware	Stoneware		
		Vitreous china	1100 °C	<5%
Construction sector	Bricks, roof tiles			
	Floor and wall tiles			
	Vitrified clay, pipes			
	Sewerage			
	Expanded clays (aggregate)			
	Abrasive	Chemotechnical		
		Electrocast	1500 °C	
		Silica, alumina		
Technical ceramics	Insulators	Porcelain		

material to form objects of required shape was clay (kaolinite, ball clay, fire clay, stoneware clay, shale clay, bentonite...). The plasticity was the required property to mould, shape, press or extrude. The piece was then dried.
- After drying, the product was fired between 900 and 1100 °C. Firing controls important properties of the finished ware (mechanical strength, abrasion resistance, dimensional stability, resistance to water and chemical and fire resistance).
- A clay body can be decorated.
- Mixing clay, quartz and flux gives rise to a large variety of products and their associated processes. They are used for:

  o Building bricks (clay blocks, facing bricks, engineering bricks, light weight bricks).
  o Roof tiles.
  o Paving bricks.
  o Chimney pipes.

10.3.2.2 Earthenware, Faience and Majolica

- Earthenware is the earliest form of pottery fired at 1150 °C to produce non-translucent porous body (porosity of 10–15%). They are usually covered with transparent and translucent glaze.
- Earthenware has lower mechanical strength than bone china, porcelain or stoneware.
- There are several types of earthenware; terracotta, faienza and majolica:

  ○ Faience (named after the Italian town, Faenza) designs a decorated earthenware having a transparent glaze.
  ○ Majolica designs any decorated earthenware having an opaque glaze.

- Table 10.2 shows the average composition of the mixture used to produce whiteware ceramics.

10.3.2.3 Stoneware for Domestic Use and Sewerage for Chemo-Technique Use

- Stoneware qualifies as a denser earthenware, impermeable and hard enough to resist scratching by a steel point. It differs from porcelain because it is not translucent (only partially vitrified) with low porosity (3%).
- The basic raw material components are Quartz, Feldspar, china clay and ball clay. It is produced by the once firing process at a temperature of 1250 °C.
- Stoneware finds application in the manufacture of household ware, drain pipes and special laboratory equipment.
- Five basic categories of stoneware have been proposed:

  ○ *Traditional stoneware*: with a dense and inexpensive body that can be used to shape large pieces.
  ○ *Fine stoneware*: used to produce tableware.
  ○ *Chemical stoneware*: used in chemical industry when resistance to chemical attack is needed.
  ○ *Thermal shock resistant stoneware*.
  ○ *Electrical stoneware*: used for electrical insulator but now replaced by porcelain.

10.3.2.4 Porcelain

- CaO–SiO_2–Al_2O_3 accounts for 90% of the chemical composition of porcelain (table 10.2).
- Porcelain is defined as a ceramic product of dense white, translucent character (properties of porcelain to transmit light). This property which depends on the structure, gives porcelain its special character.

10.3.2.4.1 Production of Porcelain
- The body of standard porcelain consists of a mix of Kaolinite (50 wt %), Feldspar (25 wt %) and Quartz (25 wt %) fired at 1380–1460 °C for 'Hard porcelain'.

TAB. 10.2 – Average compositions of mixtures used to produce whiteware ceramics (weight %) and average chemical compositions (weight %).

Constituent	Kaolin	Ball clay	Feldspar	Quartz	Bone-ash	Talc	Limestone		
Earthenware	25	25	15	35					
Porcelain	60	10	15	15					
Bone china	25	25			50				
Vitreous	20	30	20	27		3			
Wall tile	20	30	35				15		
Chemical composition	CaO	SiO_2	Al_2O_3	Fe_2O_3	TiO_2	MgO	Na_2O	K_2O	P_2O_5
Earthenware	0.12	67.48	29.71	0.60	0.60	0.22	0.13	1.14	
Porcelain	0.09	62.27	34.68	0.75	0.44	0.24	0.23	1.29	
Bone china	28.83	34.79	16.84				2.35		17.18
Vitreous	0.14	63.09	32.62	0.73	0.76	1.25	0.10	1.31	
Wall tile	9.58	59.21	28.99	0.52	0.51	0.23	0.11	0.85	

- 'Soft porcelain' in contrast to hard porcelain contains more Feldspar and is fired at 1200 °C. The decrease in the firing temperature is possible using flux (Nepheline, Syenite, Talc...).
- The firing process can be categorised into the following sections:

  ○ Heating period.
  ○ Oxidising period.
  ○ High firing period with a reducing atmosphere.
  ○ Cooling period with a neutral to slightly oxidising atmosphere.

- The reactions which take place during the firing of porcelain can be summarized as follows:

  ○ At 573 °C, quartz is converted to its high temperature modification.
  ○ At 500/600 °C, kaolinite loses its chemical bounded water and change to metakaolinite.
  ○ At 950 °C a Spinel forms that changes to Mullite at 1150 °C (primary Mullite) that remains stable at high temperatures. Primary Mullite originated from a solid phase is present as scaly aggregates. The aggregates of primary Mullite are clearly separated from the surrounding liquid phase during the initial stages of firing.
  ○ The liquid phase takes place at 985 °C (eutectic temperature of potassium feldspar and Quartz).
  ○ The potassium feldspar first reacts with silica originated from Kaolinite during the thermal decomposition whereas the free Quartz (from the batch) will only react with the liquid phase and becomes dissolved at higher temperatures.
  ○ The complete melting process is finished at about 1150 °C (melting temperature of pure potassium feldspar).
  ○ Sodium feldspar forms a eutectic with silicate at about 1062 °C. The melting point of pure Albite is 1118 °C.

- The reactions described above result in the primary phases in porcelain. Further reactions between these phases take place during the continuing firing process. The liquid phase of eutectic composition formed in the first state is not saturated by silica and alumina with increasing temperature and therefore able to absorb silica and Mullite.
- The dissolution of quartz is of great importance for the firing technique of porcelain. It is known that silica increases the viscosity of the liquid phase which means the increase in the tenacity of the liquid phase with an increasing content of silica.
- The decrease in viscosity of the liquid with increasing temperature is compensated by the dissolution of silica. Therefore, porcelain is characterized by a wide firing range which ensures sufficient stability against distortion of the items during firing.
- The Mullite aggregates are penetrated by alkalis which causes dissolution of the boundary material in the liquid phase.
- The initial composition of the liquid phase or feldspar and silica is close to the eutectic point of K_2O–Al_2O_3–SiO_2 system and thus near the phase boundary of Mullite.
- Therefore, the loss of small amounts of alkali in the initial phase will shift the composition of the liquid phase to the field of primary crystallisation of Mullite. This results in the separation of Mullite that crystallises in the shape of idiomorphic long and predominantly thin needles. It is formed in the presence of a liquid phase and called 'secondary Mullite' whereas the 'primary Mullite' results from the decomposition of clay minerals in a solid-state reaction. Secondary Mullite leads to considerable amounts found in porcelain body fired at 1200 °C.
- With increasing temperature secondary Mullite is dissolved as a consequence of the enrichment of silica in the liquid phase. Thus, in order to restore the balance of the entire system, the liquid phase must be fed with alumina. This is achieved by the dissolution of Mullite.
- The reactions during the firing of porcelain which have been mentioned so far can all be explained by the equilibrium reactions of the K_2O–Al_2O_3–SiO_2 system and the local concentrations of the individual components.
- Porcelain containing less quartz or no quartz at all will show different reactions:

    o Mullite is only slightly absorbed if at all with increasing temperature.
    o There is no shift in the chemical composition of the liquid phase of porcelain free of quartz which would act as a moving force for that reaction.

- In the first level of firing, the dissolution of Quartz continues until dissolved completely or equilibrium has been achieved and the liquid phase is saturated with silica.
- The remaining Quartz is then converted to Cristobalite in the further course of firing.
- In porcelain with a high content of silica, the conversion of Quartz to Cristobalite can already be observed before equilibrium to the complete system is reached. Due to the poor diffusion capability of silica enrichment of silica around the

Quartz particles occurs. With this, the rate of transformation of quartz into Cristobalite becomes higher than one of dissolution of silica in the liquid phase.
- At this stage of the firing process, a reaction between the liquid phase and primary (scaly) Mullite takes place. The alkali ions increasingly infiltrate primary Mullite and gradually cause its recrystallisation into secondary (needle) Mullite.
- Later on, the primary Mullite disappears completely which results in a porcelain body containing Mullite needles only which are embedded in glassy phases.
- The penetration of liquid phase into primary Mullite does not occur in a porcelain free of quartz. The primary Mullite remains unchanged even during long soaking periods at high firing temperatures close to the softening point of the body.
- Free alumina as it is present in alumina porcelains shows only poor reaction in porcelain bodies.

10.3.2.4.2 Properties and Use of Porcelain
- The thermal expansion is low and lies between 4×10^{-6} and 6×10^{-6} K^{-1} at temperatures between 20 and 600 °C.
- Quartz with its high thermal expansion will cause a high thermal expansion of the body. When it dissolved during firing, its influence diminishes and results in a decreasing thermal expansion of the glassy phase.
- Tensile and flexural strengths of glazed porcelain are, respectively, between 30–60 N/mm^2 and 60–100 N/mm^2.
- Porcelain is used in different applications: tableware, sanitary porcelain, chemical technical porcelain and electro-porcelain.
- Tableware is preferably glazed with translucent glaze to bring out its whiteness and translucence. For sanitary ware, white opaque or colour glazes are used. The glaze should have a good flow during firing, be resistant to wear and chemical influence. Glazes used in tableware have a high resistance to scratching. A high brilliance is required in art ceramics.
- Porcelain for chemical technical application contains much less quartz than standard porcelain and the resistance of thermal shocks is improved. For special electro technical purposes, feldspar is replaced by a mixture of alkaline earth carbonate in alkaline earth porcelain.

10.3.2.4.3 Bone China
- Bone china is a vitreous transparent pottery made from a body of the following composition: calcined bone 45–50%, China clay (Kaolinite) 25–30% and China stone (Cornish stone) 25–30%.
- $CaO-SiO_2-Al_2O_3-Fe_2O_3$ accounts for almost 80% of the chemical composition.
- It contains a minimum of 15% of phosphate derived from animal bone in the fired body.
- With bone china, high translucency and high mechanical strength occur simultaneously: the high strength being due to the small size and great number of

crystals, the high translucency arising from a good match of refractive indices of the two types of crystal and glass.
- Bone china has a lower thermal shock resistance than hard porcelain and is not suitable for cocking ware but met the conditions for being used in tableware.

10.3.2.5 Vitreous China

10.3.2.5.1 Tableware Application
- Vitreous china is a product covering the boundary between a dense earthenware, fine grained stoneware and soft porcelain.
- The main constituents are Ball clay 10%, China clay 45%, Quartz 25%, Nepheline and Syenite 20%.
- The maximum value for water absorption is 3%.
- It finds its application as tableware.

10.3.2.5.2 Sanitaryware (Vitreous China)
- Sanitaryware includes domestic equipment such as; wash basins, toilets, urinals, bidets, and ceramic drinking fountains...
- They are made from a mixture of clays and other minerals that are rendered impervious to water by firing at high temperatures. These products are mainly made of vitreous china (semi-porcelain) or earthenware.
- A traditional slip-casting is composed of 26% ball clay, 24% china clay, 26% quartz and 26% of flux (feldspar or nepheline-syenite).
- The raw meal is mixed with water to produce a slip casting. The product is released from the moulds, dried, glazed and fired at 1280 °C for vitreous china.
- The water absorption should not exceed 0.5% of the weight of the piece when dry.
- Vitreous china sanitaryware is coated on all surfaces exposed to view in normal use with an impervious glaze giving a smooth shiny white or coloured finish which is durable and easy to clean.
- Vitreous china is impermeable to water. It is proof against crazing in use and remained hygienic even if the glaze becomes damaged and chipped. It is also stronger than earthenware if made to the same fineness. It has the double firing shrinkage of earthenware and also sagged during its firing to a glassy state.
- Introduction of calcium in the slurry reduces the porosity of the body (fluxing action).
- The addition of Wollastonite to a ceramic mix introduces:

  ○ Insoluble calcium in the slurry.
  ○ Silica and calcium combined in the form of a single compound.
  ○ This addition stops lamination, gives good die filling characteristics, and makes for easy drying and pressing with improved bonding. The product in both green and fired stages has high strength.

- Vitreous china articles must be beautiful, functional and easy to make.

10.3.2.6 Glazes, Glaze Frits, Fluxes, Colour Stains

- Glazes are thin vitreous coating formed in place on a ceramic body. Glazes are usually applied to make the body non-porous smooth, glossy or opaque, mechanically stronger and chemically more resistant. Moreover, they improve the aesthetic appearance of the ceramic ware.
- Glazes are required to combine with different ceramic bodies. The thermal expansion of the glaze must be near but never higher than that of the body to eliminate the tendency towards crazing.
- The use of Wollastonite produces a high compression glaze, but not as high as those causing peeling outcomes.
- The glazes are similar to industrial glass' but there is a greater variety of glazes than glass'.

10.3.3 Ceramics in the Construction Sector Ceramics

- CaO–SiO_2–Al_2O_3–Fe oxides account for more than 80% of the chemical composition of the following major ceramic products in the construction sector.

10.3.3.1 Vitrified Clay Pipes

- Drain pipes are used for the removal of effluents and for the construction of sewerage systems. They can be vitrified or made with stoneware. The fired stoneware body has a slight open porosity and is watertight. It can be glazed, at least on the inside to improve the flow properties. Stoneware pipes are fired in a tunnel kiln between 1050 and 1250 °C.

10.3.3.2 Expanded Clay Aggregates

- They are porous ceramic products with uniform pore structure of fine closed cells and with a densely sintered firm external skin obtained by firing the clay between 1100 and 1300 °C.
- Aggregates have grain densities between 0.15 and 1.7 kg/dm^3 and thermal conductivity ranging from 0.07 to 0.18 w/mK.
- They are used as loose or cement bound material for the construction industry (loose fillings, lightweight concrete, blocks, prefabricated lightweight concrete component, structural lightweight concrete for onsite processing...) and also loose materials in garden and landscape (in lightweight substrates for green roofs, filter-beds and drainage fillings, and also insulating layer between floors in building, often part of underfloor heating systems, etc....).

10.3.3.3 Refractory Products (See Chapter 8)

- Refractory products are ceramic materials capable of withstanding temperature above 1500 °C.

Application of Ceramic Products 313

They are used in many industrial applications such as steel, iron, cement, lime, glass, ceramic, aluminium, copper, petrochemistry, incinerators, power plants and house heating system (storage heater blocks).

10.3.3.4 Wall and Floor Tiles

- Ceramic wall and floor tiles are widely used as decorative covering for walls and floors and include mosaics. They are typically used in bathrooms, kitchens, swimming pools but also in commercial kitchens and food processing areas where grouted with impervious resins.
- They are prepared from a mixture of ball clay, sand, fluxes, colours and other mineral raw materials (such as Wollastonite...).
- The chemical composition is given in table 10.2.
- Recipes are based on local raw materials and may contain mainly ball clay associated with grogs and rejects.
- They are shaped by extrusion, casting and dust pressing at room temperature. They are then dried and fired at temperatures sufficient to develop the required properties. Open flame tunnel kilns are used for biscuit firing. White earthenware is fired at 1000/1230 °C for cottoforte, cotto and majolica at about 1000 °C.
- Wall and floor tiles can be glazed, unglazed or engobed. Glazes are used and fired at around 1200 °C (temperature of the single firing of stoneware).
- They are classified according to their method of shaping (extrusion, pressing, casting...) and their water absorption (linked to their porosity); less than 3%, between 3 and 6%, between 6 and 10% and higher than 10%.
- The addition of Wollastonite ($CaSiO_3$) to the body gives buff to grey whiteware which can be glazed in a short single firing at a fairly low temperatures.
- The body dries rapidly without cracking or shrinkage and the finished product can be easily sawn, nailed and drilled with ordinary tool. This is due to the acicular crystal constituents having good cleavage. They break down easily partly because the bond is uniform.
- Example of composition; ball clay 31%, Talc 33%, Wollastonite 36%, Cone 01–1% (Orton) and temperature from 1117 to 1154 °C.
- More than half the total world output of Wollastonite goes into tile making (if requirement for body, glaze and frit are included).
- The industrial applications of natural Wollastonite and synthetic monocalcium silicate are based on basic properties: the supply of calcium ions as insoluble form, silicon combines, low thermal expansion, fibrous skeleton (20/1 ratio).
- Surface treatment or surface coating improves the mineral filler performance by increasing the adhesion between the matrix and the fibre.

10.3.4 High Technology Ceramics

- Technical ceramics are applied in many industries and cover both established products like insulators and new applications. They supply elements for the aerospace and automotive industries (engine part, catalyst carrier), electronics (capacitors, piezo electrics), biomedical products (bone replacement), environmental protection (filters) an many others.

10.3.4.1 Electrical Porcelain

- Electrical porcelain is used as insulator and part of electrical appliances, including soapstone (the mineral Talc, or source rock talc-schist – *also called steatite*) with small amounts of bonding clay. Small amounts of feldspar and alkaline carbonate produce vitrified body at 1400 °C.
- Porcelain is an outstanding insulating material. The dielectric strength is in the range 300–350 kV and its specific resistance is between 10^4 and 10^5 Ω/cm.

10.3.4.2 Inorganic Abrasive

- Inorganic abrasives can be fused with alumina, synthetic corundum, silicon carbide, cubic boron nitrate and Diamonds.
- Some inorganic abrasives are bonded (grinding wheels) or coated (abrasive paper and tissues).
- A principal characteristic of grinding is the effect of numerous not orientated cutting materials in the workpiece.
- Abrasive products are used for grinding, cutting off, polishing, pressing, sharpening for metal, wood, glass, stone, and plastic.

10.3.4.3 Chemo-Technical Stoneware

- Chemo-technical stoneware belongs to the group of clay-based ceramic material with impermeable bodies that is a branch of stoneware.
- The raw materials are based on stoneware clay containing Illite and/or Serizite and Kaolinite leading to a chemical composition SiO_2 40–70% and Al_2O_3 25–50%.
- In the glassy phase (45–60%), some crystalline components such as Mullite and Quartz are present.
- Chemo-technical stoneware has:

    ○ Corrosion resistance to media, such as acids (other than hydrofluoric), dyes, solvents and salt solutions irrespective of their concentration and temperature.
    ○ High compression strength and low tensile strength.
    ○ Abrasion resistance. The hardness is in the range 7–8 of the Mohs hardness scale. Corundum stoneware with a fine distribution of corundum grains has optimum abrasion resistance.

- The special applications come from the combined corrosion and abrasion.
- The applications are closely associated with the chemical industry and with every chemical process technology. Laboratory equipment are in the field of tanks, vessels, storage jars, bowls, tubes, sinks, bench tops...

Chapter 11

Application as Fillers

11.1 General Introduction

- The potential for formulation blended cement can be seen in the system CaO–SiO$_2$–Al$_2$O$_4$–Fe oxide system in relation with depletion of natural resource and rapid growth of by products from increased industrial and agricultural activity.
- Fillers are finely divided inorganic substances which are added to mineral or organic constituents providing the principal useable properties of the material.
- Fillers contain one or many constituents.
- The total CaO–SiO$_2$–Al$_2$O$_3$–Fe oxide accounts for at least 95% of the typical bulk composition of the most commonly used fillers.
- Fillers can be considered and termed 'mineral admixtures', 'mineral additions', 'cement extender', 'supplementary cementitious materials' when the major constituent is cement or concrete. When applied to medicine or domestic use the term 'excipient' is used.
- Fillers can be classified by their origin (natural, industrial or domestic by products) or by the mechanism of their interaction.
- The mechanism of interaction between the filler and the other constituents can be associated to two mechanisms (figure 11.1):

  (1) Mechanical properties improve by the optimization of the granulometry. Filler is used to fill the space between the particles of the material. Consequently, it increases the compacity of the material. Filler can also facilitate the mixing of the material.
  (2) Fillers modify the chemical properties of the material and thus may change the thermodynamic phase assemblages of the material depending on its reactivity. Indeed, fillers can be active, innocuous, inert or nearly inert and even deleterious such as clay in concrete. The reactivity is a key parameter to

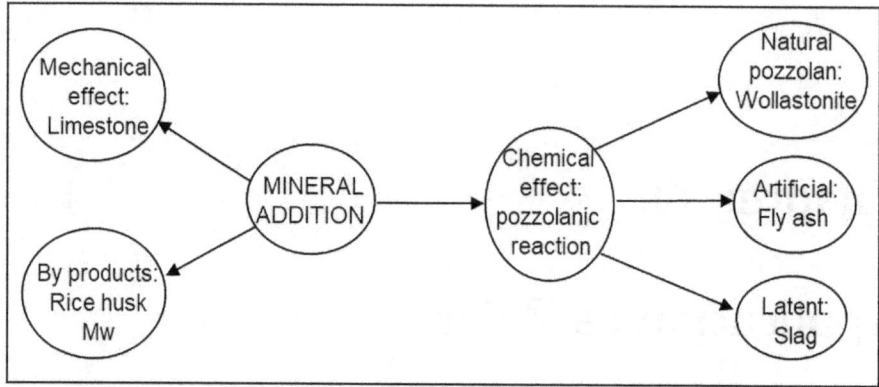

FIG. 11.1 – Various benefits of the uses of mineral additions.

determine the use of filler. For example, blast furnace slag activated by alkalis can be used as a hydraulic binder, whereas Wollastonite ($CaSiO_3$) that is difficult to activate, is used as an aggregate.

- The reactivity of filler in the specific case of cement or concrete can be quantified by a pozzolanic activity that estimates its ability to react with calcium hydroxide in the presence of water, to form hydrates possessing cementitious properties. During Portland cement (PC) hydration in the presence of slag or pozzolana, both slag and pozzolana react with Portlandite ($Ca(OH)_2$) developed by cement hydration to form hydrates having high specific surface. The phase assemblage depends on the amount of filler and temperature that is an important parameter regarding the kinetics of pozzolanic reaction.
- Fillers find applications in the field of cement and concrete in the following ways; addition to the raw materials or substitution of traditional constituents in cement manufacturing, addition to cement or concrete to improve their properties (cement extenders). For example, limestone addition as filler impacts the shrinkage, crack propagation, diffusion of chloride ions, sea water resistance, sulphate bearing water resistance and freeze–thaw resistance. The main industrial products are slags from iron and steel industry (BF, LD, EAF, ladle, ferroalloys stainless steel slags), from metallurgy (copper, zinc...), ashes from power station (fly ash, bottom ash...) natural or artificial pozzolana (metakaolin, rice husk...). Some additives require special post treatment (granulation of slag...) while some others can be improved during the manufacture of the material.

11.2 Mono-Component

11.2.1 Calcium Oxide

11.2.1.1 Lime

- Lime is used in the environment sector to neutralise flue gas pollutants, capture heavy metals, and treat organic and mineral sludges.
- Lime is used:

  o In sugar industry to precipitate impurities (10–15%).
  o In chemical industry, in leather tanning in the paper industry, to precipitate $CaCO_3$.
  o In agriculture, to adjust the pH of soil by chemical effect to give optimum growing conditions (30%).
  o In steel and non-ferrous metal industry to form slags.
  o In the glass industry, as fluxing agent (30–40%).
  o In civil works, to stabilize soils and roadways.
  o In construction as lime-based mortars.
  o In the production of cement, bricks and structural blocks.

- Lime can be used as refractory (MP 2613 °C), as long as it cannot hydrate through the presence of water.

11.2.1.2 Limestone

- Limestone is used in the pharmacy industry in tooth paste formulation.
- Limestone filler is frequently added as a constituent in Portland cement (EN 197 CEM II) and concrete as it modifies the particle size distribution, improves the workability, shrinkage, density and strength by a physical effect. It also modifies the mineral phase assemblage precipitated during cement hydration leading to improve the durability; decrease in the diffusion of chloride ions and improvement of the sea water, sulphate bearing water and freeze–thaw resistance:

  o Limestone, finely ground can complement the PSD of cement in such a way that the rheological properties (water demand, workability) and compactness are improved.
  o A partial substitution to gypsum by finely ground limestone is possible without adverse effects on the set, shrinkage and compressive strength.
  o The reaction between calcite and C_3A leads to the formation of calcium carboaluminate inducing an accelerating effect on C_3A hydration but also a decrease in the thickness of interfacial transition zone between the cement paste and the aggregates. Moreover, the addition of $CaCO_3$ accelerates the hydration of C_3S especially at early age as limestone surface can act as heterogeneous nucleation site for C–S–H. Thus, compared to an inert substance, limestone fillers can have a positive contribution on the compressive strength that depends on several

parameters; the limestone filler content, its nature and fineness, the amount of cement, the reactivity of C_3A, the grading curve of the aggregates and their mineralogical nature.

- Limestone filler can also be used in the case of formulations using calcium aluminate cement. Finely ground limestone has been used to compensate the loss of strength occurring during the so-called conversion in high alumina cement that occurs at temperatures below 30 °C. The formation of calcium carboaluminate and AH_3 in addition to CAH_{10} when CA reacts with $CaCO_3$ is expected to be at the origin of this behaviour.
- The influence of the nature of limestone must be taken into account. The limestone filler has about the same behaviour in concrete provided the $CaCO_3$ amount is high enough and it does not contain large amounts of clay, dolomite and pyrite:

  ○ The presence of pyrite in limestone filler is undesirable because of the late formation of ettringite by its oxidation and combination with calcium aluminate.
  ○ The presence of clay leads to an increase in the water demand and reduces the positive effect of the limestone.

11.2.2 SiO_2

11.2.2.1 Silica

- Silica shows a low solubility in water, acidity and viscosity at high temperatures.
- First level applications:

  ○ SiO_2 as fine microcrystalline particles is used as foundry mould.
  ○ Amorphous SiO_2 is used as additive in concrete (silica fume, rice husk).
  ○ Synthetic silica are produced from sodium silicate; fumed silica, precipitated silica, colloidal silica, silica gel:

    ▪ Precipitated silica is amorphous silica prepared by precipitation from sodium silicates solution.
    ▪ Silica gel is produced by acidification of sodium solution with an acid mineral or carbon dioxide.
    ▪ Colloidal silica (silica sol) is produced from a solution of sodium silicate; the solution, is passed through an ion exchange to partially remove the sodium ions forming a silica suspension and pH is adjusted to control the particle size.

  ○ SiO_2 is used as:

    ▪ Additives in pharmaceutical drugs where silica aid powder flow when tablets are formed.
    ▪ Extender in paint.
    ▪ Functional filler in plastic and rubber.

Application as Fillers 319

- Lumps if used as aggregate for road construction.
- High temperature thermal protection in fibber form.

  ○ Vitreous silica glass is used for high temperature application refractory.

- Second level applications:

  ○ SiO_2 after reduction is a precursor for microelectronics (photovoltaic).
  ○ SiO_2 is present in association with other oxides in cement, ceramic, refractories, metallurgy, electronic components and glass (Aluminosilicate glass) industry.

- End level applications:

  ○ Furnace tube, melting crucible and thin film of silica grow.

11.2.2.2 Silica Fume

- Silica fume (SF) is a by-product of the manufacture of metallic silicon and ferrosilicon by reducing quartzite with carbon at high temperatures. Gassified SiO is deposited as SiO_2 during cooling. SF contains 85–98% of silica that is mainly amorphous.
- The specific surface is about 20 000 m^2/kg. It is composed of fine spheres with a mean diameter of 0.1–0.2 μm.
- SF is available as bulk, densified powder or slurry.
- The addition of finely divided silica fume to Portland cement improves the mechanical properties of the mortars by optimizing the particle size distribution (which increases the compaction of the mix) and by chemical effect (which decreases the Portlandite content, increases C–S–H content and lowers its Ca/Si ratio by pozzolanic reaction thus reducing the amount of capillary porosity) leading to greater compressive strengths. SF is used in rather small amounts (compared to other additives) and its homogeneous dispersion in concrete often requires the use of superplasticizers. Consequently, SF is used to design high performance concrete. The thickness of the interfacial transition zone between the paste and aggregates is reduced by the addition of SF. The incorporation of SF reduces permeability and improves durability in various aggressive environments. The sulphate resistance is due to the lower content of free CH and to the incorporation of alumina into the C–S–H, leaving less alumina available to interact with sulphates to produce ettringite.
- In the case of calcium aluminate cements, the addition of silica fume with admixtures (mainly sodium polyphosphates) reduces the gauging water, acts on the kinetic of the hydration and changes the phase assemblages of the hydrated products.

11.2.3 Alumina

Application: mechanics (hardness), thermal (high melting point), insulating properties, chemical properties (resistance to chemical attack), ability to sintering.

- First level of applications:

    ○ Al_2O_3 is used as technical ceramic devices at high temperatures (refractory, coating).
    ○ Corundum is mainly used as abrasive (mixed with other oxide emery).
    ○ Al_2O_3 is used for electrical devices (spark plug insulator) or in jewellery (Sapphire).

- Second level of applications:

    ○ It is the precursor of the aluminium industry and is present in association with other oxides in ceramic, refractories, metallurgy, glass and ceramic industry.

11.2.4 Fe Oxides

Application: catalyst, sorbent, pigment, flocculants coatings, gas sensor, ion exchanger, lubrication.

- First level applications:

    ○ FeO is used as a pigment in cosmetics.
    ○ Fe_3O_4 is used as a black pigment.
    ○ The iron oxides such as Magnetite, Hematite and Maghemite are used as pigments for black red and brown colours, respectively, in the construction industry (roof tiles, paving slabs...).
    ○ Iron oxides-based materials are efficient catalysts. They are used as catalyst in the Haber process and in the water gas shift reaction.
    ○ For oxidation–reduction and acid-base reactions, the most applied iron oxides as catalyst are Magnetite and Hematite.

    - Fe_2O_3 exhibits good sensing characteristics towards hydrocarbon gases carbon monoxide, and alcohol.

    ○ Many toxic anions (Ca, Zn, Pb...) or anions (CrO_4^{2-}) are removed using various phases containing iron oxide. The high specific surfaces and surface charge of iron oxide regulate free metal and organic matter concentrations.

    - Fe_2O_3, as nano ferromagnetic particles, is used as a magnetic storage media in audio and video recording and magnetic optical devices. Iron oxide finds application in magnetic recording, magnetic storage devices, toner and ink, magnetic resonance imaging, wastewater treatment, bioseparatin and medicine.

    ○ In therapeutic application, iron based magnetic nanoparticles are used for hyperthermia and drug delivery. In *in vitro* application, the main use is in diagnostic application (separation-selection and magneto-relaxometry). Nano particles of Fe_3O_4 are used as contrast agents in MRI (Magnetic resonance imaging).

- Second level applications:

    ○ Iron oxides are used as raw feedstock of the steel industry.

11.3 Multi-Component By-Products from Industrial Process

11.3.1 Slags

11.3.1.1 Introduction/Characterisation

- Blast furnace slag (BFS) is a by-product in the production of iron. The role of slag in this metallurgical operation has been discussed in chapter 7 (metallurgy). The total of CaO, SiO_2, Al_2O_3, Fe oxides account for >95% of the typical bulk composition of BFS (TiO_2, MgO are present at a small quantity). The slag after cooling can be added to lime or Portland cement and used as filler. It forms slag-lime binder or Portland-slag cement. BFS can be activated by alkalis and used as binder (see chapter 6 – hydraulic binder). BFS can be also added directly to the constituents of concrete. Thus, as cement addition BFS cements are used in all branches of concrete and reinforced concrete construction, in civil engineering and building construction, in hydraulic engineering and in the concrete block industry. However, Portland slag cement may be employed in applications in which a high short-term mechanical strength is not required. The main area of their use is applications in which a high corrosion resistance or slow hydration heat liberation of the binder are required. Cements in which slag is the dominant product (60–70%) are used for the construction in sulphate environment and are suitable when low heat of hydration is needed.
- The range of chemical composition is as follows (table 11.1): SiO_2 27–40%, Al_2O_3 5–33%, CaO 30–50%, MgO 1–21%, Fe oxide <1% and alkalis 1–3% while a representative composition is given in table 11.1. The amplitude of the range is due to the diversity of the materials involved in the manufacture of iron (iron ore, flux, refractories...). It exists owing to an adverse influence of MnO with slags of low basicity and low Al_2O_3.

TAB. 11.1 – Range of chemical compositions of BF slags.

BF slag type	Super-acidic	Acidic	Basic
SiO_2	47.4–60.4	33.3–45.8	25.1–37.3
Al_2O_3	21.9–23.9	8.9–28	10.2–21.2
Fe_2O_3	0.1–0.6	0.1–8.1	0.1–1.3
FeO	5.7–21.5	1.1–8.9	0.5–3
CaO	1.9–4.4	33.1–46.5	46.6–51.6
MgO	0.8–2.8	0.2–4.5	0.2–1.1
Basicity modulus $Mb = CaO + MgO/(SiO_2 + Al_2O_3)$	0.05–0.09	0.6–0.9	0.1–1.1
Activity modulus $Ma = Al_2O_3/SiO_2$	0.37–0.50	0.26–0.80	0.3–0.8
% Glass	95–98	85–98	15–85

- Depending on the cooling rate BF slags contain glassy phase (through rapid cooling) or polycrystalline material (through slow cooling). When BF slag is cooled rapidly (granulation) the percentage of glass can vary from 85 to 95%; it is called ground granulated blast-furnace slag (GGBS). The glass phase depends on the chemical composition and the rate of cooling. With high basicity, higher speed of cooling is required. The crystalline phases are solid solutions of Melilite (C_2AS–C_2MS_2), and other minor phases like Merwinite (C_3MS_2), Larnite (C_2S), and Wollastonite (CS). SiO_2 is present in the glassy phase in the form of SiO_4 tetrahedra, the monomer and dimmer being the predominant silicate species. Calcium and alkalis are present in the glass in six-fold coordination. Merwinite occurs as primary phases when the $CaO + MgO/SiO_2$ is greater than 1.3. Five percent of Merwinite have a beneficial effect on mechanical strengths at 2 and 28 days which are attributable to enrichment of Al_2O_3 in the glassy fraction (WAN 2004). Due to the good correlation between the Gehlenite concentration and the reactivity of the BFS the Gehlenite concentration can be regarded as an indicator for the reactivity of the BFS.
- Basicity ratio or any parameter alone is not sufficient to predict the reactivity of BFS with water (DEM 1980).

11.3.1.2 Portland – BFS Cement and BFS Concrete

- Ground granulated blast-furnace slag GGBS hydraulic activity is weak and requires activation to proceed at an acceptable rate. Activation can be chemical, mechanical or thermal.
- When GGBS is mixed with water, a dissolution process gets under way resulting in the formation of a thin SiO_2–Mg enriched and Ca deficient hydrated layer at the surface. The presence of this hydrate layer inhibits further migration of water toward the slag surface and ion dissolution, thus resulting in a discontinuation of the hydration reaction. Addition of alkaline activator increases the pH and allows the dissolution to continue; calcium hydroxide, Portland cement, NaOH, KOH and Na_2CO_3 are a suitable alkaline activator.
- The rate at which the heat of hydration is released is slowed down with increasing amounts of slag in Portland cement. Thus, cements containing GGBS are characterized by a lower heat of hydration than PC. Setting times are consequently higher than those of Portland cement.
- The addition of different amounts of slag to cement (CEM II, CEM III and CEM V as defined by CEN – EN 197-1) leads to different hydrate assemblages. At low slag contents, the products of hydration and the microstructure are similar to those of Portland cement; CH, C–S–H, ettringite, calcium monosulfoaluminate and calcium monocarboaluminate. However, CH content is lower and C–S–H has a lower average Ca/Si ratio. When greater contents are used, other phases may become stable such as Stratlingite (C_2ASH_8) is often present. If alkali hydroxides or their salts are used as activators, C–S–H with a low Ca/Si ratio is produced as the main hydration product. It contains larger amounts of Al_2O_3 and MgO than C–S–H phase formed in the hydration of PC and contains alkalis.

- The strength development of the Portland slag cement is slower than that of Portland cement and is influenced by the curing conditions. At 7 days, a decrease in strength is generally recorded and at later age, strengths are similar or higher than those of PC.
- Curing under water yields the highest mechanical strengths. Moreover, the mechanical strength increases with increasing temperature. Grinding to high fineness may be especially indicated in Portland slag cement with very high slag contents to compensate the limited reactivity of the slag. Small amounts of sodium silicate reduce the setting time and are effective as a grinding aid and have a positive effect on strengths.
- There is a wide variation in the reactivity of granulated BF slags characterized by their different readiness to undergo hydration. Thus, reactivity depends on the rate of cooling. An increase in the amount of structural defect produced by rapid cooling increases the readiness of the slag to react. The reactivity increases with increasing the glass content of the slag and with increasing fineness. The reactivity depends on the oxide composition of the slag:

  ○ It increases with increasing alkalinity of the slag and in particular with increasing content of CaO.
  ○ Increasing contents of Al_2O_3 tend to increase early strength, mainly because the amount of ettringite formed is increased under these conditions.
  ○ The hydraulic modulus ($HM = CaO + MgO + Al_2O_3/(SiO_2)$) has been determined to assess the reactivity of slags. The minimum value should not be less than 1.
  ○ It is also convenient to use the basicity modulus $(CaO + MgO)/(SiO_2 + Al_2O_3)$ which varies from 0.05 to 0.9 for acid slags and from 0.9 to 1.9 for basic slags.
  ○ The hydraulic index, $Ih = CaO + 1.4\ MgO + 0.56\ Al_2O_3/SiO_2$, was also proposed (CHER 1968).

- Under hydrothermal conditions at 175 °C, α-C_2SH is formed as the main hydration product in cement with high slag content resulting of mechanical strengths under prolonger autoclaving.
- Cement compositions containing granulated BF slag are more disposed to carbonation than are plain Portland cements. The rate of carbonation is influenced by the curing conditions and generally decreases with increasing degree of hydration and compressive strengths. However, the response of concrete mixes made with BFS to corrosive agents such as acids or sulphate, is qualitatively better than that of mixes with plain PC. This is largely due to the reduced amounts of calcium hydroxide (Portlandite). Indeed, cement with high contents of slag contains little or no CH and their C–S–H are more compact. Expansion caused by silica reactive aggregates decreases when slag content increases.
- Across the age range, there is a ternary blend of limestone filler, GGBS and PC that present an optimum strength development, better than binary LF or BFS and plain PC. It is attributable to the complementary behaviour of both admixtures, LF improves early strengths whereas GGBS improves later strengths by its cementing reaction.

- The addition of reactive fillers (*i.e.* Silica Fume) has an impact on the strengths development, partially due to the occurrence of other types of hydrates.

11.3.2 Fly Ashes

11.3.2.1 Introduction/Characterisation

- During the combustion of pulverized coal, two solid products are formed; the coarser ash collected at the bottom is known as bottom ash and the finer ash carried in the flue gases is termed 'pulverized-fuel ash' shortened PFA. The value of specific surface area of PFA ranges between 2000 and 5000 m^2/kg.
- With the demise in industrial use of coal as a fuel, these have become increasingly scarce by-products, so the use of fly ash will cease to be commonplace.
- PFA derived from anthracite and bituminous coal is low calcium and defined as class F. PFA derived from lignitic or bituminous coals is high calcium and named class C with a lime content ranging from 10 to 30%.
- Typical chemical composition ranges of PFA class F and C are given in table 11.2. The total CaO–SiO_2–Al_2O_3–Fe oxides account for at least 95% of the typical composition of fly ash. Fly ash contains a glassy phase (60% and more of the ash). The main crystalline constituents are Quartz (SiO_2), Mullite ($3Al_2O_3 \cdot 2SiO_2$), Magnetite (Fe_3O_4), and Hematite (Fe_2O_3). High calcium ashes contain free lime, Periclase, tricalcium aluminate and dicalcium silicate. Fly ash contains some residual carbon leading to their grey colour.
- Bottom ash contains glass, Quartz, alumina Mullite and Anorthite.
- Fly ashes produced from fluidized bed combustion (FBC) are enriched in sulphate because of the need to remove SO_3 from the flue gas. Their high sulphate content prevents their use in the production of OPC (MAJ 1993).
- Lignite fly ash self-hardens when compacted in moist state. At ambient temperature in the presence of calcium hydroxide and water, the glass phase of the fly ash undergoes a pozzolanic reaction. In addition to Portland cement, fly ash acts as a reactive filler.
- Fly ashes can be utilized in the following way:

    o As a raw materials in the manufacture of Portland cement clinker.
    o As filler for the manufacture of blended cement.
    o As finely divided mineral admixture for concrete as partial replacement of fine aggregates or/and of cement.
    o For the manufacture of sintered light weight aggregates.
    o For the manufacture of aerated concrete block.

11.3.2.2 Portland Fly Ash Cement and Fly Ash Concrete

- Portland fly ash cement (CEM II, CEM III and CEM V as defined by CEN – EN 197-1) is produced by intimate blending of PC with PFA. The amount of fly ash is limited to 30–40% but an addition of about 15–25% is typical.

TAB. 11.2 – Typical chemical composition ranges of PFA class F and C.

Oxide (wgt %)	Bituminous	Lignite
SiO_2	20–60	15–45
Al_2O_3	5–35	20–25
Fe_2O_3	10–40	4–15
CaO	1–12	15–40
LOI	0–15	0–5

- The reactivity of FA is dependent upon the loss on ignition (unburned coal), the specific surface area, mineralogical and chemical compositions. Sulphate and MgO have to be controlled for soundness and alkalis for use in concretes containing reactive aggregates.
- Fly ashes show a pozzolanic activity and slowly hydrate when alkalis and Ca(OH)$_2$ are added. Thus, the hydration of clinker minerals is mainly responsible of the setting and initial mechanical strength development as the reaction rate of fly ash is very slow. The hydration of FA contributes to strengths only at longer hydration times but also affects other properties of the hardened material. Indeed, it modifies the assemblages of hydrates formed during cement hydration depending on the FA content and type but also temperature.
- At room temperature, FA cements develop lower early strengths, but higher long-term strengths if the degree of replacement does not exceed about 20–30%.
- The rate of pozzolanic reaction accelerates with increasing temperature of curing.
- Fly ash also affects the rheology of the fresh concrete mix, related to the spherical shape of the ash particles. The increase in water demand due to the fineness of the additions and the decrease in the rheological properties are compensated by the addition of admixtures. The utilization bituminous fly ash reduces the amount of water required for a given degree of workability. The spherical particles smaller than 45 μm contribute to the reduction in water requirement.
- The microstructure of the hardened paste of PC containing FA differs from that of pure PC. The amount of Portlandite is reduced at later stages of hydration especially in mixes with high ash contents. Moreover, the volume of pores of Portland-fly ash cement pastes tends to be greater than that of similar pastes from pure PC but the pore structure tends to be finer as long as the cement/ash replacement ratio is not above 30%.
- The resistance to chemical attack including that to sulphate is improved by the presence of fly ash in concrete mixes. Reduced concrete deterioration due to alkali silica reaction is observed in mixes of cement fly ash. However, at higher fly ash contents, the resistance to carbonation is significantly reduced.
- The use of PC-fly ash cement is indicated in application where reduced evolution of the heat of hydration and/or a high chemical resistance is required and where slower strength development is acceptable.

TAB. 11.3 – Typical chemical composition of red mud (weight %).

CaO	SiO$_2$	Al$_2$O$_3$	Fe$_2$O$_3$	MnO	MgO	Na$_2$O	K$_2$O	TiO$_2$	P$_2$O$_5$	LOI	Σ C/S/A/F
13.9	7.2	15.6	37.2	<0.1	0.3	3	<0.1	9.5	0.6	12	99.3

11.3.3 Red Mud (KUR 1997)

11.3.3.1 Introduction

- Red mud is the by-product of the manufacture of alumina from bauxite by the Bayer process. Bauxite is washed, ground and dissolved in basic soda at high pressure and temperature.
- The total CaO–SiO$_2$–Al$_2$O$_3$ accounts for 95% of the typical bulk composition of the red mud. The chemical composition shown in table 11.3, indicates that red mud contains a large amount of iron oxide (40%) and alkalis (10%).
- The mineralogy of the red mud is composed of mineral coming directly from the bauxite [Anatase, Boehmite (AlOOH), Calcite, Gibbsite, Goethite (FeO(OH), Hematite, Calcium ferrite and Kaolinite] and phases generated by the process:
  - 3(Na$_2$OAl$_2$O$_3$SiO$_2$)Na$_2$X with X = CO$_3^{2-}$, 2OH$^-$, SO$_4^{2-}$ or 2Cl$^-$.
  - 3CaO(Fe$_2$O$_3$)$_y$(Al$_2$O$_3$)$_{1-y}$ kSiO$_2$ (6–2 k)H$_2$O.
  - Na$_2$Ti$_3$O$_7$3H$_2$O (Kassite, Perovskite, Portlandite).
- The density is around 2.6 g/cm^3.

11.3.3.2 Applications

- The large amount of iron oxide bearing compounds gives the red colour to the mud.
- The quantity of alkalis and a weak pozzolanic activity are the main characteristics which yield potential applications that are summarized in figure 11.2.
- Red mud can be a source of Fe$_2$O$_3$ for cement raw mix. A maximum of 10% is incorporated in the raw mix. The clinkering temperature is lower by 100–150 °C due to the fluxing effect of the iron oxide.
- For the production of bricks, the red mud is pressed with a binder and fired at a temperature depending to the proportion in the mix (500–1000 °C).
- Red mud contains components that in combination with lime produce calcium ferro-aluminate hydrates able to produce strengths. A mix of 70% red mud and 30% CaO develops a compressive strength of 7 MPa.
- Red mud is mixed with limestone and sintered in a rotary kiln for the simultaneous production of calcium aluminate and belite cement (C$_2$S).
- When calcined between 650 and 750 °C red mud develops interesting pozzolanic properties and red hematite is obtained with very good tinting properties. It can be used at higher amounts than usual pigments (11% instead of 4%). As CH is quickly consumed, the development of efflorescence is limited.

Application as Fillers

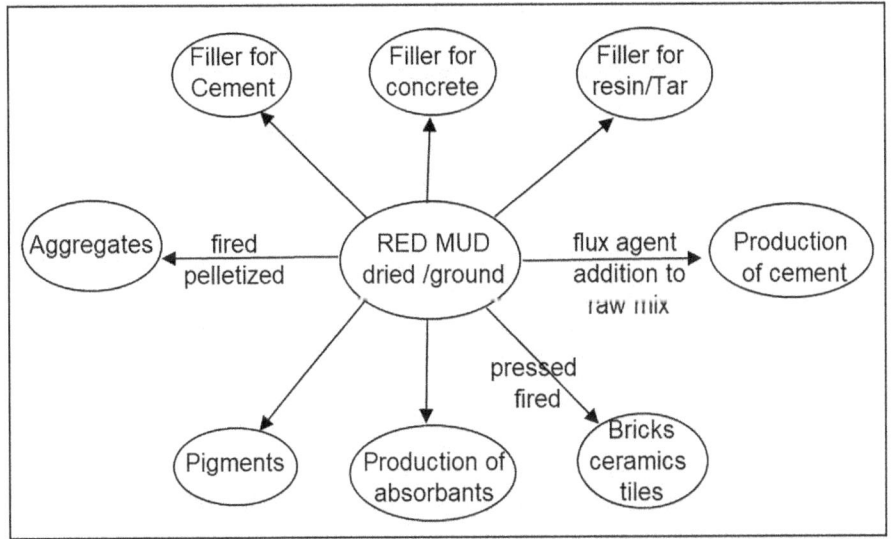

FIG. 11.2 – Applications of red mud.

- Red mud addition to cement is used for mortar and coloured concrete blocks, filling depleted mine shaft and dam structure.
- Aggregates for concrete, bitumen or resin are obtained by drying pelletizing and firing.
- A large part of red mud remains unused due to the amount of alkalis and the volume of the residue. Every ton of alumina produced leaves one ton of solid residue in suspension and 4 tons of slurry.

11.3.4 Cement Kiln Dust (CKD)

11.3.4.1 Introduction/Characterization

- Cement kiln dust is an industrial by product generated in PC manufacture from electrostatic precipitators. It is a fine grained and powdery material.
- CKD consists of four major components: unreacted raw feed, partially calcined feed and clinker dust, free lime, and enriched salts of alkali sulphates, halides and other volatile compounds. This means a high percentage of glassy phase (up to 65%) and reactive silica and CaO but also a highly alkaline material.

11.3.4.2 CKD with Other SCM in Cement/Concrete

- Cement kiln dust so generated is partly reused in cement plants and landfilled. Due to lack of landfilling space and ever-increasing disposal cost, utilization of CKD in highway uses, waste treatment, soil stabilization, cement mortar/concrete, etc., has become an attractive alternative to its disposal.

- CKD can be used with fly-ash and granulated blast furnace slag: up to 15% by mass of cementitious material.
- In ground granulated blast-furnace slag cements, CKD can be used up to 10% by mass of the cementitious material (DET 1996).
- At least 3 advantages can be observed, due to introduction of CKD:
  - Decreased initial and final setting time.
  - Increased strengths.
  - Pore refinement.
- More generally, the properties developed by building materials including CKD and the beneficial use of CKD have been evaluated and reviewed (KUN 2012).
- On the contrary, its use requires attention:
  - As its composition may greatly vary with respect to plant, CKD must therefore be evaluated on a plant-by-plant basis.
  - The leachate obtained from cement kiln dust may contain hazardous compounds and its caustic nature poses harmful effects to the environment. So, it is essential to know the characteristics of leachate obtained from CKD for beneficial utilization towards solid waste management.
- Many attempts have been made to design and develop novel blended cements based on CKD + another SCM (Fly ash or BFS) (CHA 2013).

11.4 Multi-Components from Natural Origin

11.4.1 Natural Pozzolans

11.4.1.1 Introduction/Characterisation

- The total $CaO-SiO_2-Al_2O_3$-Fe oxides account for at least 95% of the typical composition of the most commonly used pozzolana.
- Specific surface (BET) of pozzolans is high: 4–12 m^2/g.
- The term Pozzolan includes all the materials which react with lime or lime containing products and water giving calcium silicate hydrates and calcium aluminate hydrates possessing cementitious properties; so called 'pozzolanic' reaction. Thus, pozzolan hardens under water and acts as active filler when added to cement and concrete.
- Natural pozzolan forms as a result of volcanic explosive type eruption, consisting mainly of silica and alumina (pyroclastic rocks). The violent ejection of the magma in the atmosphere is the reason for the quenching and the presence of glass. The deposit of the pyroclastic material can be submitted to diagenesis cementation phenomena transforming into compacted rocks called tuffs. Lithoidal volcanic tuffs are the zeolitized counterparts of the true pozzolans. Only the pyroclastics deposits rich in silica and in glass or containing zeolitic materials present pozzolanic activity.

TAB. 11.4 – Chemical composition (wgt %) of a series of Pozzolans.

	LOI	SiO_2	Al_2O_3	Fe_2O_3	CaO	MgO	Na_2O	K_2O	SO_3
Rhyolite pumicite (USA)	3.43	65.74	15.89	2.54	3.35	1.33	4.97	1.92	
Roman pozzolan (Italy)	3.05	53.08	18.20	4.29	9.05	1.23	3.08	7.61	
Santorin earth (Greece)	4.80	63.80	13.0	5.70	4.0	2.0	3.80	2.50	
Bavarian trass (Germany)	7.41	62.45	16.47	4.41	3.39	0.94	1.91	2.06	
Ratka trass (Hungary)	6.34	73.01	12.28	2.71	2.76	0.41			0.10
Diatomaceous earth (Czech Republic)	7.62	65.22	19.23	2.76	0.80	0.25			0.31

- The chemical and mineralogical composition of the pozzolans vary in a wide range due to the origin such as Rhyolites Pumicites (USA), Roman pozzolans (Italy), Santorini earth (Greece), Bavarian Trass (Germany), Ratka trass (Hungary) and diatomaceous earth (Czech Republic). The chemical composition varies between the following limits: SiO_2 45–60, $Al_2O_3 + Fe_2O_3$ 15–30, CaO + MgO + alkalis 15 and LOI 10%; the table 11.4 gives the chemical composition of some specific pozzolans.
- Natural pozzolans have a varying amount of glass (55–70%) and minerals; opal in trass from Austria, Quartz, Feldspar, Chabazite, Analcime, Mica, Illite and Montmorillonite in Rhine Bavarian Trass, Augite, Analcime, and Leucite in Italian pozzolan. The structure and the composition of the volcanic glass favour the progress of the zeolitization process generating the volcanic tuff.

11.4.1.2 Pozzolan-Containing Cement and Concrete

- Pozzolan added to Portland cement (CEM II, CEM III and CEM V as defined by CEN – EN 197-1) reacts with the calcium hydroxide released by the calcium silicate of the clinker during the hydration to form C–S–H having a low Ca/Si ratio and greater alumina content as compared to those found in plain PC pastes.
- Portlandite persists for a long time in hardened cement paste also when the pozzolana/Portland cement ratio should be sufficient to combine all the calcium hydroxide formed during the hydration of PC. Thus, replacement of pozzolan for Portland cement causes some changes in the microstructure, in the phase composition as well as the engineering properties of the plain Portland cement.
- The pozzolanic reaction is slower than the hydration of clinker so that the expected reduction of calcium hydroxide content becomes perceivable only after a certain time.
- Replacement of pozzolan for PC causes a decrease in early mechanical strengths but the decrease is less than that foreseen on the basis of the dilution factor since pozzolan promotes hydration of PC. The ultimate strength of pozzolanic mortar is higher than that of a Portland one having the same 28 day strength because

the pozzolanic reaction goes on for many years. Tensile strengths of Pozzolanic cement are generally higher than those of PC.
- The water demand of natural pozzolan cement than that of the parent cement depends on the high specific surface of pozzolan. The higher water demand of pozzolanic cement does not affect mortar shrinkage since mortars are gauged with the same water/cement ratio but it does affect the shrinkage of concrete which are generally compared on the basis of the same workability.
- Creep is related to the strength of concrete at the time of loading so no significant differences appear when concrete having the same strengths are compared.
- When concrete undergoes thermal gradients, pozzolanic cements behave better than PC having the same compression because they have a lower heat evolution and a higher tensile strength.
- Pozzolan significantly increases the chemical resistance of cement due to the lower content of Portlandite in the paste and the lower porosity of the paste. Carbonation of concrete is severe for porous concrete but beneficial for the dense and compact one since calcite precipitating in capillary pores reduces porosity and permeability. Thus, aged pozzolanic concrete has a good resistance toward carbonation, leaching and chemical attacks (acids, ground water bearing sulphate and sea water) but also chloride ingress that promotes steel rebar corrosion. The replacement of cement by 30–40% volcanic pozzolan prevents expansion due to alkali silica reaction.
- The use of pozzolanic cement is suggested when a high chemical resistance or a heat of hydration is a primary need. Other conditions being equal, pozzolan containing cement behaves like the other one in terms of resistance to physical actions.

11.4.2 Metakaolin

11.4.2.1 Introduction/Characterisation

- Metakaolin (MK) is a dehydroxylated form of Kaolinite. Kaolinite is a clay mineral with the chemical composition $Al_2Si_2O_5(OH)_4$. It is a layered silicate with one tetrahedral sheet of silica linked through oxygen atoms to one octahedral sheet of octahedral alumina (AlO_6). Kaolinite is formed by the alteration of a variety of crystalline and amorphous rocks such as feldspars and volcanic ash. Deposits rich in Kaolinite are known as china clay or kaolin. MK can be made in a pure or impure form depending on the mineralogy of the clay.
- Disordered Kaolinite dehydroxylates between 530 and 570 °C and ordered Kaolinite between 570 and 630 °C. Dehydroxylated disordered Kaolinite shows higher pozzolanic activity than ordered (KAK 2001). Dehydroxylation causes the clay to become chemically reactive. The rate of calcining does not influence the pozzolanic properties provided that dehydroxylation is complete and that Kaolin has not been over calcined. Much of the aluminium in MK becomes tetrahedrally coordinated.

- Above about 800 °C kaolin begins to convert to relatively inert ceramic materials such as Spinel, Silica and Mullite.
- The particle size of MK is small; the N_2 BET surface is 2000–20 000 m^2/g.
- MK is also used in the manufacture of porcelain, and its formation is an integral part of the process for the manufacture of ultramarine pigment.
- MK presents pozzolanic activity and thus can be used as filler in cement and concrete.

11.4.2.2 Metakaolin-Containing Cement and Concrete

- Metakaolin, manufactured for use in concrete is milled to a fine powder, before calcining.
- MK can be activated by alkali metal, hydroxide, and alkali metasilicate or by calcium hydroxide. MK reacts with up to 1.6 times its mass of CH according to the reaction $AS_2 + 5CH + 5H \rightarrow C_5AS_2H_5$. MK reacts with aqueous alkali to give a variety of products with a zeolitic structure.
- MK changes the pore structure and modifies the composition and structure of the hydrates. MK reacts with CH formed during PC hydration to give C–S–H and 'C–A–S–H' phases. To remove all the CH from the cured PC paste approximately 20–25 mass % of the cement should be replaced by MK.
- Cement containing MK evolves approximately the same total quantity of heat as normal concrete with the same binder content; MK addition slightly reduces the initial and final setting times of cement paste.
- MK increases the compressive strength of concrete and can increase the rate at which concrete gains tensile strength during the first six hours after mixing depending on the amounts used (JON 2001).
- MK is used in concrete where its improved durability in aggressive environment is important in relation to the removal of all or parts of CH and the lower permeability. MK can be used to prevent alkali-silica reaction and to improve resistance to sulphate attack (high proportions of MK are required; about 20 mass %). The diffusion rate of chloride in cement and concrete is reduced by a factor 3 or more when 10% of the PC is replaced by MK.
- MK is used for its engineering and durability properties to repair bridges, to construct foundation and also in concrete for decorative applications such as paving slabs, floor screeds, renders and sculptures (KUR 2011).

11.4.3 Rice Husk Ashes (RHA)

11.4.3.1 Introduction/Characterisation

- Rice husks (RH) are by-products of rice paddy milling industry and are considered as agricultural waste.
- RH is composed of cellulose (40–50%), lignin (25–30%), ash (15–20%) and moisture (8/15%).
- After burning in defined conditions (T°/time), the rice husk ashes (RHA) are composed of 80–90% SiO_2, 1–3% K_2O and <5% unburnt carbon. The thermal

treatment is achieved in an industrial furnace (Mehta process, Ashmoh process, Shresta process, Black Meal process, CBRI process, CGCRI process).
- RHA is highly pozzolanic with a PSD of 160 m²/g and a bulk density between 96 and 160 kg/m³.

11.4.3.2 RHA-Containing Cement and Concrete

- RH produced annually in large quantity (100 million tons) is highly resistant to natural degradation. The use of RHA as a complementary addition on cement and concrete is a proper disposal (HWA 1997).
- Hydration of cement blended with RHA shows an optimum content of 30%. The excess of RHA has no contribution to the pozzolanic reaction. Thus, large amounts of RHA have an adverse effect and reduced strengths (TUA 2011).
- RHA tends to shorten the setting time.
- Significant reduction in the porosity and refinement in the pore structure are obtained with RHA addition. Consequently, durability is improved as the permeability is significantly reduced after 28 days of curing for the cement paste with 10–30% RHA.
- Addition of 2–3% RHA improves the stability and workability of concrete by reducing the tendency towards bleeding and segregation. RHA absorbs a large amount of water due to its high PSD and thus reduces bleeding and segregation characteristics of concrete mixture. This will lead to a more impermeable and durable concrete.

11.4.4 Wollastonite

AND 1970

11.4.4.1 Introduction/Characterisation

- As a reminder Wollastonite is a calcium metasilicate ($CaSiO_3$) and has a theoretical composition of 48.3% calcium oxide and 51.7% silicon dioxide but may contain trace to minor amounts of aluminium, iron, magnesium, manganese, potassium, and sodium. It occurs as prismatic crystals that cleave into massive-to-acicular fragments. It is usually white, but also may be grey, brown, or red depending on its composition (AND 1970).
- Major uses for Wollastonite are in plastics, ceramics, metallurgy, paint, and friction products and as asbestos substitutes.

11.4.4.2 Wollastonite Used as Asbestos Substitute

- Wollastonite is used as mineral reinforcing agent for many applications that were satisfied by Asbestos, such as mineral wool. This is because it does not cause any health issues associated with Asbestos. Its applications as tensile reinforcement are somewhat limited as the length of the Wollastonite fibres is less than the one

Application as Fillers 333

of asbestos; aspect ratio of 20/1 for Wollastonite compared to 44 000/1 for Asbestos.
- It has been used successfully to improve impact resistance of concretes, for instance in roof tiles where it has been shown to greatly increase the lifespan of the tile.

11.4.4.3 Wollastonite Used in the Paint Industry (20 000 t/year)

- Wollastonite is most widely used in the types of paint employed for exterior house coatings including primers and under coating namely flat and emulsion paints.
- Its use as an extender in paints is largely based on its colour, acicular nature and low oil absorption.
- Wollastonite is used as anti-corrosion paint for industrial coating including flat and emulsion oil or water-based paints.

11.4.4.4 Wollastonite Used in Plastic and Rubber Industry (60 000 t/year)

- Wollastonite is used as reinforcing filler for the manufacture of numerous components manufactured mainly from nylon and also from other resins (polyurethane, polypropylene, polyester, polystyrene...).
- By means of its low hydro-absorptivity, dielectric constant and a thermal expansion, it reduces the quantity of resin, increases the mechanical properties and the resistance to heat and reduces shrinkage.

References Part II

AND 1970 – Andrews R.W. (1970) Wollastonite, monograph, Her Majesty's Stationery Office, pages 114.
BAN 1982 – Ban-Ya S., Iguchi Y., Honda H. (1982) Heat of mixing of liquid FeO-SiO_2 slag, Proc. Int. Symp. Phys. Chem. Iron and Steelmaking, Toronto, Met. Soc. CIM, pp. II: 39–44.
BAN 1995 – Banfill P.F.G. (1995) Superplasticizer for ciment fondu. Part 2: effects of temperature on the hydration reactions, *Adv. Cem. Res.* **7**, 151.
BAR 1974 – Barret P., Ménétrier, D., Bertrandie D. (1974) Contribution to the study of the kinetic mechanism of aluminous cement setting I - latent periods in heterogeneous and homogeneous milieus and the absence of heterogeneous nucleation, *Cem. Concr. Res.* **4**, 545.
BAR 1983 – Barnes P. (1983) *Structure and performance of cements*, 1st edn. Spon Press.
BEN 2002 – Bensted J., Barnes P. (2002) *Structure and performance of cements*, 2nd edn. Spon Press.
BES 1980 – Bessbertnykh A.V., Khokhlov V.K., Cheloudko V.V. (1980) Étude thermodynamique de la cinétique de la formation de l'alite, 8th Int Congress Chemistry Cement, Paris 1980, Vol I, pp. 42–46.
BIE 1909 – Bied J. (1909) British patent 8193, French patents 320290, 391454, 1909.
BOD 1972 – Bodsworth C., Bell H.B. (1972) Physical chemistry of iron and steel manufacture, 2nd edn. Longman Group, p. 529.
BOY 1984 – Boyd D.C., Macdowell J.F. (1984) *Commercial glasses, advanced in ceramics*, The American Ceramic Society, Columbus, OH, Vol 18.
CAM 2005 – Camilleri J., Montesin F.E., Brady K., Sweeney R., Curtis R.V., Ford T.R.P. (2005) The constitution of mineral trioxide aggregate, *Dent. Mater.* **21**, 297.
CHA 2013 – Chaunsali P., Peethamparan S. (2013) Influence of the composition of cement kiln dust on its interaction with fly ash and slag, *Cem. Concr. Res.* **54**, 106.
CHE 1968 – Cheron M., Lardinois C. (1968) The role of magnesia and alumina in the hydraulic properties of granulated blast-furnace slags, Proc. 5th ICCC Tokyo, sup. Paper IV-121 pp. 277-285.
CON 1981 – Conjeaud M., George C.M., Sorrentino F.P. (1981) A new steel slag for cement manufacture: mineralogy and hydraulicity, *Cem. Concr. Res.* **11**, 85.
DAR 1948 – Darken L.S. (1948) Melting points of iron oxides on silica; phase equilibria in the system Fe-Si-O as a function of gas composition and temperature, *J. Am. Chem. Soc.* **70**, 2046.
DAV 2015 – Davidovits J. (2015) *Geopolymer chemistry and applications*, Publ. Saint-Quentin, Institut Géopolymère, France, 623 pages.
DAW 1983 – Dawson P.R., Ostwald J., Hayes K.M. (1983) Calcium ferrite in the new low silica sintering technology, *BHP Tech. Bull.* **27**, 47.
DEM 1980 – Demoulian E., Gourdin P., Hawthorn F., Vernet C. (1980) Influence de la composition chimique et de la texture des laitiers sur leur hydraulicité, *8th ICCC Paris* **III**, 89.
DET 1996 – Detwiler R.J., Bhatty J.I., Bhattacharja S. (1996) Supplementary cementing materials for use in blended cements. Research and Development Bulletin RD112T, Portland Cement Association, Skokie, Illinois.
EDS 1956 – Edström J.O. (1956) The phase CaO.2 Fe_2O_3 in the system CaO- Fe_2O_3 and its importance as binder in ore pellets, *Jernkontorets Ann.* **140**, 101.
EGU 1989 – Egundebi G.O., Whiteman J.A. (1989) Evolution of microstructure in iron-ore sinters, *Ironmak. Steelmak.* **16**, 379.

FUK 1992 – Fukuda K., Maki I., Ito S. (1992) Remelting reaction within belite crystals during cooling, *J. Am. Ceram. Soc.* **75**, 2896.

FUK 1993 – Fukuda K., Maki I., Ikeda S., Ito S. (1993) Microtextures formed by the remelting reaction in belite crystals, *J. Am. Ceram. Soc.* **76**, 2942.

GAN 2011 – Gandolfi M.G., Taddei P., Siboni F., Modena E. (2011) Development of the foremost light-curable calcium-silicate MTA cement as root-end in oral surgery. Chemical-physical properties, bioactivity and biological behavior, *Dent. Mater.* **27**, 134.

GAR 2001 – Gartner E.M., Young J.F., Damidot D.A., Jawed I. (2001) Hydration of portland cement, p. 91, *Structure and performance of cements* (J. Bensted, P. Barnes, Eds). 2nd edn. Spon, 2002, pp. 57–113.

HEW 1998 – Hewlett P.C. (1998) *Lea's chemistry of cement and concrete,* 4th edn. Arnold London.

HUG 1967 – Hughes H., Roos P., Goldring D.C. (1967) X-ray data on some calcium-iron-oxygen compounds, *Mineral. Mag.* **36**, 280.

HWA 1997 – Hwang C.L., Chandra S. (1997) The use of rice husk ash in concrete, *Waste materials used in concrete manufacturing* (S. Chandra, Ed). Noyes publications, 1996, Chap. 4, pp. 184–234, 672 pages.

IMA 1968 – Imaoka M., Yamazaki T. (1968) *Report of the institute of industrial science,* The University of Tokyo, Vol 18, Serial No 118, pp. 15–17.

JON 2001 – Jones T.R. (2001) Metakaolin as a pozzolanic addition to concrete, *Structure and performance of cements* (J. Bensted, P. Barnes, Eds). 2nd edn. Spon Press, chap fifteen, pp. 372–398, 2002.

KAK 2001 – Kakali G., Perraki T., Tsivilis S., Badogiannis E. (2001) Thermal treatment of kaolin: the effect of mineralogy on the pozzolanic activity, *Appl. Clay Sci.* **20**, 73.

KAL 1960 – Kalyanram M.R., Macfarlane T.G., Bell H.B. (1960) The activity of calcium oxide in slags in the systems $CaO-MgO-SiO_2$, $CaO-Al_2O_3-SiO_2$, and $CaO-MgO-Al_2O_3-SiO_2$ at 1500 °C, *J. Iron Steel Inst.* **195**, 58.

KUN 2012 – Kunal, Siddique R., Rajor A. (2012) Use of cement kiln dust in cement concrete and its leachate characteristics, *Resour. Conserv. Recycl.* **61**, 59.

KUR 1997 – Kurdowski W., Sorrentino F. (1997) Red mud and phosphoric gypsum and their fields of application, *Waste materials used in concrete manufacturing* (S. Chandra, Ed). Noyes publications, 1996, pp. 290-351, 672 pages.

KUR 2011 – Kurtis K.E. (2011) Benefits of metakaolin in HPC, *HPC bridges views*, Vol 67, pp. 6–9.

LEA 1956 – Lea F.M., Desch C.M. (1956) *The chemistry of cement and concrete,* 2nd edn. Arnold, London.

LEA 1970 – Lea F.M. (1970) *The chemistry of cement and concrete,* 3rd edn. Edward Arnold & Co, London.

LI 2012 – Li L., Hu J.-H., Wang H. (2012) Smelting oxidation desulfurization of copper slags, *J. Iron Steel Res. Int.* **19**, 14.

LUX 2000 – Luxán M.P., Sotolongo R., Dorrego F., Herrero E. (2000) Characteristics of the slags produced in the fusion of scrap steel by electric arc furnace, *Cem. Conc. Res.* **30**, 517.

MAJ 1993 – Majling J., Roy D.M. (1993) The potential of fly ash for cement manufacture, *Am. Ceram. Soc. Bull.* **72**, 77.

MAK 1979 – Maki I. (1979) Phase constitution of the Alite in Portland cement clinker from modern manufacturing processes, *Il Cemento*, **76**, 167.

MAN 1990 – Mangabhai R. (1990) *Calcium aluminate cements: proceedings of the international symposium on calcium aluminate cements.* E&F.N. Spon, London.

MAN 2008 – Fentiman C., Mangabhai R., Scrivener K.L. (2008) *Calcium aluminate cements: proceedings of the centenary conference.* IHS BRE Press.

MOR 1954 – Morey G.W. (1954) *The properties of glass,* 2nd edn. Reinhold, New York.

OLU 2007 – Oluwadare G.O. (2007) Roles of alumina and magnesia on the formation of SFCA in iron ore sinters, *Trends Appl. Sci. Res.* **2**, 483.

PAR 2004 – Parr C., Simonin F., Touzo B., Wöhrmeyer C., Valdelièvre B., Namba A. (2004) The impact of calcium aluminate cement hydration upon the properties of refractory castables, Technical paper TP-GB-RE-LAF-043 presented at TARJ meeting, Ako, Japan, pp. 1–17.

PAR 2010–1 – Parirokh M., Torabinejad M. (2010–1) Mineral trioxide aggregate: a comprehensive literature review-Part I: chemical, physical, and antibacterial properties, *J. Endod.* **36**, 16.

PAR 2010–2 – Parirokh M., Torabinejad M. (2010–2) Mineral trioxide aggregate: a comprehensive literature review-Part III: clinical applications, drawbacks, and mechanism of action, *J. Endod.* **36**, 400.

PER 1999 – Péra J., Ambroise J., Chabannet M. (1999) Properties of blast-furnace slags containing high amounts of manganese, *Cem. Conc. Res.* **29**, 171.

ROB 2008 – Roberts H.W., Toth J.M., Berzins D.W., Charlton D.G. (2008) Mineral trioxide aggregate material use in endodontic treatment: a review of the literature, *Dent. Mater.* **24**, 149.

SCH 1951 – Schuhmann Jr R., Ensio P.J. (1951) Thermodynamics of iron-silicate slags: slags saturated with gamma iron, *J. Metals Trans AIME.* **3**, 401.

SCH 1953 – Schuhmann R., Powell R.G., Michal E.J. (1953) Constitution of the $FeO-Fe_2O_3-SiO_2$ system at slag-making temperatures, *J. Metals Trans. AIME*, **197**, 1097.

SCH 1980 – Scholze H. (1980) *Glass science and technology*, 3rd edn. Elsevier, Paris.

SCR 1990 – Scrivener K.L., Taylor H.F.W. (1990) Microstructural development in pastes of a calcium aluminate cement, *Calcium aluminate cements: proceedings of the international symposium on calcium aluminate cements.* E&F.N. Spon, London, pp. 41–51.

SEI 2008 – Seifert S., Neubauer J., Goetz-Neunhoeffer F., Motzet H. (2008) Characterisation of the crystalline phase distribution within self-levelling compounds (SLC) using 2-dimensional XRD (GADDS), *Calcium aluminate cements, proc. of the centenary conference 2008, Avignon* (C.H. Fentiman, R.J. Mangabhai, K.L. Scrivener, Eds). pp. 581–592.

SOR 1984 – Sorrentino F., George C.M., Vershaeve M. (1984) Patent FR 25413100 A1 "Method for manufacturing steel desulfurisation slag", 1984.

SOR 1995 – Sorrentino D., Sorrentino F., George C.M. (1995) Mechanism of hydration of calcium aluminate cements, *Materials science of concrete* (J. Skalny, S. Mindness, Eds). Vol IV, pp. 41-91.

TAN 1974 – Taneja C.A. (1974) Role of magnesia on hydration of high alumina slags, VI ICCC Moscow, Supl. Paper Section III-2, 1974.

TAO 2008 – Tao D. (2008) Predication of component activities in the molten aluminosilicate slag $CaO-Al_2O_3-SiO_2$ by molecular interaction volume model, *J. Mater. Sci. Technol.* **24**, 797.

TAY 1990 – Taylor H.F.W (1990) *Cement chemistry.* Academic Press Ltd., London.

TOR 2010 – Torabinejad M., Parirokh M. (2010) Mineral trioxide aggregate: a comprehensive literature review-part II: leakage and biocompatibility investigations, *J. Endod.* **36**, 190.

TUA 2011 – Van Tuan N., Zhang Qi, Guang Ye (2011) Modeling the hydration of cement blended with rice husk ash, *13th ICCC at Madrid* **6**, 353.

TUR 1961 – Turkdogan E.T. (1961) Activities of Oxides in $CaO-FeO-Fe_2O_3$ melts, *Trans. AIME* **221**, 1090.

WAN 2004 – Wang P.Z., Trettin R., Rudert V., Spaniol T. (2004) Influence of Al_2O_3 content on hydraulic reactivity of granulated blast-furnace slag, and the interaction between Al_2O_3 and CaO, *Adv. Cem. Res.* **16**, 1.

WHI 1974 – White J. (1974) Slag control in basic steelmaking processes: an examination of the possibility of eliminating fluorspar, *Ironmak. Steelmak.* **2**, 115.

WOE 1960 – Woermann E. (1960) Decomposition of alite in technical Portland cement clinker, *Chemistry of cement, proceedings of the fourth international symposium*, Washington, paper II-S8, Vol I, 1960, pp. 119–129.

YOU 2006 – Young F., Skalny J., Eds (2006) Materials science of concrete VII - materials science of concrete series, *Publ. Am. Ceram. Soc.* 306.

YU 1968 – Yu X., Butt M., Timashev V.V., Kaushanski V.E. (1968) Properties of tricalcium silicate and its solid solution, *Proc. 9th Conf. Silicate ind.* Budapest 59.

ZAR 1982 – Zarzycki J. (1982) *Les Verres et l'état vitreux.* Barcelone, Masson, Paris, New York.

www.ingramcontent.com/pod-product-compliance
Lightning Source LLC
Chambersburg PA
CBHW050155230526
45470CB00001B/100